LOW DOSE RADIATION

THE HISTORY OF THE U.S. DEPARTMENT OF ENERGY RESEARCH PROGRAM

LOW DOSE RADIATION

THE HISTORY OF THE U.S. DEPARTMENT OF ENERGY RESEARCH PROGRAM

ANTONE L. BROOKS

WSU PRESS

Washington State University Press
Pullman, Washington

WSU PRESS
WASHINGTON STATE UNIVERSITY

Washington State University Press
PO Box 645910
Pullman, Washington 99164-5910
Phone: 800-354-7360
Fax: 509-335-8568
Email: wsupress@wsu.edu
Website: wsupress.wsu.edu

© 2018 by the Board of Regents of Washington State University
All rights reserved
First printing 2018

Printed and bound in the United States of America on pH neutral, acid-free paper. Reproduction or transmission of material contained in this publication in excess of that permitted by copyright law is prohibited without permission in writing from the publisher.

Library of Congress Cataloging-in-Publication Data

Names: Brooks, Antone L., 1938- author.
  Title: Low dose radiation : the history of the U.S. Department of Energy research program / by Dr. Antone L. Brooks.
Description: Pullman, Washington : Washington State University Press, 2018.| Includes bibliographical references and index.
Identifiers: LCCN 2017047624 | ISBN 9780874223545 (alk. paper)
Subjects: LCSH: Radiation--Measurement. | Brooks, Antone L., 1938- | United States. Department of Energy. | Radiation--Dosage. | Radiation--Toxicology--Research--United States.
Classification: LCC QD117.R3 B76 2018 | DDC 612/.01448072--dc23
LC record available at https://lccn.loc.gov/2017047624

DISCLAIMER

This report was prepared as an account of work sponsored by an agency of the United States Government. Neither the United States Government nor any agency thereof, nor any of their employees, makes any warranty, express or implied, or assumes any legal liability or responsibility for the accuracy, completeness, or usefulness of any information, apparatus, product, or process disclosed, or represents that its use would not infringe privately owned rights. Reference herein to any specific commercial product, process, or service by trade name, trademark, manufacturer, or otherwise does not necessarily constitute or imply its endorsement, recommendation, or favoring by the United States Government or any agency thereof. The views and opinions of authors expressed herein do not necessarily state or reflect those of the United States Government or any agency thereof.

I would like to dedicate this book to my lovely wife Janet. She has been my beautiful California movie star from the first day I met her. I must acknowledge her unfailing love and support through all my life. Without her this book, my science, and our family would not have been possible. Thank you so very much, love ya.

Contents

Preface	1
Foreword, by Gayle Woloschak	3
Summary	5
Acronyms and Abbreviations	7
Introduction	9
1. Life and Times of a Radiation Biologist	15
2. A Brief History of Radiation Biology	33
3. The Birth of the DOE Low Dose Radiation Research Program	41
4. Early Observations and New Technology	49
5. Paradigm Shifts in Low Dose Radiation Biology and Application of Data	75
6. Biomarkers of Radiation Exposure and Dose	119
7. Mechanisms of Action	131
8. Modeling	173
9. Taking a Systems Biology Approach to Risk	193
10. Program Communication and Monitoring	199
11. Current and Potential Impact on Standards	205
12. Applying Lessons Learned to Future Direction	211
Epilogue	231
Appendices	
A. BERAC Report Program Plan	233
B. First Call for Proposals	253
C. Dose Range Charts	256
References	259

Preface

While writing and preparing this book for publication, I worked closely with two communication specialists at Pacific Northwest National Laboratory (PNNL) whose enthusiasm and expertise I appreciated. By the time they got involved in the process I had already written a full-length manuscript, but I'm a scientist, not a writer. I have always valued collaboration with communications people, and I had an especially good relationship with one of these guys—a giant, long-haired fellow who shared my enthusiasm for the outdoors and telling jokes. When we weren't talking about how to best represent the science so someone might actually want to read this book, we talked about fishing and other outdoor pursuits in the American West. He showed what seemed like genuine interest in some of my stories, so I decided to share with him an autobiography I had written for my family. He read it, showed to his colleague, and the next thing I knew they were trying to edit that book too. Actually what they wanted to do was include some of my life's details in this history of the DOE Low Dose Radiation Program. Why? Radiation and the pursuit of a mechanistic understanding of its effects on health has been a driving theme in my life since childhood in southern Utah growing up amidst and eventually studying fallout clouds from nuclear tests conducted in neighboring Nevada.

While I have always listened to my communication specialists, I was less sure about the enthusiasm these two collaborators had for including the details of my life in a science book. With some initial reluctance, I relented and allowed them to reduce my book-length manuscript down to chapter length. You'll find the result in chapter 1—a summary of my life up to the moment when I took the job as Chief Scientist for the Low Dose Radiation Program.

Radiation and its effects have consumed my professional life. Even my personal life from an early age has been affected mightily by my drive to truly know what radiation does to our bodies. I hope chapter 1 provides some context for the work described in the rest of this book.

I would like to acknowledge the financial support of the U.S. Department of Energy (DOE), especially Marvin Frazier for developing the program and Dr. Noelle Metting for her continued support of this project. I would also like to acknowledge Pacific Northwest National Laboratory (PNNL) and Drs. William Morgan and Katrina Waters for providing work space, supplies, and funding to complete the project. I must also recognize the excellent editorial help provided by PNNL. Early in the project Julie Wiley helped to get an initial draft of the manuscript on the Web and put me in touch with communication specialists Jeffrey Holmes and Andrew Pitman. That duo provided not only editorial help but made serious suggestions on revising content, addressing technical aspects of document production, and providing the communications skills to make it more readable and interesting without compromising the science. Without their generous help and interaction the book would never have made it to the publisher. Finally, I heartily thank Washington State University Press for publishing the book and for their commitment to shining light on the important work completed during the history of the DOE Low Dose Radiation Program.

This book is the work of a single individual, with considerable technical help, and reflects the author's view on many different research projects. The opinions and interpretation of the published data compiled in this book are from the author and do not reflect those of the U.S. Department of Energy or any of the investigators that conducted and published the research.

Foreword

The Department of Energy's Low Dose Radiation Research Program, conducted from 1998 to 2008, was a period of discovery and excitement. Dr. Antone (Tony) Brooks served as Chief Scientist of the program, and his history captures that enthusiasm. Most radiation researchers believed that effects of radiation at high doses could (simplistically) be extrapolated to low doses, that similar mechanisms were at work regardless of the dose. Thus, the greatest attention was initially given to well-defined biological endpoints such as cell death and cancer. What was shocking about the findings from DOE's Low Dose Radiation Program was that biological responses at low doses are unique. Tony's history reviews the work conducted by the program, as well as the release of the resulting data, papers, and conclusions of that work. His involvement in the project was intimate, and he describes the outcomes of the work done, its interfaces with the scientific findings of the time and of the future, and the potential impact on radiation protection, regulatory agencies, and radiation science as a whole.

In many cases low dose responses were unrelated to what was known to occur at high doses. Such unexpected findings included, for example, the bystander effect. Un-irradiated "bystander" cells in a culture sharing only growth media with the irradiated cells show similar responses including induction of mutations, cell death, chromosomal injury, and others. Delayed effects were also discovered—consequences of radiation (other than mutations) were identified in the daughter cells of irradiated cells sometimes as distant as 13 cell generations. Adaptive responses were also uncovered where a "tickle" low dose of radiation could protect a cell from the damaging effects of a subsequent high dose exposure. These low dose responses were quirky, expressed in some cells and not others, and most were identified not only in cells in culture but also in whole animal systems.

The late twentieth and early twenty-first century were also an exciting period for science in general. Genomes were being sequenced, systems biology was just beginning, single cell biology was impacting molecular techniques, and the resolution of microscopic techniques was improving. Low dose radiation biology was enmeshed in that milieu and took advantage of many of the discoveries of the time. To some extent, some of the unusual responses identified at low radiation doses were identified because of the emphasis on single cell biology. What was occurring in the broad scientific community impacted the Low Dose Program studies. The development of the microbeam, for example, a technology that allowed for irradiation of single cells, was part of the movement to understand single cell responses.

Dr. Brooks provides a compendium of information that resulted from the Low Dose Radiation Program: he traces the origins of the program and its birth from DOE's long-standing interest in radiation effects through its discoveries and findings and into the impact on regulatory agencies. To this day, investigators from the Low Dose Program are heavily involved in agencies involved in establishing regulations for the U.S., including the National Council on Radiation Protection and Measurements, NASA, Nuclear Regulatory Commission, and others. The program resulted not only in new research, but also in training of graduate students, post-doctoral fellows, and others who have seeded a new generation of investigators who think about sensitive approaches for detection of radiation-induced effects.

The value of this book is not only in the science covered or the vast array of references offering a real resource to the community—but it is also, and perhaps most importantly, the story of the significant contribution this program made in identifying and understanding aspects of the low dose response to radiation. Finally, by putting this research into context, Dr. Brooks points out that new understanding will come with the development of techniques yet to come. There is a true need for continued investigation into this area and continuing the legacy of DOE's low dose radiation program.

Gayle Woloschak, PhD
Professor of Radiation Oncology,
Radiology, and Cellular Molecular Biology
Robert H. Lurie Comprehensive Cancer Center
Feinberg School of Medicine
Northwestern University

Summary

What follows is an overview of the research progress made by the U.S. Department of Energy Low Dose Radiation Research Program over the ten-year period from 1998 to 2008, offering a useful literature review and background information for anyone interested in conducting research on the effects of low-dose or low dose-rate radiation exposure.

The first chapters describe the state of the field when the program began, the need for the program, the development of the research program, and the people that played essential roles in outlining its scientific direction and securing its funding. Chapter 4 describes the technology and molecular techniques developed and combined in research projects. Such combinations illustrate how these advances in technology and biological techniques made possible measurements in the low-dose region not previously possible.

The research resulted in a number of unique observations that have led to paradigm changes in radiation biology. These are described in detail in chapter 5 and include bystander effects, genomic instability, and adaptive protective responses. The discussion also illustrates the importance of genetic background on all these observed responses. The observations evaluated in chapter 5—and further discussed in chapters 6 and 7—provided a biological mechanistic basis for the observations. This mechanistic approach will make the data more useful for understanding the impact and implications associated with low-dose responses. Such understanding is essential if any of this research is to be useful in radiation protection and the formation of radiation standards.

Chapter 8 shows how models have been developed to help understand the observations at all different levels of biological organization. Without useful models, the basic biological information cannot be applied to radiation standards. Chapters 9 and 10 illustrate that without the communication of the information, not only to scientists but to the larger community, the information has limited value. Adequate communication is thus essential before scientific information can be useful. Chapter 11 discusses the impact of the program on standard setting and radiation protection. Chapter 12 looks to the future to determine how such complex and abundant data can be integrated and interpreted to understand and predict radiation risks in the low-dose and dose-rate region.

To date, these data have had major influence on understanding the biological processes triggered by low doses of radiation, but additional research, development of methods of using the data, and communication are required before such data can impact radiation standards.

Acronyms and Abbreviations

AEC	Atomic Energy Commission
ALARA	as low as reasonably achievable
ALS	Advanced Light Source
ANL	Argonne National Laboratory
ATM	ataxia telangiectasia mutated protein
BEIR	Biological Effects of Ionizing Radiation
BER	biological and environmental research
BERAC	Biological and Environmental Research Advisory Committee
BNL	Brookhaven National Laboratory
CHO	Chinese hamster ovary
DDREF	dose and dose-rate effectiveness factors
DIE	death-inducing effects
DOE	U.S. Department of Energy
DSBs	double-strand breaks
EGF	epidermal growth factor
EGFR	epidermal growth factor receptor
EPA	Environmental Protection Agency
ERDA	Energy Research and Development Administration
FISH	fluorescent in situ hybridization
FNA	French National Academy
GCI	Gray Cancer Institute
Gy	Gray
HFC	human fibroblast cells
HMEC	human mammary epithelial cells
HRS	hyper-radiosensitivity
HZE	high Z
ICRHER	International Consortium for Research on Health Effects of Radiation

ICRP	International Commission on Radiation Protection
ICRP	International Council on Radiation Protection
IRR	increased radioresistance
ITRI	Inhalation Toxicology Research Institute
LANL	Los Alamos National Library
LBNL	Lawrence Berkeley National Laboratory
LC-MS/MS	liquid chromatography/tandem mass spectrometry
LET	linear energy transfer
LLNL	Lawrence Livermore National Laboratory
LMDS	locally multiply damaged sites
LNT	linear-no-threshold
MnSOD	manganese superoxide dismutase
NAS	National Academy of Sciences
NASA	National Aeronautics and Space Administration
NCI	National Cancer Institute
NCRP	National Council on Radiation Protection and Measurements
NRC	National Research Council
OMB	Office of Management and Budget
ORNL	Oak Ridge National Laboratory
PCC	premature chromosome condensation
PNNL	Pacific Northwest National Laboratory
RAC	Radiation Advisory Council
RBE	relative biological effectiveness
ROS	reactive oxygen species
SCE	sister chromatid exchanges
TGFß	transforming growth factor beta
TRDLs	tandem repeated DNA loci
UNSCEAR	The United Nations Scientific Committee on the Effects of Atomic Radiation

Introduction

The U.S. Department of Energy's (DOE's) Low Dose Radiation Research Program was initiated in 1998 and remained one of the worlds' largest research efforts in this area for over ten years. It has currently been reduced from a high of over sixty projects to the point where only two small projects exist today. The United States thus went from the leader in this important research area to being a non-player today. The data derived from the program provided critical information on the biological responses in the low dose region and resulted in a database that is essential for elucidating the effects of low dose radiation. The program drove pioneering fundamental science discoveries that provided the foundation for setting standards to protect the public and led the way in sparking research and debate across the international scientific community. As outlined in the book, the data from the program made it necessary for the field of radiation biology to evaluate the current paradigms and suggested that biological responses in the low dose region were unique and that the risks from exposure to radiation in the low dose and dose-rate region are overestimated.

As former chief scientist of the program (from 1998 to 2004) and someone who has dedicated his entire professional life to this work, I wrote this book not just to chronicle the research as a historical footnote. I wrote it to underscore the value of the research generated toward the continued protection of human health and to share lessons learned while conducting the science. Moreover, another major motivation is the potential application of the knowledge in the event of nuclear crisis ranging from a "dirty bomb" to a full-scale detonation. The data can also be applied to many areas that impact our decisions made daily such as radiation used in medicine, nuclear industry, nuclear waste clean-up, and industry. Resources being used to protect against very small radiation doses may not represent a wise use of our limited money, and this spending is driven mostly by fear of low doses of radiation. The science we conducted could be instrumental in helping to quell the fears of the public and in the subsequent triaging of any such tragic incident. The public—even some public health officials and scientists—are often woefully misinformed about the actual risks of low dose radiation exposure.

The chapters that follow this introduction document the first ten years of the program and provide information about its development, fundamental science findings, and the scientists that made the program viable. Although the book is primarily a scientific text, chapter 1 is a narrative focusing squarely on one scientist's life in particular—mine. I've always believed that communicating scientific work to non-scientist decision makers is about as important as conducting it. So when communication specialists convinced me that my story links uniquely to the program's history, with some reluctance I agreed to share some stories about my life growing up and starting a career in southern Utah during above-ground nuclear testing. Beyond this brief digression about my experiences with radiation and my desire to study it, overarching goals for telling the program's history include the following:

- Telling the story of the program's research and subsequent impact on current thinking and low-dose paradigms associated with the radiation biology field
- Summarizing the data generated in the program to provide a scientific basis for setting radiation standards
- Highlighting lessons learned during the program's lifespan, including how those lessons might be useful in case of a nuclear event
- Stimulating research on the potential adverse and/or protective health effects of low doses of ionizing radiation and the crucial work that remains to be done

The program's research in the United States no longer exists, so it is important that this book tells not only the past accomplishments, but the real needs for future research required to define biological responses in the low-dose region. Serious challenges remain, however, and must be resolved before this information can be used to determine adequate and appropriate standards regarding risk from low-dose radiation exposures.

It is important to note that most researchers involved in the DOE Low Dose Radiation Research Program were conducting similar research with funding from other sources. This book is not a review of all the literature developed by the scientific community on low-dose radiation effects, but focuses instead on the information produced with funding from the DOE Program. I recognize that there may be some program publications that have been missed and not included in the book and apologize for these inadvertent omissions.

Historically, radiation risks in the low-dose region have been difficult to evaluate for two major reasons. The first is background radiation, which is modified by elevation, geographic location, environment, and diet, and which therefore varies over a wide dose range. Background radiation also includes many man-made exposures to low-dose radiation from a wide range of sources, including military activities, medicine, and industry. All

INTRODUCTION   11

of this background radiation influences the total dose that individuals and populations receive. Second, a high spontaneous frequency of cancer and genetic effects exists that is influenced by many factors, including lifestyle, genetic background, and environmental conditions. These make detection of any low-dose radiation-induced effects very difficult.

The relationship between specific human radiation doses and many biological responses has been reviewed extensively by national and international groups and has resulted in many publications from the National Commission on Radiation Protection (NCRP), the International Commission on

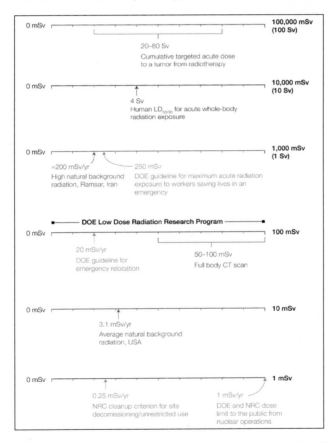

Figure 1. Simplified version of the dose range charts developed by Dr. Noelle Metting. The exposures are given on a log scale, and each line shown represents a dose 10 times greater than the previous line. It shows doses experienced in everyday life relating to current regulations, background radiation, research doses, acute radiation effects, and radiation therapy.

Radiation Protection (ICRP), the National Academy of Sciences, and other agencies and regulatory groups associated with radiation protection.

Dr. Noelle Metting, program manager of the DOE Low Dose Radiation Research Program since 2000, developed a chart that summarizes the many orders of magnitude of the radiation exposures that people were, and continue to be, exposed to and compares the level of radiation allowed in radiation protection standards to other common human exposures. Figure 1 has been simplified from an important figure developed by Dr. Noelle Metting. The original, carefully generated figure by Dr. Metting is shown in Appendix C. For those seeking detailed information, it is important to go the appendix where the information is coded in a way to address their individual questions about standards, background radiation, medical radiation, space radiation, and radiation doses that can cause death. Figure 1 summarizes a few important pieces of information for the average reader that illustrate the wide range of radiation that we deal with in our daily lives.

These two figures relate the radiation exposures from many specific human activities to some of the biological effects seen at these dose ranges. They also illustrate the doses under investigation in the program and show that in this dose region, it is not possible to detect changes in the frequency of cancer or genetic effects in human populations.

The inability to detect these changes was one of the driving forces in the program's initiation. It was felt that to ensure that radiation protection standards are adequate and appropriate, there must be a better understanding of the mechanisms of interaction of radiation with biological systems in this low-dose range, as well as of the possible impact of these exposures on radiation risk. These figures and resulting discussion illustrate the importance of detecting and understanding the biological changes induced at this very low dose, and the challenge that the program faced.

Although the radiation responses in this low-dose region have not been directly detected, they have been predicted by models, and there have been extensive discussions and questions about the current regulations in this dose region. Radiation protection standards are set (after adjustments for radiation type, dose, and dose rate) using the Linear-No-Threshold (LNT) model. The LNT model constructs a straight line from the observed health effects induced by high doses to the predicted health effects of low radiation exposures. The research developed by the program provided evidence that this model is over conservative and needs to be further revised.

These estimates are used to predict cancer risks in the low-dose range where changes in cancer frequency cannot be detected. Thus, there is a degree of confusion between measured data, real risk, and extrapolated risk—risk that cannot be supported by data. This extrapolation was based on old radiation paradigms. Research by the Low Dose Radiation Research Program has generated data that impact many of these paradigms. These data are reviewed and summarized here to provide a scientific basis for further evaluation and evolution of thinking in the fields of radiation biology and radiation protection.

When low-dose research was initiated, several well-accepted radiation paradigms existed. The program's research has demonstrated that some of these paradigms may need to be challenged. For example, it was assumed that the cell was the important biological unit for determining radiation response, and that energy deposited in an individual cell was responsible for the biological effect observed in that cell. This was called the "hit" theory. Results from recent research using microbeams developed at PNNL and other techniques developed by the program have demonstrated that this is not the case, and that "bystander" cells, which have no energy deposited in them, also can respond to radiation exposure with a wide variety of changes. Some of these biological changes seem to be damaging, while others are protective.

Another widely held paradigm was that cells act independently of each other and that cancer risk could be evaluated based on single-cell responses. It has now been well established that extensive cell-cell, cell-matrix, and cell-tissue interactions determine the outcome of any radiation exposure and demonstrate that it takes a tissue to produce a tumor. Cells respond very differently in complex tissues than they do in tissue culture, and these complex responses have been evaluated by the program and must be considered in risk estimates.

Another paradigm was that a single cell "hit" by radiation could produce a single rare event, such as a mutation, that was inherited by the daughter cells and was the most important change during the induction of radiation-induced cancer. Research has demonstrated that in addition to causing mutations, 1) ionizing radiation changes the gene expression in many genes; 2) gene expression is altered as a function of radiation dose, with unique low-dose and high-dose genes identified; and 3) changes in gene expression can change the fate of the cells in terms of many biological endpoints. This suggests that very different mechanisms of action are involved following exposure to low doses of radiation than those activated by high doses. This further suggests that it may not be possible to extrapolate radiation effects linearly from high doses to low doses, as is currently done.

A final paradigm was that a mutated cell passed on its mutation to each cell in subsequent generations and that this was the basis for the induction of cancer from radiation. It has been demonstrated that, in addition to mutations in individual cells, radiation can also produce genomic instability in cell populations, and this instability can be seen only after many cell divisions. The role of genomic instability in radiation-induced cancer, especially at low doses, is a major paradigm change and a major focus of research in the program.

It is important to know if the dose-response relationships are linear in the low-dose region or if the response is less or more than that predicted by a linear dose-response curve in the low-dose region. Thus, the shape of the dose-response relationships following low doses remains one of the most important questions in the field of radiation biology. Before the program was started, it was demonstrated that small radiation doses could decrease the response of cells to the induction of chromosome aberrations and mutations produced by a subsequent high dose of radiation. This was called the "adaptive" response. Research conducted by the program has demonstrated that adaptive responses can occur in many animal, tissue, cell, and molecular systems at low doses and not only modify the response to subsequent high doses but also decrease the background frequency of the endpoints of interest. This may suggest the possibility of a radiation-induced decrease in cancer risk in the low-dose region. The mechanisms involved in these and other responses are being carefully researched and are reviewed in this book.

As new research in the program was conducted, it became obvious that new radiation paradigms are needed to describe the response of biological systems to low doses of radiation. This research has moved beyond simple descriptions of the new phenomena of "bystander effects," "adaptive responses," "changes in gene and protein expression," and "genomic instability." It has developed the basis for a mechanistic understanding of the interaction of radiation with complex biological systems. Only with new research that helps develop this mechanistic understanding will it be possible to develop standards based more on a systems approach that considers all the biology involved in radiation-induced changes. It is hoped that this research will provide a solid scientific basis for adequate and appropriate radiation standards.

CHAPTER 1

Life and Times of a Radiation Biologist

We were out in the cool early morning, hoeing and watering the new corn before the heat of day set into the desert of southwest Utah. I enjoyed watching the water trickle down the rows, bringing life to this climate where little life was possible without irrigation. Like our forefathers, we did much of the hard work in the very early morning. For a short second, the sky lit up to the brightness of midday then returned to the pale light of the nearing dawn. "Quick," Dad said, "count and you can tell how far away the atomic bomb was on Frenchman Flat." The earth quivered as the shock wave rolled across the sleepy southern Utah town of St. George. We knew the speed of light and how fast the shock and sound waves traveled. I enjoyed such exercises and was pleased to announce that the bomb was less than a hundred miles away. We had just experienced another of the 103 above-ground nuclear weapons tests at the Nevada Test Site. This was an exciting experience, a memory engraved into my young brain.

The close interactions with the atomic era were an integral part of our lives as I grew up. While in junior high, a small truck with loudspeakers mounted on the top of it would drive around town and tell us to remember to come to the basketball game, the dance, or whatever event was planned for our small community. One day in 1953 while out playing basketball in the street, the truck came around announcing, "We would like everyone to go into their house, since there is a fallout cloud from the nuclear test coming across the town." This was the fallout from "Dirty Harry," a shot where the fallout did not go where they wanted it to go. However, being rather involved in the game and using the logic of youth, we determined that if the fallout was in the air, and air was both inside and outside of the house, it would do little good to go inside. In addition, my team was only down by two, and recess was almost over. Our level of concern for radiation and fallout was about zero. We knew the government would not do anything to hurt us since we said the pledge of allegiance every day in school. This was America.

I have lived and worked through some very interesting times in radiation biology.

1. Growing Up in St. George, Utah's Dixie

I was born on July 6, 1938, in St. George in southwest Utah and the northeast corner of the Mojave Desert. Dad was the sheriff and later the postmaster. Mom taught English at Dixie Junior College, named for the "Dixie" Cotton Mission where early Mormons had grown cotton and sugar cane during the Civil War. With a master's degree from Columbia University, Mom could be considered one of the first liberated women in our town. She was an expert on Utah history and wrote a number of books, including *The Mountain Meadows Massacre*, a study of the 1857 incident where Utah Territorial Militiamen attacked a California-bound wagon train and killed about 120 settlers. This book was not popular with the Mormon church, but it was history. She often said, "Only the truth was good enough for the church I belong to." I have always tried to follow that philosophy in science.

My concern about fallout from the nuclear tests really started when I was fifteen, working in a service station. Some "government people" came in and asked us to wash all the cars that had just passed through the Nevada desert. The cars had been in a fallout cloud and needed to be cleaned at government expense. If this fallout was bad enough that they did not want it on the cars, why should we wash it all off and concentrate it in the sump of the service station where I worked every day? We washed a lot of cars that day. I asked the "government man" about concentrating the fallout and he replied that there was not enough on the cars to be a health concern, and if we washed it off, it would not be in the air for the travelers to breathe.

Around the time I graduated from Dixie Junior College, its president was found to have leukemia, and the word was out that the atomic bomb fallout had caused the disease. There was, however, little acceptance in my own mind that the fallout had caused anything. My mother had taught me to be a "free thinker," and suggested that we really didn't know if the fallout had caused the disease. I was well and healthy. Everyone that I knew was well and healthy. I think that her philosophy became part of me as I realized that there were many different opinions on scientific topics. In science the truth is based on data and how it is interpreted. This, of course, changes with time. Each scientist needs to know and understand the data and make a call as to what the data means. What would the "truth" be about the health effects of radiation?

2. University of Utah

From Dixie Junior College I went to the University of Utah in Salt Lake City, at first majoring in education before graduating with a BS in

Physical Biology and a minor in Ecology. Since I was not going to teach, I determined I would have to go on and get a master's degree. As I started graduate school, I found that Dr. Robert C. Pendleton had just received a grant from the Atomic Energy Commission (AEC) to study the movement of fallout from the environment through the food chain with the ultimate deposition in people.

Each week we would go on a field trip in a different direction to obtain milk, grain, grass, hay, water, and soil samples from dairy farms. The isotopes of primary interest in these studies were plutonium-239 (^{239}Pu), iodine-131 (^{131}I), strontium-90 (^{90}Sr), cerium-144 (^{144}Ce), and cesium-137 (^{137}Cs). During this same timeframe, the University of Utah was also asked by the game department to help them when there was crop damage caused by deer. We shot deer, took samples, and determined how much radioactive material from the environment was being incorporated into the animals. Dr. Pendleton was very concerned about the transport of some of the isotopes through the food chain. For some radionuclides, like ^{137}Cs, there was a higher concentration in the deer than was found in their food. As predators, like a mountain lion, ate the deer, the amount of ^{137}Cs was further concentrated. Thus, there were tropic level increases in the amount of ^{137}Cs as the radionuclide moved up the food chain.

Of course, as a hungry graduate student eating whatever I could get my hands on, I turned into the top predator on the food chain. After taking the deer samples—liver, bone, paunch content, lungs, etc.—for the studies on the concentration of radioactivity in the animals, I would get out my knife and with a few carefully placed cuts remove the loin and put it in my pocket for home use. Right good eating! The farmers were also very generous with me, often saying, "Here, take a gallon of milk home." As a poor graduate student I never turned down food or any other kindness.

There were unnumbered adventures on these field trips chasing fallout which are still an important part of my memory. Once, my fellow student Bruce Church and I took a string of pack horses into the Uinta Mountains to gather samples that were contaminated with fallout. We were in the mountains for over a week, catching fish, shooting deer, getting soil samples, taking all kind of vegetation, etc. We came back out of the mountains with two very big ice chests full of fish. Our plan was to take them back to the University of Utah where we could determine the amount of radioactive fallout in them. A game department man saw the ice chests and thought he had a real set of poachers and went to check out the fish. I hollered at him, "Stay away from those fish. They are all radioactive." He jumped back and retreated to his car. We all got a good laugh out of it. Years later at a hearing in Vernal, Utah, on the potential for radiation from the Nevada fallout to induce cancer in humans, a game department employee testified that a representative of the government had told him that all the fish in the Uinta Mountains were radioactive. Bruce was at the court hearing and called to me to see if I knew who could possibly say such a stupid thing. That would be me! I learned you must be careful what you say and who you say it to.

At about this time I met Janet Lorene Poulsen and fell in love. Since I really did not have any money, our dates included visiting the library to study, Saturday night dances (which were free), or car rides. We were soon married in the Los Angeles Temple on July 24, 1962, not three weeks after the atomic test Sedan put fallout clouds over Salt Lake City. Sedan was part of a program to test peaceful use of nuclear bombs for mining, canal digging, etc., and it left the biggest manmade crater in America and accounted for the second-most human exposure of any U.S. nuclear test. Pendleton and his students worked overtime, and when Janet and I got back from our wedding and honeymoon, driving from California to Utah, I chased fallout for twenty of our first thirty days of marriage.

It was a unique experience as a scientist to start my career in the natural environment chasing fallout from nuclear bombs. The fallout was on everything, in everything, and exposed everyone in its path to ionizing radiation. During the time I was doing radiation ecology and finding radioactive material everywhere, I developed a serious fear of radiation. Fueled by the news media, the film industry, and even many scientists, this same fear still exists today in many people.

As part of my master's thesis I convinced some of the farmers living on the farms where we were testing for fallout to come into the University of Utah for whole body counts. Their participation was very useful since it demonstrated that the farmers were taking in more ^{137}Cs than they were able to clear from their bodies. This resulted in an increase in the level of ^{137}Cs in the farmers over the eight months that we counted them, triggering the question "was there any way to limit the intake of radioactive material into the human population?" This question continues today with the advent of nuclear accidents. I recently prepared a paper to compare the rate of elimination of ^{137}Cs from the farmers in Utah with that following the Fukushima nuclear accident (Brooks et al 2016).

My concern for fallout helped me determine that I needed to get in the counter and have a whole body count to see how much fallout I had in my body. Figure 2 shows some of these results. My love for venison and the large volume of fresh milk from the "hot farms" provided a direct pathway for ^{137}Cs into my body. As you can see on the graph, men have a higher concentration of ^{137}Cs than women and I (ALB) had one of the highest ^{137}Cs contents of anyone in my study. We found that the volume of milk consumed was directly related to the concentration of ^{137}Cs in people, and other sources of intake also influenced the body burden. Was this level of ^{137}Cs a hazard? Because of this, was I destined to die from cancer or some other terrible radiation-induced disease? My concern for potential adverse health effects to me, my expecting wife, my family, and other residents of Utah continued to increase. This, along with other questions that came up often in my research, prompted me to continue my education and invest my life in studies to try to determine the health risk from internally deposited radioactive materials.

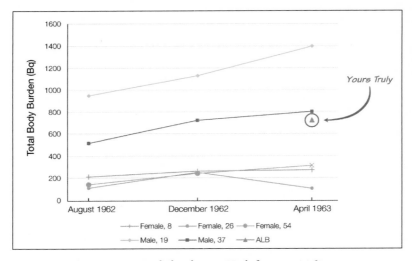

Figure 2. Body burdens in Utah farmers, 1962

Nine months and one day after our marriage, Janet and I had our first son, Mark Leavitt Brooks, a good, healthy, busy baby that could never take a nap.

Working with Dr. Pendleton, I studied fallout for a little over two years. This resulted in a master of science degree with a thesis titled *Comparative Radiation Ecology of Six Utah Dairy Farms*. Life as a radiation ecologist had been good. I liked camping, the outdoors, and the research. I could have stayed at Utah and continued to study the field of radiation ecology for my PhD, but I felt that it was time for me to move on and address my real concern. Do low levels of radiation from internally deposited radioactive material increase the risk for cancer? I needed to conduct laboratory-based research on the health effects of radiation where I could control the variables. Only through this type of science could I develop an understanding of the "true" health impact of fallout and low dose radiation. I would be able to conduct studies at all levels of biological organization—molecular, cellular, tissue, organ, and whole animal—and carefully document the responses of each.

3. Cornell University

I applied to a number of different graduate schools, but Cornell University offered the best opportunity: the most money! It was a tax free grant and more money than I had ever made working in the service station. More money was extremely important to a poor boy. Cornell best fit my ultimate goal of understanding the health effects of radiation.

Cornell had a world class radiation biology program with Dr. Cyril Comar as the head of the department, known for his study of ^{90}Sr uptake, distribution, and dose; Dr. Alison P. Casarett, a radiation biologist, had just written Radiation Biology, one of the first books on the subject; and Dr. Fred Lengermann, an expert in metabolism of radioactive materials and radiation biology. Dr. Lengermann agreed to be my major professor, and I took my radiation biology classes from Dr. Casarett. I was able to move out of radiation ecology, with its multiple and uncontrollable variables, and go full bore into the laboratory to study the health effects of radiation. During that first year at Cornell, we had our second son, Kriston Poulsen Brooks, born on November 20, 1964.

I determined that my studies for my PhD thesis would focus on the induction of radiation induced chromosome breakage and rearrangements in somatic cells (bone marrow) and genetic cells (spermatogonia from the testes). The next task was to select the proper biological system to study. I wanted to work on animals and not cells in culture. In the 1960s, no human or animal epithelial cell lines had been developed. All tissue culture was done with fetal calf serum in the media, and the only cells that would grow in this environment were fibroblasts, the most common cells of connective tissue in animals. Since most cancers occur in epithelial cells, not fibroblasts, I did not, and still do not, think that results derived in the wrong cell lines are meaningful to evaluate radiation-induced cancer.

At this time radiation risk was understood to be equally split between the induction of mutations and cancer. As I studied the literature, there were a number of papers on the induction of chromosome aberrations in the Chinese hamster (Bender and Gouch 1961, Brewen 1962, Brewen 1963), which had only twenty-two chromosomes compared to forty six chromosomes in humans, forty in mice, and forty-two in rats. The morphology of the Chinese hamster chromosomes was well defined, making it easy to determine if each chromosome was normal or had been altered by radiation exposure. As I continued to work with this animal, it turned out to be very attractive as an experimental subject. Since it was not inbred, the Chinese hamster developed a wide range of different tumor types, similar to the human (Benjamin and Brooks 1977). Later I discovered that the retention of ^{239}Pu and cesium-144-praseodymium-144, both of which clear rapidly from rats and mice, were retained in the liver of the hamster with a similar half-life as seen in humans (Brooks et al. 1974). It would be essential to develop techniques that would make it possible to measure the induction and repair of radiation-induced chromosome aberrations in both somatic cells (bone marrow) and reproductive cells (spermatogonia in the testes). Thus, my task was set and the direction of my research established. My PhD thesis (1963–1967) was to determine the dose, dose-rate, and time relationships between the frequency of radiation induced chromatid type aberrations in the bone marrow, a target for cancer induction, and the mitotic spermatogonial cells in the testes, the target for genetic damage in the next generation.

I used a Cobalt-60 (^{60}Co) gamma source to expose the animals to graded levels of radiation and sacrificed the animals at different times after the radiation exposure. I isolated cells from both the bone marrow and testes, prepared them for chromosome scoring, and recorded the frequency and type of chromosome damage induced by the radiation in each tissue. Thus, I generated chromosome data as a function of tissue type, dose, and time after the exposure. This proved to be a useful experimental design, since dose, dose-rate, and time are all essential variables to control in any radiation experiment. My research resulted in publications in radiation-related journals (Brooks and Lengermann 1967, Brooks and Lengermann 1968).

For radiation-induced chromosome aberrations, the frequency, distribution, and types of aberrations induced were known and changed as well-described functions of the dose. When staring down the microscope to study radiation induced chromosome aberrations, it was necessary to first determine the number of chromosomes present. This made it possible to know if you were scoring a whole cell or a cell that was broken during chromosome preparation. Next, each chromosome had to be examined to determine if it was normal or whether it had been broken or if it had attached to either itself or another chromosome to form an exchange, which in some cases would leave fragments. The frequency and types of abnormalities induced by radiation exposure were then recorded for each cell. I invested many, many hours staring down the microscope.

Extensive research demonstrated that there is no increased risk of mutations in offspring of irradiated parents. By twenty-four hours after the radiation, the aberration frequency was almost back to the background level. This hallmark study indicated a repair or loss of radiation damage that was almost complete by one day after radiation exposure.

By conducting these studies in the whole animals (*in vivo*), and not just in tissue culture (*in vitro*), it was possible to make these useful observations and help understand the risk for the induction of chromosome aberrations. With the *in vivo* techniques for study of chromosome damage in place for single acute exposures, showing the amount of damage they produced and the rapidity with which it was repaired, I felt prepared to tackle my real lifetime goal: to study the effects of low dose and dose-rate exposure from fallout. The radiation exposure from fallout gives its dose over months and years, which is a long period of time compared to the research that I was conducting.

My fear of radiation was still alive, but I now had tools in hand that would be essential as I moved on from graduate school to the real world of science.

4. Lovelace Inhalation Toxicology Research Institute

After I got my PhD in 1967, my next task was to find a real job. Dr. Roger O. McClellan had just left the U.S. Atomic Energy Commission and been appointed to head up a young inhalation toxicology laboratory in Albuquerque, New Mexico. At this time, the laboratory at the Lovelace Foundation for Medical Education and Research was focused on health effects of fallout.

Traveling to New Mexico for my interview, I liked Albuquerque's desert environment and the wide open spaces around the laboratory. I was also impressed with the research being conducted. They had an active research program on the health effects of internally deposited radioactive material. This was a perfect fit. I was offered a post-doctoral position in the Lovelace Fission Product Inhalation Program. I was pleased with the job and the offer, and excited to make the move.

A short time after I got to Albuquerque, the name of the institute changed to the Inhalation Toxicology Research Institute (ITRI). We were a young and vigorous group of investigators that got along well. Dr. McClellan was very good at getting money for us to do our research, and our job was to do carefully planned and executed science.

Dr. McClellan was an exceptional science manager. As the cell toxicology group leader and project coordinator of several research projects, I was high enough in the chain of command to observe his workings, feel the heat of his wrath, and bask in the pleasure of his praise.

At performance reviews Dr. McClellan often asked two questions: The first was "What have you done for ITRI during the past year?" The easy response was an iteration of scientific papers published, meetings attended, presentations given, grants received, etc. However, the second question was the tough one. "What have you done to increase your value to the Institute?" This really made me think. What have I done? It stimulated me to take classes in cell and molecular biology, a newly developing field. It was a good question, one I tried to ask each year during my self-appraisal.

My first research at Lovelace was to address concern that internally deposited ^{90}Sr, one of the materials I had found in milk in Utah, was more hazardous than chronic external radiation exposure. I remember my shock when I determined I would have to inject Chinese hamsters with up to five microcuries (µCi) per gram body weight to result in radiation doses that were capable of producing measurable increases in chromosome aberrations in the bone marrow. Wow! This was about one billion times higher than the concentrations of ^{90}Sr we were finding in the milk from fallout in southern Utah (5 picocuries [pCi] per liter). This was the first time I really related the amount of activity in the fallout to the amount required to produce a health effect. There was a huge activity and dose difference.

What I concluded from my studies of ^{90}Sr was that there were no unique biological changes or potential effect from this internally deposited radionuclide. In fact we determined the bone is a very radiation resistant organ, and the frequency of cancer in the bone increased only after very large doses. The regulation and control of external exposure was appropriate and adequate for the control of exposure to ^{90}Sr. Perhaps I should not have been so afraid of the ^{90}Sr in the milk. With the help of Dr. McClellan, this research resulted in a paper in *Nature* (Brooks and McClellan 1968). This publication was a real feather in the cap of a young scientist and helped move me along my scientific journey.

During this time, March 20, 1968, we had our third child (Jana Lee Brooks), a beautiful, easy baby.

In addition to the fission products in the fallout produced by the bombs, there was some ^{239}Pu released into the environment because of incomplete fission. At this time it was suggested that since this radionuclide was an alpha emitter and was taken up and retained for long periods of time in bone and liver, it was "the most hazardous substance known to man."

Liver cells divide infrequently, and several of the radionuclides of interest from fallout or nuclear power (^{144}Ce, ^{239}Pu, Americium-241 [^{241}Am]) concentrate in the liver and deliver large doses over long periods of time. Using the Chinese hamster liver as a model system, it would be possible to measure both liver cancer and chromosome aberrations in the same species to help determine the relative hazard from ^{239}Pu. We conducted multiple studies using the different internally deposited radioactive materials listed: californium-252 (^{252}Cf), ^{239}Pu, ^{241}Am, ^{144}Ce-^{144}Pr and tritium (^{3}H). The responses to the internally deposited radioactive material could be directly compared to external exposures given as a single short dose or as protracted chronic exposure to gamma rays.

These studies generated some highly important data and resulted in a number of open literature publications (Brooks et al., 1974). They were summarized in an article published in Science. The data made it possible to determine the Relative Biological Effectiveness (RBE) of internally deposited alpha and beta particles in producing chromosome damage in the liver relative to external exposure to ^{60}Co gamma rays, delivered at both high and low dose-rates (Brooks 1975).

There were a number of important scientific discoveries from these studies. First, ^{239}Pu was not "the most hazardous substance known to man," but had the same effectiveness in producing chromosome damage as any other alpha emitter. Next, in the same system, alpha particles were twenty times as effective in producing chromosome damage as beta-gamma emitting radiation. Finally, at low dose-rate, the biological response was the same whether the dose was delivered from an external source or from internally deposited radioactive material. Thus, there was no unique hazard from the internally deposited radioactive material.

The "Hot Particle Hypothesis" suggested that a single ^{239}PuO$_2$ particle deposited in the lung would result in a very high dose to cells that were near the particle and would be very effective in causing lung cancer. It assumed that the response to produce cancer was a single cell event and that the cells that had received the high dose would then progress to produce the cancer. At the time this was a credible hypothesis with a huge impact on nuclear safety. If true, current radiation standards for ^{239}Pu were too low by a factor of more than ten thousand.

Because this was really a question about non-uniform dose distribution vs. the lung specifically, we were able to develop the Chinese hamster liver as a model system to relate chromosome aberrations and cancer produced by uniform radiation vs. "hot particles." I approached Dr. Otto Raabe, the head of our aerosol science department, to have him make me ^{239}PuO$_2$ particles of known size and activity. If I could have such particles, it would be possible to hold the total dose constant and vary the local dose by either injecting a small number of large particles or a lot of small particles into animals. He looked at me and said something like, "Brooks you come in here and act like you are ordering a loaf of bread. This is not an easy or simple request!" I replied that I knew he was clever enough to do it.

Otto and a number of other scientists at ITRI played important roles in the development of an aerosol particle separator. This invention made it possible to isolate mono-dispersed plutonium particles with well-defined sizes and activity. I injected known amounts of radioactive ^{239}PuO$_2$ particles of different sizes into the jugular sinus of the Chinese hamster. The liver was the first organ that the particles came in contact with and retained over ninety percent of the particles for extended periods of time. Ionic plutonium citrate, which also deposited in the liver and resulted in uniform exposure of liver cells, served as a positive control. Thus, we could have the same total activity and dose in the liver with very different particle sizes and micro-energy distributions. The large particles resulted in less than 1% of the liver cells "hit" by an alpha particle, with 99% of the liver cells not being traversed by an alpha particle. The same total activity distributed in the small particles would "hit" all the cells. This protocol made it possible to test the Hot Particle Hypothesis and determine if a constant level of activity from large particles resulted in a higher hazard than small particles or a uniform dose distribution from plutonium citrate.

We studied the frequency of chromosome aberrations and the induction of cancer in the liver as endpoints to define the role of dose-distribution on biological damage. If the Hot Particle Hypothesis was confirmed, the animals with the large particles would have a much higher cancer frequency than the animals with the small particles. We found that regardless of the micro-distribution of the energy in the liver or the form of the plutonium, the same number of chromosome aberrations were present per cell per unit of radiation dose, and the incidence of cancer was the same. Thus, the energy distribution did not influence the number of chromosome aberrations or cancer produced.

The observation that the particles and the citrate produced similar amounts of chromosome damage and cancer was not what I expected. The data only made sense if the total dose to the organ was responsible for the chromosome aberrations and cancer incidence, not the localized dose. In retrospect, this seemed to be an example of the "bystander" effect in the 1970s before it was formally defined using alpha sources (Nagasawa and Little 1992) and microbeams (Geard et al. 2002). This result provided strong evidence that the hot particle hypothesis was not supported by research. This again reduced my fear of hot particles.

On April 25, 1974, son Andrew Hafen Brooks joined our family in Albuquerque. Andrew was my most creative little person, always wearing a wild outfit and keeping us constantly entertained.

During this same time I was able to get acquainted with Arland "Red" Carsten at Brookhaven National Laboratory. He had a large colony of Hale-Stoner-Brookhaven mice that were maintained on tritiated water (HTO) at 3 billion pCi/l (110,000 Bq/l), more than 100,000 times higher than the EPA's drinking water maximum of 20,000 pCi/l (740 Bq/l). I arranged to study the frequency of chromosome aberrations in the liver using a subpopulation of these animals. In my study we sacrificed the animals after 90, 330, 500, and 700 days of drinking the HTO water. We studied the frequency of chromosome aberrations in their liver and determined that the response was not significantly different from that predicted from chronic exposure to ^{60}Co. The data suggested that per unit dose HTO was about 1.5–2.0 times as effective as chronic gamma ray exposures (Brooks et al. 1976). This experience helped me to understand the relationship between biological damage and regulatory levels of tritium allowed in the drinking water. The conservative nature of the regulations for tritium include a large safety factor (almost a million) that exists between levels of tritium required to produce chromosome damage and the amount allowed in the water for human consumption (740 Bq/liter) (Brooks et al. 2013).

5. Washington, DC, and DOE Headquarters (1976–1978)

Dr. McClellan suggested that it would be worthwhile for me to go on assignment to Energy Research and Development Administration (formerly AEC, soon to be DOE) headquarters to learn about the funding of research and develop a better feeling for the interactions between science and regulations. In December of 1976, my family and I moved to Washington, DC. I was a technical representative in charge of reviewing and providing scientific guidance on the radiation studies on animals, mostly beagles with internally deposited radioactive materials. I also was involved in reviewing all the radiation biology programs being conducted at that time.

The dog studies were a very well-coordinated program with open communication and regular meetings to review progress and compare notes. At the time I started with the program it was funded at a level of about twenty million dollars per year. In the 1970s, this was an especially large research investment directed to address real health concerns related to the release of radioactive materials into the environment from fallout and nuclear energy production. The studies were focused on several well-defined questions. For example, was acute radiation exposure more hazardous than chronic low dose-rate exposure? Was internally deposited radioactive material more hazardous than external radiation exposure? Did the dose-distribution within and between organs influence the cancer risk? What was the dose-response relationship between the level of radioactive materials and the causes of death (cancer and non-cancer)? These large animal studies were conducted at a number of universities, two national laboratories, and the specialty laboratory, Lovelace Respiratory Research Institute, where I worked. Each of these had a well-defined mission. The objective of these studies was to give a ranges of doses where the highest doses would produce cancer from the internally deposited radioactive materials. The experimental designs of these studies was outlined in a book by Dr. Roy Thompson (1989), and the history of the results was compiled in a huge work by Dr. Newell Stannard (Stannard and Baalman 1988).

My experience at DOE headquarters was both educational and interesting. It provided me with a great overview of the research being conducted during the late 1970s, and how science and regulations interact. I had the opportunity to visit and review each of the laboratories where radiation biology research was being conducted. This included not only the dog-study laboratories, but also all the national laboratories with research units in basic radiation biology: Brookhaven National Laboratory, Oak Ridge National Laboratory, Lawrence Livermore National Laboratory, Lawrence Berkley National Laboratory, and Los Alamos National Laboratory (LANL).

The distance between the public and the scientific community was a lesson I needed to learn. We got a Freedom of Information Act Request from an activist requesting all the information we had on the health effects of ionizing radiation. Such letters generally get kicked down the chain of command, and being on the bottom of the food chain, this one landed on my desk for a response. It was obvious that the person or persons asking for the information had no idea what they had requested. There were, even at that time, literally tons of information. I went to my boss and said, "Why don't we do what they have asked? We could get several big dump trucks, fill them with books and scientific publications and dump them on the front lawn of their house." He just looked at me and said, "Brooks, go and write up a two-page response." This is one of the problems that we have in communicating the health risks of radiation with the public. The public doubts that the scientific community understands the risks and biological responses to radiation. In reality, we really know more about the health impacts of ionizing radiation than any other substance in the environment. We have data on the response to radiation all the way from humans to the most basic molecular studies, which we have used to establish the radiation standards and limit the exposures.

There is always competition for funds for any scientific project. Since the "dog projects" were very expensive, about twenty million dollars per year, they represented the biggest single chunk of the DOE budget for research on radiation effects. However, since there was no useful human data on the health effects of many of the radionuclides, the routes of uptake, the dose, dose-rate, and dose distribution, the dog data was the only data available to set standards in humans and control uptake and radiation dose to humans from radioactivity in the environment. We were able to make the case that it was important to complete the studies that were underway and finish them with publications that would summarize this valuable information.

While working with the DOE in the late 1970s, we welcomed our second daughter. Lara Kay Brooks was born July 14, 1977, in Maryland.

Toward the end of my stay at DOE, I was offered a "full-time-job" in Washington, DC, and was tempted to stay on. We enjoyed the area, the work, and the ability to be involved in providing direction and advice to the research community, but I thought I was a better scientist than I was a bureaucrat. I decided to get back to the laboratory bench and fire up my research efforts again.

6. Back to New Mexico and Lovelace ITRI

Back in New Mexico, I was the head of the cell and molecular science group. Using techniques and methods developed to study radiation effects, we invested a couple of years doing research on the health effects of energy production from fossil fuels.

Soon our sixth and last child, Evan Swainston Brooks, was born, arriving on April 20, 1980. A happy baby, he was always active and very involved in everything that we were doing.

My years at ITRI helped me gain a good respect for radiation, and an understanding of how much contamination was required to produce adverse health effects from this and other sources of environmental pollution. This experience decreased my fear of environmental radiation exposure. It was obvious to me that compared to a large number of environmental contaminants we studied, radiation is not a strong mutagen or carcinogen.

7. Pacific Northwest National Laboratory

In 1989 I was offered a job at Pacific Northwest National Laboratory (PNNL) in Richland, Washington, by Dr. William J. Bair. I started work on July 24, 1989, Janet's and my 27th wedding anniversary.

PNNL and the Northwest was a good fit for me. The laboratory was located by the Columbia, Snake, and Yakima Rivers, a good environment for wildlife of all kinds, and as an avid outdoorsman it was a wonderful location for my family and for me.

The director of the cell and molecular biology at PNNL was Dr. Marvin Frazier. Within about a year after I started at PNNL, Dr. Frazier went to DOE headquarters in Washington, DC, to do the same job I had done during my stay there. We had a number of good discussions on what it was like to work at DOE. As seen later in the book, Dr. Frazier was instrumental in the development and initiation of the DOE Low Dose Radiation Research Program.

At PNNL they had an extensive research program on health effects of radiation, which was a good fit for my background and skills. Here I continued research on interaction between internally deposited radioactive materials and external exposure. That work went well. I had a good group of dedicated technicians and scientists, and moved into a major effort to conduct research on the health effects of radon. Radon was the major contributor to environmental radiation exposure. Much of this research focus was driven by Dr. David Smith. I remember a conversation I had with Dr. Smith, where he asked me what problems remain in the field of radiation biology. My response was, "The lack of information on the potential health effects on inhaled radon." I do not know if this or other conversations on this subject had any impact, but DOE started a major research program on the health effects of radon in homes.

One day I was talking to my colleague, Dr. Fred T. Cross, about all the studies he had conducted on the lung cancer production associated with inhalation of radon and its daughter products. As I looked at the data on lung cancer in rats, I noted that there were a lot of cancers associated with the airways of the respiratory tract and the deep lung, but there were no cancers in the trachea. Since all the radon and its daughters had to pass down the trachea to get to the deep lung, these radioactive materials should deposit energy from the alpha particles in this region. I asked Dr. Cross, "Why were there no cancers in the trachea?" He told me that they probably were not getting any radiation dose. So being a simple biologist, I determined to measure the amount of chromosome damage in different regions of the respiratory tract to determine if they were getting the same dose. This simple decision led me to several years of radon research.

Because of this research, I was selected to be on the National Academy of Sciences (NAS) committee to evaluate the health effects of radon and to produce an extensive report. Helping produce this report, "The Health Effects of Radon Exposure" BEIR VI, was a useful experience for me. It helped me understand not only the scientific impact of these reports but also the importance and impact of communicating the results to the public. Public perception of radiation risk is communicated by one-liners, since it often seems that no one ever reads the report. After working very hard to produce a very well-balanced scientific report, the only thing the public saw was a one-liner: "Radon in homes is the second leading cause of lung cancer." The report was written by scientists, and the summary and one-liners were written by a communications specialist. Other good science in the report didn't see the light of day or reach the public. For example, the report showed that most (70%) of the radiation dose used to calculate the radon risk was from homes below the EPA action limit of 4.0 pCi/liter or 150 Bq/m^3. Thus, remediation of homes to remove radon would only change the sum of the radon dose by a small amount and have little impact on the "calculated" total lung cancer risk. The general public was seemingly oblivious to the fact that most of the "calculated" radon risk was from smokers and the interaction between radon and cigarette smoke as detailed in the report.

My research on radon and my participation in the BEIR VI committee were major factors in my being asked to testify before a congressional committee on the risk of radon in homes. This committee was headed up by Representative Waxman of California to determine what role EPA should take in controlling and removing radon in homes. Again, this was an interesting experience on how politicians view science in making policy that influences us all. The EPA was pushing for the regulation of radon, and environmental groups were there to show how hazardous radon or any radiation was. Also interested were the building trades that were concerned about the costs of remediation, and scientists were there to provide some basic data on which they could base decisions.

When it was my turn to speak to the committee, Representative Waxman asked me something like, "Dr. Brooks we bring you in here to tell us about the science associated with the risk from radon in our homes. I really don't want to hear much about that. I have a wife and four children, that I dearly love, and I want to know if I found radon in my home at the level recommended by EPA for remediation what would you recommend that I do?"

"I have a wife and six children," I responded.

He asked, "Do you dearly love them?"

"Well," I said, "Three of them are teenagers."

This got a good laugh out of the committee. I went on to explain that the risk from radon at the EPA action level is so low that it is similar to many risks I accept easily. I would not do anything to change the radon level if it was only at the action level. I would be sure to stop smoking if I was a smoker, and if I was not a smoker, I would put a fan in my basement and leave a window open. Only if the level of radon was much higher than the action level, perhaps ten times, would I take any further action.

Ultimately, the committee took the most conservative approach and decided to set regulations protecting the public against the smallest doses, which research suggested produced insignificant biological change.

With radon or other alpha emitters, it was not possible to study individual cells and their responses to exposure. The microbeam, developed at PNNL, made it possible to place an alpha "hit" on a specific cell. Using biology we could then determine the response of the cell hit as well as its neighbors. It was determined that not only hit cells but their neighbors were responding to the radiation exposure—the "bystander effect." As we saw during my plutonium research at Lovelace, the whole organ was responding, not individual cells. This research opened up a whole new area of biology on cell-cell and cell-matrix communication.

8. Move to Washington State University

Quite abruptly, the research funding for radon and microbeam research both vanished. For the next year I invested my evenings and weekends writing proposals for funding. I was lucky, good, and blessed since I got a proposal funded by the National Aeronautics and Space Administration (NASA) on the effects of space radiation (high Z particles) on cell killing and chromosome damage in the rat. In space there are heavy ions (high Z particles) like iron traveling at close to the speed of light that have very high energy and are damaging as they hit biological tissues. Soon after this, an additional proposal was funded by the National Institutes of Health (NIH) to study basic mechanisms of chromosome aberration production using chromosome painting techniques which were just coming into the mainstream of cytogenetics.

With these two funded proposals in hand, I left PNNL and walked across the street to Washington State University Tri-Cities. I had my own funding, so they were happy to hire me. I could again do something that I felt I was good at. This put me back in the laboratory and at the microscope.

I was involved conducting research with a good new technician, Lezlie Couch, a new laboratory, and adequate funding. Lezlie worked with me as a colleague for the next fifteen plus years and always provided a great deal of scientific input. She was exceptionally bright, and could make a rough draft into a publishable scientific paper in English rather than in the southern Utahan that I spoke and wrote. My recognition later in life was in a large part due to her excellent work.

9. DOE Low Dose Research Program

In 1998 the DOE Low Dose Research Program needed a chief scientist and put out a call for proposals to fill this position. It sounded like exactly what I wanted to do, as I had spent my whole scientific career studying low dose radiation effects. My expertise fit perfectly and I was fortunate enough to get the position. This book documents the history of that program and underscores the need to continue that important work. Particular emphasis is given to the achievements and lessons learned. In 2017 as I worked and was paid by Pacific Northwest National Laboratory to complete the book, the editors determined that I should include part of my history as a forward to the book. This was a tough decision for me since the book is about the research program, not me. My only hope is that this will provide some background and entertainment for the readers. I'm proud to have been a part of the program and honored to have this opportunity to tell its story.

CHAPTER 2

A Brief History of Radiation Biology

In 1901, only five years after radiation was formally discovered, Pierre Curie used a radium source to give his arm a high dose in order to investigate its effects. The exposure produced a burn about the size of a quarter that lasted over a month before healing to a grayish spot, and also generated a fine paper for *l'Académie des sciences*. Over the next decades, researchers developed a partial understanding of the biological effects of radiation, mainly through animal studies. They had a good idea of what many of the acute effects of radiation exposure would be when the United States dropped atomic bombs (A-bombs) on Hiroshima and Nagasaki in 1945. That event was a turning point in many ways, bringing acute public awareness to the issue of potential dangers associated with both high doses from acute radiation and lower chronic doses from fallout. Certainly it was a turning point for radiation biology.

Radiation dose or amount is measured as energy deposited per unit mass of tissue. In the past dose was expressed as rads but now the international units are Gy (gray). One Gy is equal to 100 rads. For reference, an acute whole body exposure of x-rays or gamma rays of about 4.0 Gy (400 rads) results in the death of half the population from acute radiation syndrome within thirty days. The acute radiation syndrome induced from high acute doses of radiation delivered to the A-bomb victims followed a similar pattern to that observed in animal models. However, there was limited information on the long-term effects of radiation, especially as the radiation dose decreased below about 0.5 Gy. There was also very limited information on the deposition, distribution, dose, or health effects from the radioactive materials deposited in the body from A-bomb fallout. These concerns resulted in the development of extensive radiation biology programs in the 1950s at several national laboratories in the United States—Oak Ridge National Laboratory (ORNL), Brookhaven National Laboratory (BNL), Argonne National Laboratory (ANL), Lawrence Livermore National Laboratory (LLNL), Lawrence Berkeley National Laboratory (LBNL), and Pacific Northwest National Laboratory (PNNL)—as well as national laboratories in England, Germany, France, and Japan.

In addition to the national laboratories, extensive research was funded by the Atomic Energy Commission at several specialty laboratories to examine specific problems such as the inhalation of radioactive materials (e.g., Lovelace Inhalation Toxicology Research Institute). Research on the health effects of radiation also began in several U.S. universities. Major universities with long-term radiation research programs involved in this early research included the University of Utah, Cornell University, Colorado State University, Columbia University, the University of Wisconsin–Madison, the University of Rochester, the University of California–Berkeley, Case Western Reserve University, Harvard University, the University of Tennessee, and the University of Texas. Many other universities had smaller research programs. With the initiation of the DOE Low Dose Research Program in 1998, many more universities were involved in research on the effects of radiation.

I do not intend this chapter to cover the details of all of this research or cite specific studies, but to broadly describe what we knew prior to the Low Dose Program. It will outline the extensive research that had been conducted on the impact of radiation on living organisms and the very large data base that existed. Just as important is what we did not yet know—those gaps in the science that led to the need for the Low Dose Radiation Research Program and, along with policy and regulatory goals, spurred its development.

1. Research on the Early Effects of Acute Radiation Exposure

Early research focused on the acute effects of radiation. This research defined the radiation damage to organs and tissues that resulted in early deaths from acute radiation syndrome. Studies were conducted on large numbers of different animal species, and it was determined that the amount of radiation required to kill half of the animals in 30 days ($LD_{50/30}$) depended on dose-rate and the species. The studies determined which organs were responsible for the deaths as a function of dose and time after exposure. This resulted in the definition of the prodromal (or initial) acute radiation syndrome, which was further evaluated to reflect the tissues affected.

At very high doses, radiation destroyed the animals' nervous systems, and death resulted in a matter of hours and days. As the dose decreased, the cause of death became cell death in the gastrointestinal system. As cells lining the gut died, fluid loss, infections, and death occurred in the first few weeks after radiation exposure. At still lower doses, the cells in the blood system were depleted, resulting in death in one to two months. Researchers studied deaths from failure of each of these systems extensively, and they are well defined.

Several *in vivo* tests were also developed to evaluate cell killing in whole animal systems. These different systems showed many similarities among species as well as some interesting differences that helped develop understanding of radiation sensitivity and resistance.

Human data came from A-bomb victims and victims of radiation accidents. Many of the animal studies were predictive of the systems failures induced by the radiation in humans, and thus the type of early damage that resulted in deaths. An interesting observation that came out of the high-dose, acute-effects research was that some animals and people were much more radiation resistant and able to survive these early radiation effects, and others were very sensitive. This was related to their genetic background. The influence of genetic background on radiation-related genetic sensitivity and resistance was one of the major areas of focus for research in the Low Dose Radiation Research Program.

2. Late Effects Induced by Radiation Exposure

Concerns about late effects of radiation, especially cancer and genetic effects, led to extensive research in Japan. That research followed the A-bomb survivors to determine the role of radiation dose in inducing late effects. In addition to the human studies, several carefully controlled animal studies using a range of species were also conducted in which the animals were exposed to graded doses of radiation that did not result in acute lethality. They were then followed for their lifetimes to determine induction of both genetic effects and late effects such as cancer and heart problems. These animal studies supported the human data, helping advance understanding of the mechanisms involved in the development of late effects, and analyzing these endpoints under carefully controlled conditions.

All of these high-dose studies, especially the study of the A-bomb survivors, provide the major information used in calculating health risk effects from radiation exposure. These studies have been extensively published, and they are well summarized in reports from the National Academy of Sciences (NAS), National Council on Radiation Protection (NCRP), the French National Academy (FNA), and the International Council on Radiation Protection (ICRP).

Research was also conducted to determine the health effects of fallout from nuclear weapons and reactor accidents. Early studies focused on radiation ecology, the movement of radioactive materials through the environment. These studies determined that different radionuclides were concentrated or discriminated against as they moved through the food chains. Such studies provided the information needed to determine radiation dose and dose-distribution in humans from the many radionuclides in radioactive fallout.

These studies were important because of the extensive aboveground atomic weapons testing from 1945 to 1962 that resulted in global contamination from fallout. By the time that growing public concern about fallout led the United States and the Soviet Union to sign the Partial Test Ban Treaty in 1963, the United States had conducted just over one hundred aboveground nuclear weapons tests at the Nevada Test Site alone, and the former Soviet Union had tested close to three hundred at sites including the Semipalatinsk Test Site in Kazakhstan. In all, there were over two thousand nuclear tests around the world. The location of the tests, atmospheric or underground, is the first major parameter to consider in estimating the fallout with most underground tests yielding little fallout; the second parameter to consider is the type of test and the yield. Most of the world would be contaminated with these tests.

Thus, nuclear weapons tests resulted in world-wide fallout which increased the background level of radiation. Nuclear tests continued but were conducted underground, which limited the release of radioactive materials to the environment.

Apart from fallout, radiation ecology was also necessary for the study and safe development of nuclear power. The first commercial nuclear power plant came online in 1957, and developments in radiation ecology helped to define the release and the movement of radioactive material through the environment and ultimate deposition and risk in humans.

Researchers at multiple government-funded programs in the United States conducted extensive studies on health effects from internally deposited radionuclides that resulted from the fallout. This research used several different animals, with a focus on the beagle. The dog studies have been summarized in several books discussed in chapter 3 and in multiple proceedings of scientific meetings. Each laboratory that conducted these studies with government funding was required to produce an annual report. Reports from these research laboratories provided a large database: PNNL, ANL, BNL, the University of California–Davis, University of Utah, and Lovelace Inhalation Toxicology Research Institute (ITRI). Extensive research on the retention, distribution, dose, and health effects of internally deposited radioactive materials was also conducted in England, France, Japan, and Germany.

3. Mechanistic Cell and Molecular Studies

3.1 Cell Killing

Beginning in the early 1950s, animal studies were supported by cellular and tissue studies that evaluated the mechanisms involved in induction of these early and late effects. National laboratories, specialty laboratories, and universities throughout the world supported this new avenue of research. The new ability to grow mammalian cells in tissue culture enabled researchers to focus on radiation's role in cell killing. Two basic types of studies were conducted: (1) colony formation assays, or the influence of radiation on the ability of cells to survive, divide, and form colonies; and (2) dye exclusion assays, or the ability of cells to stay alive regardless of their ability to divide. Most of the early studies used colony formation assays.

At this time, the cells used in the culture systems were fibroblasts. By supplementing the media with fetal calf serum, these cells could grow in culture, divide, and form colonies that were easy to measure and provided a direct way to determine the ability of radiation to kill cells, or at least limit their ability to grow and form colonies. However, this cell type was not ideal for understanding cancer, because most cancers arise from epithelial cells. As research progressed, media was developed which made it possible to grow epithelial cells in culture. This resulted in more realistic dose-response relationships for cell killing, chromosome damage, and mutation induction.

Many well-characterized immortalized cell lines were developed during this time that could be used by many laboratories. Despite their limitations, cell survival studies were very useful in defining the influence of radiation type (alpha, beta, gamma, and neutron exposure), exposure characteristics (dose-rate, dose fractionation, or Linear Energy Transfer [LET]), and chemical protection on cell survival. The slopes of the dose-response relationships and the shape of the dose-response curves were defined as a function of all these variables. This provided very useful information in understanding the mechanisms involved in radiation-induced cell deaths following large radiation exposures.

In these *in vitro* studies, the sensitivity to detect changes in the very-low-dose region (<0.1 Gy) was limited, so most of the studies used a dose range of 0.5 Gy up to many Gy, or to the point where most of the cells were killed. The cell-killing curves developed for low-LET radiation showed an apparent plateau in the low-dose region with an exponential decrease in cell survival as the dose increased. On the other hand, exposure to high-LET radiation showed an exponential decrease in cell survival over the whole dose range. Many studies were conducted to determine factors such as dose rate, dose-fractionation, and genetic background of the cells that would influence the shape of the dose-response curves.

The DOE program-funded research played an important role in the development of new techniques to measure cell killing in the low-dose region. These new data will be reviewed in chapters 4 and 5 and the implications for risk discussed in chapter 9.

3.2 Mutation Induction

In the early days of radiation biology research, a primary concern about radiation exposure was the potential for that exposure to increase the mutation frequency. Early research on mutations in *Drosophila* (fruit fly) and other test organisms suggested that there was a linear dose response for the induction of mutations over a wide range of doses and that there was little repair of this radiation-induced genetic damage. This implied that there was a *linear no-threshold* increase in genetic damage as a function of radiation dose, i.e., that no dose threshold exists below which radiation produces less damage per unit; each unit of radiation linearly increases genetic risk. The very early data suggested that genetic damage from radiation would accumulate across generations and eventually have a marked impact on the health of human populations. However, extensive research in mammalian systems showed that there was significant repair of radiation-induced genetic damage, and that the damage was dependent on dose-rate, sex, and many other factors. Surprisingly, the studies on A-bomb survivors and their offspring did not detect radiation-induced genetic effects.

Research at ORNL used mice (mega-mouse studies) to evaluate the induction of mutations in mammals. They irradiated male mice with large acute doses (3.0 Gy), just below the level to induce lethality, let them recover, and gave a second 3.0-Gy dose. They then mated these mice and evaluated the induction of mutations in the offspring in specific genetic loci. The frequency of mutations transmitted to the offspring at these loci could be related to the radiation exposure and dose. Many other studies were conducted on female mice, as well as studies on mice exposed to different radiation types, dose-rates, and dose fractionation. Additional genetic endpoints were used in many of these studies. These studies have provided a valuable data set on which the genetic risks from radiation exposure were based.

Studies were conducted at LANL to determine if the frequency of genetic damage increased after irradiation of many generations of mice. These studies determined there was no buildup of genetic mutations over the many generations exposed and that the reproduction process seemed to limit the transmittal of genetic damage to offspring. They also determined that there were differences in radiation sensitivity between sexes, strains, and dose patterns. The studies also quantitatively measured the repair of genetic damage and helped to explain the lack of buildup of genetic damage following radiation exposure over many generations.

Animal research provided a usable dose-response relationship for the induction of mutations. The data base developed in these studies still provides one of the major inputs for estimating the genetic risk in humans. Because of the low risk of induction of genetic disease, it was concluded that the radiation risk for genetic damage was small relative to the risk for radiation-induced cancer, and most research focused on radiation-related cancer.

Because of the low frequency of mutations detected per unit of radiation dose, it was not possible to make many measurements in the low-dose region. Developments of techniques to measure DNA damage and mutations in the low-dose region have since been developed and will be discussed. The implications of radiation-induced DNA damage, mutations, changes in gene expression, and other genetic alterations will be evaluated extensively in this book.

3.3 Chromosome Aberrations

Another indication that radiation impacts could instigate changes in genetic material was the observation in the late 1930s that radiation causes chromosome breakage and rearrangements in plant cells. Techniques became viable to observe chromosome damage in human cells in the 1950s. These changes were also present in many types of cancer. It was predicted that the frequency and types of chromosome aberrations provided a good measure of cancer risk following radiation exposure. Techniques were developed to culture human blood lymphocytes and to measure the frequency of chromosome aberrations in these cells when they were exposed in either tissue culture or individuals. The response was the same. Chromosome aberrations thus provided the most sensitive biological change that could be used to detect radiation-induced damage. The frequency of chromosome aberrations was carefully related to radiation dose and became a useful biodosimeter. The frequency of chromosome aberrations in blood lymphocytes still remains the gold standard to estimate radiation exposure in human populations exposed in radiation accidents where little measurement or no other type of dosimeter is available.

The literature on radiation-induced chromosome aberrations is extensive and includes measurements made in humans and experimental animal systems. Techniques have been developed that make it possible to stain each chromosome a different color so that the frequency, location in the genome, and type of radiation-induced damage can be carefully measured. Development of molecular and cellular techniques to measure radiation-induced changes in the type and frequency of chromosome aberrations following low doses of radiation in a number of different tissues will be discussed in this book. Such research has resulted in increased efforts to determine the usefulness of chromosome aberrations in estimating radiation dose and predicting human cancer risks from radiation exposure.

3.4 DNA Damage and Repair

After the discovery of DNA by Watson and Crick in 1953, much scientific opinion coalesced around the idea that DNA was the most important target molecule modified by radiation in radiation-induced cancer and genetic effects, resulting in a very large number of studies focused on the ability of radiation to induce DNA damage. Studies were also designed to explore the mechanisms and repair of that damage. In the early days of research on DNA damage, the endpoints measured depended on large amounts of DNA being damaged. This resulted in the early studies being conducted *in vitro* following very high doses of radiation. These basic mechanistic studies determined the types of DNA damage induced by radiation and defined different repair types and pathways. This information still provides one of the best links between radiation-induced DNA damage and the induction of cancer and genetic effects.

New research suggested that radiation results in different types of DNA lesion than that produced during normal endogenous oxidative metabolism. Exploring the differences between radiation-induced DNA damage and damage produced by normal oxidative metabolism was initially identified as a major research focus and remained one of the important elements of the program. Techniques were developed to measure DNA damage in individual cells that made it possible to study the damage following the lower radiation doses relevant to the Low Dose Radiation Research Program.

In the following two chapters, I will discuss recent developments that link the radiation-induced DNA damage to signaling pathways. This modern research has demonstrated that not only are the DNA alterations, breakage, base substitutions, and rearrangements important in cancer, but that these radiation-induced DNA alterations trigger many signaling pathways. These pathways regulate the cell's response to radiation, such as cell cycle changes, differentiation pathway changes, and alterations in gene induction and expression. Many of these changes can be detected, even following low doses of radiation. The impact of these unique alterations in signaling pathways triggered by DNA damage was an important area of research for the program.

CHAPTER 3
The Birth of the DOE Low Dose Radiation Research Program

1. Background

Public concern about the potential health risks associated with the nuclear age prompted extensive research. At the time the DOE Low Dose Radiation Research Program began, we knew more about the health effects of ionizing radiation (at high doses) than any other environmental risk factor.

Radiation effects had been measured following exposure to all different types of radiation, delivered over a range of radiation dose rates and by many exposure modes. The uptake, deposition, distribution, dose, and biological effects of the non-uniform distribution of radiation from internally deposited radioactive materials were also well known in both animal models and humans. Finally, the large number of people exposed to high and low radiation doses following the Hiroshima and Nagasaki bombings, and to local fallout world-wide from nuclear testing, had been studied carefully. Additional human exposures from accidents associated with the development of nuclear energy and in the medical field provided important information on health effects of ionizing radiation. The impact of radiation on these large human populations had been extensively reviewed and evaluated in many epidemiological studies and the risks from the exposure characterized (NRC 2006).

However, despite the breadth and thoroughness of that research, there was and is a lower limit to the level of energy deposition, exposure, and dose that can be related to radiation-induced disease using human epidemiology methods. This limit is related to several factors that prevent the detection of radiation-induced cancer and genetic effects in humans, including:

- Variable levels of background radiation
- High and variable human cancer frequency
- Multiple environmental factors and lifestyles that influence cancer frequency
- Problems associated with determining dose and exposure in human populations
- Difficulty in defining appropriate exposed and control populations

The cost and scientific effort required for long-term follow-up of exposed human populations to evaluate the health effects produced by radiation are also obstacles.

Additional late-effect problems in linking radiation to cancer relate to the long period between radiation exposure and development of cancer. Finally, it is not possible to determine if any cancer observed in either the exposed or control population was induced by the radiation or from other causes.

These problems made it impossible to experimentally determine the relationship between cancer incidence and low dose radiation, but predictions still had to be made for regulatory and other purposes. Scientists developed models to make these predictions—but they could not be experimentally verified. This uncertainty about accurately predicting cancer risk in human populations was recognized by the scientific community and the public, and was a major factor in establishing the DOE program to fund additional research.

2. Development of the DOE Low Dose Radiation Research Program

As DOE started to phase out research with dogs and radon, money became available to address the concern about the potential health impact of exposure of large populations to low doses of radiation. This concern was elevated by the high cost of nuclear waste cleanup, the expansion of radiation use in medicine, the potential for nuclear terrorist events, and the concerns about potential release of radioactive materials from nuclear power. At this time several important scientific advances, such as the sequencing of the genome, could be combined with new technology and applied to address the old problem of the impact of low doses of radiation on risk.

The need to research risk from low-dose radiation was recognized at the highest levels of government, especially in relation to the expenses associated with radiation protection, environmental cleanup, and nuclear waste storage. Senator Peter Domenici, a Republican from New Mexico and chairman of the Senate Appropriations Subcommittee on Water and Energy for many years, was a strong supporter of basic radiation research and made many presentations where he outlined these expenses and the uncertain nature of the risks in the low-dose region. He suggested that the use of the LNT model for estimating these risks resulted in much of the expense of environmental cleanup. He was one of the government leaders who recognized the need to base radiation protection standards on the best possible science.

The DOE Low Dose Radiation Research Program needed and received strong support at the scientific, political, and government agency level. Supporters such as Sen. Domenici and those working in the DOE were essential to getting the program started and funded.

In 1990, Dr. Marvin Frazier, a Pacific Northwest National Laboratory scientist who had a very strong background in radiation biology as well as molecular biology, was hired by DOE Headquarters as a technical representative for the Office of Biological and Environmental Research (BER). This position provided scientific input on important political and program decisions. One of his early assignments from BER Director Dr. David Smith was to review the status of the research being conducted by DOE in the field of radiation biology and to determine what parts of the research could be justified to continue to receive funding. This justification was to be based on research that would contribute new knowledge that could be applied to setting radiation standards.

At that time, much of the funding for radiation biology research was invested in lifespan studies on the health effects of radiation on dogs. These studies had been conducted at national laboratories, universities, and in specialty laboratories for more than twenty years. DOE's review of these programs found that they provided valuable information on the risk and health effects of internally deposited radioactive materials. Because limited information existed on these effects on humans, the dog study data were essential and were used in setting standards for these effects.

The studies provided understanding of the risk associated with changes in dose-rate, dose distribution, LET, physical and chemical properties of the radionuclides, and the route of administration and uptake. They also established the relationships between uptake, distribution, and retention used to calculate radiation dose and dose-rate from a wide range of internally deposited radioactive materials. This research created a valuable data base that made it possible to relate the dosimetric variables to the biological effect in each individual dog. Thus, it was possible to treat individual dogs as clinical subjects, follow the development of pathology related to radiation in each dog over its lifetime, and relate the biological changes induced in the animals to dosimetric parameters. The dog studies helped establish factors used to set standards such as the influence of non-uniform dose distribution, the risk per unit of dose for different types of radiation exposure (w_D), a wide range of different radionuclides and exposure types, tissue weighting factors (w_j), and the dose and dose-rate effectiveness factors (DDREF).

The DOE assessment of the dog programs suggested that, although these studies had been valuable, continuing the high level of funding would have limited promise for future mechanistic research or impact on standards. The programs were given two years to wind down and publish data. For several years after these projects were terminated in the mid-1990s, DOE provided limited radiation biology funding. The funding was decreased because the LNT models were thought to be conservatively protective, and the scientific tools and methods available in epidemiology and toxicology were inadequate to address questions associated with cancer risk following low doses of radiation.

The initiation of the DOE Low Dose Radiation Research Program was the major area of focus for renewed research in DOE on the health effects of radiation. To understand how the program received funding, it is important to briefly discuss the players and interactions involved in getting political and agency support for the program.

In our discussions Dr. Frazier emphasized that it was important to understand how developing credibility for BER with the Office of Management and Budget (OMB) was needed to initiate the program. Dr. Frazier's experience with OMB started at hearings on air quality standards, for which he reviewed documents. He suggested that the expert review group being used by OMB seemed to be picking data to meet their preconceived ideas. He wrote a paper and presented it at their meeting, which caused the committee to take a recess and discuss his suggestions. As a result of this meeting and interactions with OMB, particularly with Dr. Gary Venethum, OMB Branch Chief, Dr. Frazier and BER gained the needed credibility from OMB to help get the funding for the DOE Low Dose Radiation Research Program.

In the early 1990s, two of Sen. Domenici's staff, Peter B. Lyons and Alex Flint, began working to determine the need and direction for a new research program in radiation biology to help define the risk in the low-dose region. With the information they had derived, they came to BER and worked to help develop the program. Dr. Ari Patrinos, Division Director of BER, was given the charge to draft a scientifically sound program. He and Dr. Frazier wrote a draft description of what the program would be and presented it to OMB, which agreed that the program would be worthwhile and had them proceed with its development.

They suggested that a major goal of the new program would be to generate data to determine if the use of the LNT dose-response models, the use of collective dose, the concept of "as low as reasonably achievable" (ALARA), and the current DDREF effects used in standard setting were supported by and could be tested using the modern genomics approaches and new tools only recently made available.

After much work between the DOE staff, Sen. Domenici's staff, and the OMB, the Low Dose Radiation Research Program began in 1998. Sen. Domenici's continued support of the program was evident in a quote from a talk at Harvard University October 31, 1997: "In this year's Energy and Water Appropriation Act, we initiated a ten-year program ($13 million/year) to understand how radiation affects genomes and cells so that we can really understand how radiation affects living organisms. For the first time, we will develop radiation protection standards that are based on actual risk."

This formed the basic philosophy for the DOE Low Dose Radiation Research Program, which focused on using developments in technology and biology to evaluate changes in the low-dose region. Sen. Domenici continued to provide support at a Gordon Research Conference on August 6, 1998, where he said, "I feel very strongly that we need the best possible standards for radiation risks, based on the best science we can produce."

DOE determined that by taking advantage of new technologies and techniques and combining them with the rapid advances in molecular and cell biology produced by the Human Genome Project, it might be possible to measure radiation-induced biological changes in the low-dose region.

Recent scientific advancements indicated that detecting changes at the cellular and molecular level in the previously undetectable range below 0.1 Gy could now be done. For example, the extensive biological advances associated with sequencing of the genome, the development of gene expression arrays, and the expansion of information on cell-cell and cell matrix communication might be combined with technologies such as microbeams, systems that could expose individual cells to known types and amounts of radiation and measure biological responses in the low-dose region. This approach could provide mechanistic data toward the development of a scientific basis for radiation standards in the low-dose region. Such studies would make it possible for the scientific community and the public to evaluate the current standards to ensure they were adequate and appropriate to control radiation exposure during clean-up, waste storage, and use of nuclear power.

3. Development of the Program's Scientific Direction

Staff within BER drafted and developed the scientific directions for the program. Dr. Frazier assigned Dr. David Thomassen as the program manager. In 1997, both Dr. Frazier and Dr. Thomassen were heavily involved in planning the first scientific meeting to get input on the needs and scientific direction from the scientific community. Leading radiation biology scientists were chosen to form a subcommittee of the Biological and Environmental Research Advisory Committee (BERAC), a standing committee to provide advice to the DOE. This subcommittee was charged with developing recommendations for DOE to use to develop the request for proposals for the new Low Dose Radiation Research Program. The subcommittee was chaired by Dr. Robert Ullrich, and the committee members were:

- Dr. Robert Ullrich, Chair, Department of Radiation Oncology, University of Texas Medical Branch, Galveston, Texas
- Dr. David Brenner, Columbia University, Center for Radiological Research, New York, New York
- Dr. Antone L. Brooks, Washington State University Tri-Cities, Richland, Washington
- Dr. Richard J. Bull, Pacific Northwest National Laboratory, Richland, Washington
- Dr. Eric J. Hall, Radiation Oncology Center for Radiological Research, Columbia University, New York, New York
- Dr. William F. Morgan, Professor of Radiation Oncology, University of California, San Francisco, San Francisco, California
- Dr. Julian Preston, Chemical Industry Institute of Toxicology, Research Triangle Park, North Carolina
- Dr. James Flynn, Decision Research, Eugene, Oregon
- Dr. Henry N. Wagner, Jr. Director, Division of Radiation Health Science, Johns Hopkins Medical School, Baltimore, Maryland
- Dr. Susan S. Wallace, Chair, Department of Microbiology and Molecular Genetics Director, Markey Center for Molecular Genetics, University of Vermont, Burlington, Vermont
- Dr. Gayle E. Woloschak, Center for Mechanistic Biology and Biotechnology, Argonne National Laboratory, Argonne, Illinois

The executive summary of the BERAC report to BER (Appendix A) illustrates the general directions recommended by the subcommittee during the scientific development of the program. As the result of this and other meetings, a document was developed called "DOE-BER Low Dose and Dose-Rate Program" that provided the initial scientific questions, possible research areas, and preliminary suggestions for a potential budget. This document was helpful as DOE staff drafted the first call for proposals from the scientific community and developed a budget for the program.

The first call for proposals was then developed by BER's Drs. Ari Patrinos, Marvin Frazier, David Thomassen, and Arthur Katz. The doses to be investigated were set at levels below the exposure levels where risk can be derived using standard epidemiological methods. Thus, the program focused only on the biological responses to low doses (<0.1 Gy) of low-LET ionizing radiation. Initially the focus was on the low-LET radiation of primary concern during waste clean-up and nuclear power production. Later the program was expanded to include high-LET radiation, as DOE and the National Aeronautics and Space Administration (NASA) combined programs to address problems of concern to both agencies. This expansion was to study high-LET radiation that would be encountered from high-Z particles and other high-LET radiation during space travel and the potential for exposure to high-LET radiation during waste clean-up.

The primary goal in this first request for proposals was to develop a research program that built on advances in modern molecular biology and instrumentation not available during the previous fifty years of radiation biology research. The program was to concentrate on understanding the relationships between normal endogenous processes that deal with background oxidative damage and processes responsible for the detection and repair of low levels of radiation-induced damage. The research focused on understanding cellular processes responsible for recognizing and repairing normal oxidative damage and radiation-induced damage. The Summary and Supplementary Information from the first call (Appendix B) illustrate the early scientific directions that the program established as a basis for evaluating the program's success over the past ten years.

Research proposals responding to this request were received and scientifically reviewed, and projects addressing all five major areas of concern were funded. I was selected as the lead scientist, and a committee was formed to provide direction and overview of the program. Dr. Thomassen was the program manager until 2001, when he received another assignment in DOE. Dr. Noelle Metting, a radiation biologist and radiation physicist from Pacific Northwest National Laboratory who had joined DOE, was assigned as the new Low Dose Research Program manager. Dr. Metting managed the program and provided guidance and direction over many years.

Low dose research has provided extensive scientific data on the responses to low doses of ionizing radiation and needs to continue to explore the mechanisms behind these responses. Current research is directed toward important questions that remain to be addressed to make the data more useful in the regulatory arena.

CHAPTER 4

Early Observations and New Technology

The DOE Low Dose Radiation Research Program was founded on the idea that it would be possible to use techniques and technologies developed as part of the Human Genome Project to measure, characterize, and understand biological responses produced by exposures to low doses of radiation. This was not completely straightforward; in some cases, we had to develop entirely new technologies of our own. Two major technological obstacles stood before us:

First, existing dosimetry techniques were too imprecise to work at this unprecedentedly small cellular and molecular scale and at low doses. It was impossible to determine which cells were "hit" and which were not at low doses, where only a small fraction of cells had energy deposited in them. A technique was needed not only to know which cells had energy deposited in them, but what their responses were, down to single traversals from ionizing radiation, the lowest dose possible to a cell.

Second, measuring cellular and molecular responses to radiation required larger doses than we needed for our work, and cost more time and money than our available funding allowed. Here, the Human Genome Project (and other NIH and DOE programs) was useful; it had developed a number of methods that were very rapid and had the sensitivity to measure radiation-induced biological changes in the low-dose region. These techniques made it possible to generate very useful data on how molecules and cells respond to low doses of ionizing radiation.

1. Development of the Microbeam

The answer to our dosimetry problem was the microbeam, one of the most important tools developed by the program. A microbeam was an instrument that made it possible to focus the alpha particles or radiation beam on single cells and to know which individual cell had energy deposited in it. Several different types of microbeams were developed by the program. The first developed was an alpha particle microbeam where these particles could be generated, focused, and known numbers of alpha particles delivered to individual cells. Other microbeams developed included an electron gun that focused beta particles, an X-ray microbeam which could "hit" known individual cells, and new type of a microbeam using an X-ray microprobe at the Advanced Light Source (ALS). This X-ray microprobe could irradiate individual cells as well as specific regions in cells without damaging neighboring cells. Strips of cell could be exposed and the response of these cells and their neighbors evaluated. The development of these machines made it possible to do the ultimate low-dose studies in which single cells could have energy deposited in them, and their responses and those of their neighbors could be studied. Our program played a central role in the development and use of microbeams, which combine physics and biology to better understand the response of individual cells to radiation.

Using microbeams, the type of radiation, the total energy deposited, and the number of particles deposited in identified cells can be altered. The response of the "hit" cell and of the neighboring cells can then be studied. From such studies it has been demonstrated repeatedly that the "hit" cells as well as "non-hit" neighboring "bystander" cells respond to ionizing radiation. Such observations have caused a shift away from the major paradigm that only cells with radiation deposited in them respond to radiation.

Microbeams were designed to accurately and rapidly expose individual cells to known amounts of energy and to a wide range of different radiation types (alpha particles, X-rays, protons, and electrons) (Nelson et al. 1996, Folkard, Schettino, et al. 2001, Randers-Pehrson et al. 2001, Wilson et al. 2001, Resat and Morgan 2004b). The development of this equipment paved the way for many more studies on bystander effects, demonstrating the importance of both direct cell-to-cell contact and communication through the release of substances from hit cells that influenced the responses in non-hit cells.

As more institutions throughout the world developed and used microbeams, regular international meetings were held each year to compare research and further develop these important machines. Publications and summaries from these meetings indicate the wide range of important studies that were enabled by microbeam technology. A publication of the abstracts in 2006 in the journal *Radiation Research 166:* "Proceedings of the 7th International Workshop: Microbeam Probes of Cellular Radiation Response" is a good example of the variety of microbeams developed and the type of research questions that would be addressed with these new technologies.

1.1 Alpha Microbeams

1.1.1 PNNL–Texas A&M

An alpha particle microbeam was developed at PNNL (Braby 2000) by combining an accelerator, which provided helium particles of known energy, with a focusing magnet to place the alpha particles on a known spot and a microscope to locate the biological target. A diagram of this machine is shown in Figure 3. The basic parts used in this microbeam were similar to those used elsewhere.

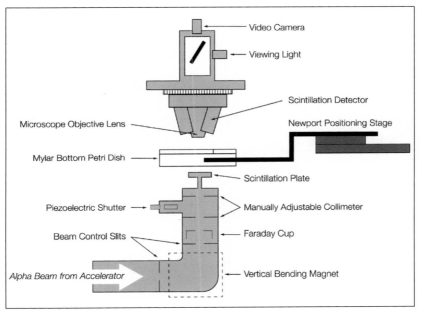

Figure 3. Schematic of an alpha-particle radiation microbeam showing how individual cells can be "hit" by known numbers of alpha particles.

The first biological studies were conducted with this microbeam at PNNL. These made it possible to relate the response of hit cells and the number of hits/cell to one biological response, the induction of micronuclei. Micronuclei are fragments of chromosomes that are not incorporated into the daughter cell nuclei so they are visible in the cytoplasm. This facilitated the generation of dose-response relationships for the induction of micronuclei (Nelson et al. 1996). These dose-response relationships were compared to the response to the same endpoint following exposure to uniform alpha irradiation from ^{239}Pu sources (NRC 1999) and to exposure to radon gas (Brooks et al. 1997). These comparisons made it possible to relate the number of hits per cell to the response to defined doses from uniform alpha sources (Brooks et al. 1994, Miller et al. 1996, NRC 1999). This demonstrated for the first time that cells hit with multiple alpha particles can survive and have no visible micronuclei induced. At the time that these studies were conducted, it was not recognized that both hit and non-hit cells could respond to alpha particle radiation.

Dr. Braby moved his microbeam from PNNL to Texas A&M University and continued his research there. At Texas A&M, the alpha particle microbeam was used to make several important scientific observations. For example, studies were conducted to determine if "hit theory" could be used to relate dose to biological response following exposure to alpha particles. Studies were designed to help define the relationships between radiation dose and hit numbers. These studies suggested that the number of cells hit may not be the most meaningful parameter to relate alpha exposures to biological responses (Braby and Ford 2000).

This microbeam was used with the well-established rat trachea model to determine how cell-cell communication is triggered by alpha particles using intact tissue. These studies tested the hypothesis that normal respiratory epithelial cells transmit signals to neighboring cells in response to a small fraction of the cells being hit or a very-low-dose radiation exposure. Energy patterns were varied and the induction of bystander effects measured in a series of studies to determine what parameters were necessary to produce bystander effects (Braby and Ford 2004).

Finally, the research team also conducted studies to determine if the bystander effects observed following alpha particle microbeam irradiation were an artifact produced by the preirradiation of the surfaces of the cell culture material (Medvedeva, Ford, and Braby 2004). These studies suggested that in some systems, bystander effects may be related to radiation-induced changes in the tissue culture surfaces and not to a cellular response. The potential for such artifacts require further investigation and must be carefully controlled to make microbeam studies meaningful. These artifacts were not present in many other systems, bystander effects do exist, cells directly communicate with each other to produce damage in non-exposed cells, and they release biologically active substances into the media that can alter the response of non-exposed cells.

1.1.2 Columbia University

Columbia University was also among the first to develop an alpha particle microbeam (Randers-Pehrson et al. 2001) and was the first to develop a microscope system to automate the location of cells. This was done by combing a scanning microscope stage with the use of a vital dye that stained the nucleus. Using this system, it was possible to locate and irradiate large numbers of cells very rapidly to a defined number of alpha particles. This made it possible to study the influence of defined numbers of alpha particles on cell transformation response in bystander C3H 10T1/2 cells (Miller et al. 1999, Sawant et al. 2001).

Columbia scientists also developed special staining techniques to identify different cell types grown in the same dish. This culture technique was combined with the microbeam. A known numbers of alpha particles could hit cells stained with one type of dye while cells of the other type did not have energy deposited in them. The number of micronuclei could be scored in both the hit cells and the non-hit cells and hit-response relationships derived in the different cell types on the same dish (Ponnaiya et al. 2004). The frequency of micronuclei was observed to increase as a function of the number of alpha particles that traversed the hit cells, as was expected. It was also demonstrated that the non-hit cells had more micronuclei than the controls, which represented a direct demonstration of the bystander effect. In Figure 4, both hit and non-hit cells are shown to have micronuclei.

The scientists (Drs. Hall, Geard, Brenner and Randers Pehrson) along with a large number of graduate students at Columbia University also pioneered a co-culturing technique in which irradiated and bystander cells were cultured on two surfaces of Mylar separated by media (Geard et al. 2002). Because the range of the alpha particles was short relative to the distance between the cells, the cells on one surface could be irradiated, and cells on the other would receive no energy deposition. Using this technique, it was possible to irradiate large cell populations to investigate the induction of chromosomal aberrations in irradiated and bystander immortalized human fibroblasts. Using this system, it could be shown that bystander effects occurred in the non-irradiated cells.

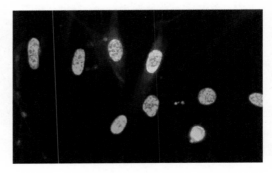

Figure 4. Stained cells showing cells hit (light blue) and not hit (darker blue) by microbeams, demonstrating that micronuclei in the non-hit cells can be damaged as a result of the bystander effect (Ponnaiya et al. 2004).

The types of chromosome aberrations change as a function of the cell cycle stage at the time of radiation. Chromosome aberrations are produced during the G_1 stage of the cell cycle before the cell has duplicated its DNA while chromatid type of aberrations are produced during the G_2/S stage of the cells after and during DNA duplication. Columbia scientists demonstrated a well-defined dose-response relationship in the cells directly exposed to alpha particles with energy deposited in the cell. The type of aberrations observed were, as would be expected, chromosomal. However, the bystander cells without direct energy deposited in them exhibited no dose-response relationship, and the level of aberrations remained elevated above that seen in the controls at a constant rate. In addition, the type of aberrations observed in the bystander cells were the chromatid type, which would not be predicted with cells that had not duplicated their DNA and were in the G_1 stage of the cell cycle used in this study. Such studies demonstrate that the bystander response is unique and may have different significance in risk analysis since chromatid type of aberations are mostly eliminated when the cell divides and seem to add little risk.

After the alpha microbeam was moved to Texas A&M University in 1997, an additional electron gun microbeam was developed at PNNL (Resat and Morgan 2004b). The electron gun microbeam is shown in Figure 5.

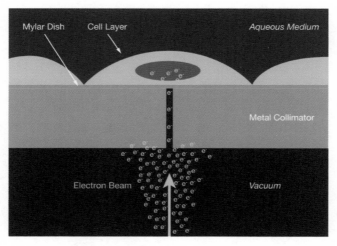

Figure 5. A spatially resolved electron gun microbeam with a cell in place to define dose distribution. Once the electrons go through the collimator, they scatter and result in a distribution of energy within the cell.

This microbeam used a focused electron beam, so it provided information on cellular responses to low-LET radiation delivered to known cells and locations (Resat and Morgan 2004b, Resat and Morgan 2004a).

Extensive research was conducted on this machine to characterize the beam size, energy distribution, penetration depth, and other characteristics of the energy deposition in cells (Wilson et al. 2001, Lynch et al. 2005, Resat and Morgan 2004b).

This device consisted of a pulsed electron beam capable of operating at energies from 20 to 80 keV. The electron gun is housed in a standard vacuum chamber pumped by a turbo molecular pump (base pressure 1×10^{-8} Tor). An electron source provides a beam with selected energy in the range of about 50 to 100 keV and fluence of about 3×10^{10} electrons cm^{-2} s^{-1} to a collimator chosen for the specific experiment. The chamber is equipped with a Faraday cup for monitoring beam current and an optical shutter to ensure no electron dark current between pulses. The spatial resolution of the device is achieved by passing the electron beam through a high-aspect-ratio hole (~20:1) fabricated in a metal foil (the collimator). Several collimators were designed with the goal of minimizing the production of X-rays while optimizing the spatial resolution of the delivered dose. The beam collimation holes (typical a few tens of microns) are formed using laser drilling. The collimator can be constructed with one hole or a series of holes depending on the biological experiment of interest (Resat and Morgan 2004b, Sowa et al. 2005).

This equipment has been very useful in measuring important track structure variables associated with the exposure of cells to electrons (Wilson et al. 2001). This instrument has also been important in determining the role of low-LET radiation in production of biological alterations in cells that did not have energy deposited in them. Such studies suggested that alpha particles were more effective in initiation of cell-cell communication than exposure to low-LET radiation. Such communication was essential for induction of bystander effects or biological alterations in non-hit cells. Some studies with low-LET microbeam radiation supported this observation as they failed to demonstrate the initiation of cell-cell communication and bystander effects (Morgan and Sowa 2005).

As the program direction changed, many of the later studies moved from the response of individual cells in monolayer tissue culture to studying the responses in complex tracheal tissues (Ford et al. 2005). At Texas A&M, tracheal tissue was irradiated with a highly collimated electron microbeam irradiator or with a single-particle, positive-ion microbeam irradiator. This made it possible to compare the responses in single-cell tissue culture systems with the responses in normal rodent respiratory tissue cells. Such studies helped provide a method to extrapolate responses in tissue culture to the responses of experimental animals and the human respiratory epithelial cells after exposure to a variety of radiation types, especially radon. These studies have been critical in linking cellular and molecular studies to human risk from inhaled radon.

1.2 Gray Cancer Institute

The Gray Cancer Institute (GCI) in England also got funding from the program to help refine and develop an alpha particle microbeam and used it to study the interaction of alpha particles with hit and non-hit cells (Prise et al. 1998). Studies from this institution were funded by the Low Dose Radiation Research Program and suggested that microbeams provide a unique tool to help understand the response of individual, identified cells to the exposure of known numbers of alpha particles (Michael et al. 2001, Folkard et al. 2002, Prise et al. 2002).

The other major development at the GCI was the development of low-LET microbeams using Xrays. The GCI pioneered the use of X-ray focusing techniques to develop systems for micro-irradiating individual cells and sub-cellular targets (Schettino et al. 2000, Folkard, Schettino, et al. 2001, Folkard, Prise, et al. 2001, Michael et al. 2001). The prototype X-ray microprobe was developed alongside the existing charged-particle microbeam to address problems specific to low-LET radiations, where very precise targeting accuracy and energy delivery are required. Figure 6 is a diagram of this microbeam.

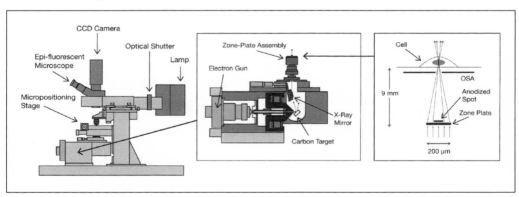

Figure 6. Schematic of a focused X-ray microbeam that illustrates spatial resolution of individual cells hit by X-rays.

The X-rays were generated by bombarding a target with energetic electrons. This generates so-called "characteristic X-rays" whose energy depends on the target material. Using this machine, it was possible to radiate one cell at a time with defined doses and with sub-micron precision. The X-rays are then focused on a very fine spot (smaller than an individual cell) using a "zone-plate," which is a small lens (less than 1 mm diameter) of a type developed initially for X-ray microscopy. Increasing the energy is simply a matter of choosing other target materials; for example, aluminum or titanium instead of carbon. It thus became possible to use very low doses approaching that of a single electron track deposited in a single cell.

This microbeam was optimized for focusing 278 eV CK X-rays; however, there are several reasons for extending the range of available energies. To do this, a variable-energy soft X-ray microprobe was developed that could deliver focused CK (0.28 keV), AlK (1.48 keV), and, notably, TiK (4.5 keV) X-rays. TiK X-rays can penetrate well beyond the first cell layer (the 1/e attenuation in tissue is 170 μm) and are therefore much better suited to studies involving tissues and multi-cellular layers. Also, from a microdosimetric viewpoint, TiK X-rays produce a spectrum of energy depositions in DNA-sized targets that more closely resemble those of conventional low-LET radiations (Schettino et al. 2000, Folkard, Prise, et al. 2001). This type of exposure is very relevant to environmental levels of exposure.

Such research made it possible to concentrate on irradiating specified individual cells within cell populations in order to identify bystander responses for low-LET radiation where non-radiated cells respond to signals from nearby radiated cells (Schettino et al. 2003). Modification of the equipment made it possible to use higher-energy X-rays to extend the studies into complex tissues and beyond experiments involving single cell layers. These types of microbeams generate types of low-LET radiation that more closely mimic the types of exposures of prime interest to DOE during waste clean-up as well as the type of radiation received by nuclear workers or during nuclear accidents.

1.3 Advanced Light Source Microbeam, Lawrence Berkeley National Laboratory

Lawrence Berkeley National Laboratory (LBNL) also developed a type of a microbeam using an X-ray microprobe at the Advanced Light Source (ALS) to precisely irradiate individual cells and specific regions in cells without damaging neighboring cells. The unique synchrotron-based source of a 12.5-keV X-ray microbeam line 10.3.1 at the ALS was used to quantitatively characterize low-dose responses of low-LET, radiation-induced bystander effects in a novel tissue-like model of human mammary epithelial cells (HMEC), normal human fibroblast cells (HFC), or in a third scenario, with both cells together in a co-culture system (Mainardi et al. 2004). Cultures were grown in microwell slide chambers and

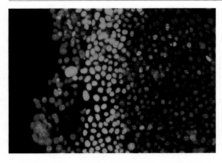

Figure 7. *Cells in culture irradiated with a strip microbeam. Green cells were hit, blue cells were not. This illustrates the bystander effect in that the signal migrates from the hit to the non-hit cells to produce damage, as proven by the scattered green cells. (Blakely et al. 2006).*

irradiated with precise stripes of radiation that were up to 100 μm wide. An example of exposed and non-exposed cells is shown in Figure 7.

With this system, a group of cells in a defined stripe could be irradiated. The response in these cells was then compared to those outside the stripe that received no radiation energy. Samples were processed for the expression of radiation-induced protein markers with fluorescent immunohistochemistry in a time course from ten minutes to several hours after exposure. Using fluorescence microscopy on a high-precision-controlled microscope stage and fiducially marked references, the physical locations of the dose stripes were mapped exactly to the location of the biological responses. Computer-based fluorescent analysis of radiation-induced signals in thousands of cells has revealed statistically significant differences in the broadening of the effects of the dose stripes to neighboring unirradiated cells with time after exposure. Such broadening of the dose stripe to involve cells not in the irradiated field represented a radiation-induced bystander effect that was quantitatively evaluated.

The sensitivity of detection in this model system is below 0.1 Gy, with dose stripes discernible after 0.05 Gy. The intensity of the fluorescence was greater in the dose stripe for larger doses (e.g., 1.0 Gy), and the fluorescence signal decreased more slowly with time after high-dose exposure than after lower doses (e.g., 0.25 or 0.1 Gy). Results from a rapid time course study show that radiation-induced signals were observed within 10 minutes after exposure in cells adjacent to, but outside of the irradiated area. The effect was apparent at 10 minutes after exposure and diminished with time, but was still significant 3 hours after exposure. A dose-dependent induction of bystander effects in several classes of radiation-induced signals was measured and the time course determined to examine how radiation exposure changes cell signaling acutely, and chronically (Blakely et al. 2006).

With the development of the microbeams and the discovery and characterization of the bystander effects, it was no longer adequate to think in terms of individual cell responses as a model for radiation risk estimates. Well-established paradigms in radiation biology had been challenged, including hit theory and the use of dose to cells as a means of predicting radiation responses. With these new observations, a broader "systems" type of thinking is required in which whole tissues or organs respond to the radiation exposure in a coordinated way.

2. Biotechnology

Other important techniques were developed that made it possible to conduct research in the low-dose region and evaluate the biological responses. These techniques are discussed briefly in this chapter, and the responses and significance of the science is covered in subsequent chapters.

2.1 Flow Cytometry and Chromosome Painting

The flow cytometer was developed primarily at LANL using a combination of rapid flow of liquids and laser technology. Cells were suspended in a suitable liquid that would fall rapidly, forming individual drops that passed in front of a laser. The laser would determine if each drop had a cell in it and measure traits associated with the cell. Using an electric field and magnets, it was possible to charge the drops with the cells of interest and deflect the drops and collect them in individual containers. This technique made it possible to sort individual cells, chromosomes, and organelles well before the Low Dose Radiation Research Program began (Wimmer et al. 1996, Cram 2002, Cram, Bell, and Fawcett 2002). This technique is an important part of research in many areas, including the research on health effects of radiation (Wilson and Marples 2007). Program researchers have used flow cytometry to evaluate many cellular and molecular changes induced by ionizing radiation in populations of cells.

An important spin-off from the flow cytometer is the development of whole-chromosome paints. Using flow cytometry, individual chromosomes could be sorted from the remainder of the genome and probes developed that were specific to each chromosome. These probes made it possible for each chromosome to be "painted" a unique color. Probes for parts of the chromosome and individual genes were further developed by the Human Genome Project, which made it possible to determine the exact location of genes on each chromosome. With the ability to mark each chromosome with whole-chromosome paints, it has become obvious that each chromosome has a unique domain in the cell nucleus during all stages of the cell cycle (Cornforth, Bailey, and Goodwin 2002, Vives et al. 2005). It was determined that at metaphase, each of the individual chromosomes involved in radiation-induced damage can be accurately identified.

Early dose-response studies conducted in the program without using chromosome painting could not determine the number and types of chromosomes involved in radiation-induced aberrations. In studies such as those in which cells or animals were exposed to high-Z particles similar to those found in space (Brooks 2001, Brooks et al. 2001), many of the complex aberrations were scored as single exchanges, and many of the complex exchanges were not detected. Later, more complete studies using these paints made it possible to determine the involvement of each individual chromosome in the chromosome aberrations (Cornforth 2001, Vazquez et al. 2002). With Fluorescent In Situ Hybridization (FISH) techniques, it became possible to do very complete dose-response curves (Loucas and Cornforth 2001) and evaluate the individual chromosomes involved (Cornforth 2001), the number of chromosomes that make up each aberration, and the location within each chromosome where the aberration is induced. Figure 8 shows a chromosome spread painted with these techniques. The figure also shows a dicentric chromosome and fragments produced by radiation (Loucas and Cornforth 2001, Cornforth, Bailey, and Goodwin 2002).

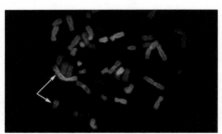

Figure 8. Chromosomes visualized with fluorescent in situ hybridization (FISH). Arrows show a simple dicentric interchange between chromosomes 2 and 8, resulting in two color junctions (Cornforth et al. 2002).

These staining techniques required the development of different systems of scoring chromosome aberrations (Tucker et al. 1995, Cornforth 2006). Chromosome painting demonstrated that many of the aberrations that were scored in the past with conventional techniques as simple exchanges involving two chromosomes actually involved multiple chromosomes (Vazquez et al. 2002, Cornforth 2001). Such studies have been extended to determine the influence of dose-rate (Loucas, Eberle, Bailey, et al. 2004) and radiation type, including gamma rays and high-Z particles, on the induction of complex chromosome aberrations (Loucas, Eberle, Durante, et al. 2004, Cornforth et al. 2002, Rithidech, Honikel, and Whorton 2007). The frequency, distribution, kinetics of repair, and type of complex aberrations have all been characterized as function dose, dose-rate, and LET.

An additional technique, Co-FISH, was developed using a combination of the staining and sorting techniques that made it possible to label one set of DNA in a cell, resulting in one chromatid being labeled, but not the other (Bailey, Meyner, and Goodwin 2001, Bailey and Goodwin 2004, Zou et al. 2004). This technique was very important in understanding the function of telomeres and their role in the formation of chromosome aberrations and DNA repair (Bailey et al. 2004, Bailey, Goodwin, and Cornforth 2004, Bailey and Cornforth 2007). These additional data on the function of telomeres may become important in the future in estimating the role of radiation on the development of disease.

The role of flow cytometry in the study of radiation biology is critical and has been carefully reviewed (Wilson and Marples 2007).

2.2 Gene Chip Technology
2.2.1 Genomics

As was expected at the start of the program, the development of gene chip technology as part of the Human Genome Project and the field of genomics proved to be very important in understanding the biological responses to low doses of radiation. This tool made it possible to rapidly evaluate the radiation-induced changes in gene expression, protein production, and metabolites following any environmental insult, including low doses of radiation. Application and modification of this technology was an important part of the program (Kegelmeyer et al. 2001, Bittner et al. 2000, Koch-Paiz et al. 2000).

With gene chip technology, it became possible to identify many radiation-induced changes in gene expression for thousands of genes at one time and to determine which genes were either up- or down-regulated as a biological response to exposures to low doses of radiation. Because of the large amount of data generated by such approaches, it was also necessary to develop additional informatics methods to handle and to interpret such data (Fornace et al. 1999). In many early studies, cells were simply exposed to low doses of radiation and the changes in gene expression determined (Amundson et al. 2001a, b, Yin et al. 2003).

The type of genes that responded to low doses of radiation could also be defined. Early studies quickly determined that many of the genes that changed their gene expression following low doses of radiation were the same genes that respond to many other forms of stress (Amundson, Do, and Fornace 1999, Amundson et al. 2001a, Amundson et al. 2002, Amundson and Fornace 2003, Amundson et al. 2003).

As with any new technology, this one has great possibilities, but there are always areas where data generated by such a broad-based approach can be misinterpreted (Amundson and Fornace 2003). Contrary to expectations, initial data did not suggest that low doses of radiation had marked impact on the expression of genes known to be associated with DNA repair (Kegelmeyer et al. 2001, Yin et al. 2003, Akerman et al. 2005). As gene chip technology was applied to evaluate the radiation response as a function of time after radiation exposure (Amundson et al. 2002, Amundson et al. 2003), radiation dose (Coleman and Wyrobek 2006, Yin et al. 2003, Ding et al. 2005), dose-rate (Amundson et al. 2003), and radiation type (Kurpinski et al. 2009), it became clear the gene responses were very complicated and were modified by all these factors (Amundson and Fornace 2003).

The take-home message of these early studies was that the genes that responded at low doses and dose rates were different than the genes that responded after high doses. This suggested a difference in the mechanisms of action following low doses of radiation compared to those following high doses (Ding et al. 2005, Coleman and Wyrobek 2006, Brooks et al. 2006, Coleman et al. 2005). Such a difference suggested that the shape of the dose-response relationship could be non-linear. It was also clear that the time course of the response and the shape of the dose-response relationships were different for many individual genes. These early gene expression studies provided the groundwork for many more mechanistic studies of the observed radiation-related processes, such as changes in reactive oxygen species status of the cells, bystander effects, adaptive responses, and genomic instability, all discussed in greater detail in chapters 4 and 5.

2.2.2 Proteomics

Gene expression and protein expression are not linked 100%, and many of the important proteins involved in biological changes do not follow the changes in gene expression. Many biological processes are altered by changes in protein expression. These may be related to time differences, differences in breakdown and up-regulation of the genes and proteins, differences in intercellular location of the proteins (Raman et al. 2007), and other factors. The use of proteomics in biology is relatively recent in the program and represented a major future direction had the program funding been continued.

Many recently developed techniques make the use of proteomics possible. Development of chips similar to gene chips was a major technical advance. With such techniques, it became possible early in the program to clone and characterize known proteins in mice and humans (Coleman, Eisen, and Mohrenweiser 2000). These protein microarrays also made it possible to define many different protein interactions with cellular components such as the chromatin in the cells (Coleman et al. 2003). As better multiplexing techniques were developed, it became possible to identify bacterial and viral proteins in mammalian protein samples (Rao et al. 2004), which could be very useful in the future to identify and diagnose diseases.

It was demonstrated that many cells could shed proteins as the result of environmental insults including radiation exposure. By applying proteomic techniques, Ahram et al. (Ahram, Adkins, et al. 2005) determined that these shed proteins could be characterized into different classes. By combining the proteomic approach with the databases that have been developed as the result of proteomic research, some of the shed proteins were identified in Chinese hamster ovary (CHO) cells following radiation exposure (Ahram, Strittmatter, et al. 2005).

Proteomics techniques have continued to improve by combining liquid chromatography/tandem mass spectrometry (LC-MS/MS) with other methods of isolating and characterizing proteins. The speed of processing samples with these techniques has been a limiting factor, but the number of samples processed has been increased by using high-intensity focused ultrasound in sample preparation in combination with LC/MS techniques (Lopez-Ferrer et al. 2008). Such techniques have also used much faster methods based on shared peptides to identify the proteins (Jin et al. 2008). These techniques hold great promise for linking the changes in gene expression to the proteins carrying out the biological functions.

The goal of all proteomic research is to link protein changes to biological function. It has been demonstrated that there are many post-translational modifications of proteins that are very important in determining the function of the proteins. Such modifications can alter the potential impact of the proteins in both positive and detrimental ways (Warters 2002). A major important change impacting the protein function is the phosphorylation of proteins. Extensive research in this area has been conducted, but only limited research in the program. However, program research has determined that it is possible to identify the phosphoproteome, which defines post-translational phosphorylation of the proteins and supplements proteomics. This research demonstrated that the phosphorylation of proteins following radiation exposure to high doses was different from that observed after low doses of ionizing radiation (Yang et al. 2006).

Yamaguchi et al. (Yamaguchi et al. 2005) identified substrate specificity for human protein phosphatase 2Cd, Wip1 as an example of how changing phosphorylation can change function. This made it possible to develop a substrate-based cyclic phosphoptide inhibitor of this protein and led to many developments in identifying protein function and its modification. This research is continuing under the program.

The ultimate goal of proteomics research is to relate molecular and cellular changes to well-defined biological changes as well as to exposure conditions. For example, Wang and Gao (Wang and Gao 2005) determined that proteomic analysis was useful in the study of neural differentiation of mouse embryonic stem cells to neurons. Such analyses can be applied to define any differentiation pathway as well as responses, as cells differentiate in a unique way as a function of radiation exposure.

Change as a function of radiation dose was the first exposure condition to relate protein profile changes with radiation exposure. These studies determined which and how much proteins changed as a function of radiation dose. Such information could then be used to estimate radiation dose where no physical dosimeters were present (Marchetti et al. 2006). This is a good example of applying program-funded research in an area that was not an emphasis of the program—in this case, biodosimetry. This is discussed in greater detail in chapter 6.

Jang et al. (Jang, Guo, and Wang 2007) used a proteomic approach to relate specific changes in calmodulin. The relationships that exist between calcium- and phosphorylation-dependent calmodulin complexes were defined using such an approach that paves the way for more extensive studies on how radiation can modify these relationships. These studies are laying the groundwork for further mechanistic studies that will be useful in defining many cellular processes triggered by radiation.

Signaling and chronic inflammation are discussed in great detail in chapter 5. These processes can be studied and provide important links between the proteome and specific molecular mechanisms that will be critical in using systems biology to better understand radiation risk (Wemer and Haller 2007).

A few studies have demonstrated that a proteomic approach can be useful in studying cancer and the processes that are important steps in cancer development. The mitochondria are the powerhouse of the cell and play a critical role in the generation of free radicals similar to those generated by radiation. Miller et al. (Miller et al. 2008) demonstrated that it is possible to use MS-based proteomic techniques to profile mitrochondrial proteins in radiation-induced genomically unstable cell lines. These unstable cell lines demonstrate a persistent oxidative stress and are thought to represent an important stage in the development of cancer.

A limited number of studies used animal models to study the role of protein changes during cancer development. Studies of radiation-induced leukemia demonstrated that proteomic techniques can suggest relationships that exist between radiation exposure, protein changes, and cancer development (Rithidech, Honikel, and Whorton 2007). Such animal-based models are essential to link radiation-induced cancer in animals to that in humans.

2.2.3 Metabolomics and Secretomics

The analysis of the many products produced as the result of metabolism or secretion stimulated by an environmental insult is called metabolomics and secretomics. Techniques similar to those used in proteomics have been used to identify these molecules. Because many products are identified using these techniques it is important to develop techniques to sort and characterize the interrelationships between these chemicals (Patterson et al. 2008).

In metabolomics, the products are sampled either in the urine or the blood. Because both body fluids are easy to obtain, metabolomics can be used as a biodosimetric technique to estimate previous radiation exposure (Tyburski et al. 2008). To date, radiation-induced changes in metabolites are not sensitive enough to detect exposure to low doses of radiation and have not been a focus of the program. However, studies using metabolomics to study radiation-induced cell killing suggest that it may be possible to identify the two major types of cell death, apoptosis and necrosis, in HL60 leukemia cells (Rainaldi et al. 2008). This type of research needs to be expanded to other cell types, especially normal cells *in vivo*.

Because the role of the microenvironment is important in the development of radiation-induced disease and the maintenance of normal organ function, it is also important to determine its role in the secretion of hormones or other substances into a tissue or organ. Chen et al. (2008) evaluated the role of the microenvironment on the "secretome." This is a new area of research that may represent an important part of the systems biological approach needed in the future for evaluating radiation risk.

2.3 Techniques to Detect and Characterize DNA Damage and Repair

DNA damage and repair play a central role in the induction of cancer. Characterization of genes, proteins, and pathways involved in repair of radiation-induced DNA damage was one of the major areas in the field of radiation biology and was an important area addressed initially by the program. Two major questions posed in the original program outline were:

- Is the damage induced by ionizing radiation and the repair of that damage different from the endogenous oxidative damage and repair present during normal life processes? Addressing this question is important because high levels of oxidative DNA damage are produced and repaired daily in every cell in our bodies.
- Does this DNA repair extend to damage from ionizing radiation? When the program began, the ability to measure DNA damage following radiation exposure was limited to very high doses of radiation (Rydberg, Lobrich, and Cooper 1994). Thus, much of the past research in this area was not applicable to the new DOE Low Dose Radiation Research Program.

Research in this area resulted in the development of techniques that could measure DNA damage following low doses of ionizing radiation and determine the similarities and differences between radiation-induced damage and repair and DNA damage produced by normal processes. The distribution of DNA damage following radiation was very non-uniformly distributed in the DNA, with local sites having multiple different types of damage (locally multiply damaged sites, or LMDS) (Ward 1994). This observation formed an important base for program research.

New techniques were developed at Brookhaven National Laboratory that made it possible to detect these multiply damaged sites following low doses of ionizing radiation (Sutherland et al. 2000a, b, Sutherland, Monteleone, et al. 2001). The basis of these techniques was to convert all of the DNA lesions to double-strand breaks (DSBs) (Georgakilas, Bennett, and Sutherland 2002), then separate the DNA according to size (Sutherland, Monteleone, et al. 2001, Sutherland, Bennett, Cintron, et al. 2003), and using single-molecule laser fluorescence sizing (Filippova et al. 2003), quantify the number of breaks and the size of the DNA strands (Sutherland, Bennett, Georgakilas, et al. 2003, Sutherland, Georgakilas, et al. 2003). Using this combination of techniques, it was possible to measure the clustered DNA damage sites and the size of the lesions following very low doses of radiation (Sutherland et al. 2002) over a wide range of different types of radiation exposure (Sutherland, Bennett, Schenk, et al. 2001, Sutherland et al. 2002, Song, Milligan, and Sutherland 2002) and under different experimental conditions (Sutherland, Bennett, Weinert, et al. 2001). It was suggested that such damage sites were unique for radiation, and the distribution of the damage depended on the radiation type (Hada and Sutherland 2006).

This early research also suggested that radiation-induced damage was formed in clusters and was different from the random distribution of DNA damage produced by normal endogenous processes. It was determined that additional research was needed on the repair and processing of radiation-induced clustered DNA damage.

An important development in understanding the relationship between cells hit by radiation and the response to the energy deposited in the cell was the development of methods to detect DNA damage and repair sites. It was determined that histones were phosphorylated in response to DNA double-strand breaks DSBs (Burma et al. 2001). This process generated sites called H2AX sites—sites of phosphorylated genes as a reaction on DNA DSBs—that could be visualized at the site of the initial energy deposition. These were thought to be a marker of the location of the DNA breakage (Al Rashid et al. 2005) and repair (Burma and Chen 2004). The γH2AX technique, developed by a number of laboratories outside the program, was incorporated into several program studies and provided very useful data.

When the γH2AX technique was combined with microbeam studies, it was possible to see which cells were hit by radiation, how many hits had occurred on a given nucleus, and how long it took the cells to repair the damage and lose the γH2AX foci. An example of this type of study and the information generated is seen in Figure 9 where single nuclei were traversed by three alpha particles (Prise et al. 2002). The γH2AX technique was an important tool used to evaluate many of the new biological phenomena seen in the program.

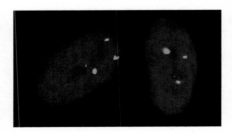

Figure 9. DNA damage is recorded as γH2AX foci in two nuclei of human fibroblasts targeted with the microbeam. Each nucleus was hit by three helium ions; a single ion in each of three locations, each helium ion delivered 100 mGy equivalent and produced four to six DNA double-strand breaks, shown in green (Prise et al. 2002).

2.4 Identification and Characterization of DNA Repair Genes

Extensive early research in radiation biology focused on the identification and characterization of many DNA repair genes, and under the Low Dose Radiation Research Program, many more DNA repair genes were identified (Cleaver, Afzal, et al. 1999, Blaisdell and Wallace 2001, Fujimori et al. 2001), characterized (Lamerdin et al. 2004, Weinfeld et al. 2001, Hirano et al. 2005), and defined. Interactions of the repair genes with other genes were evaluated (Miller et al. 2005, Wiese et al. 2002), the pathways involved in the repair were defined (Pierce et al. 1999), and their roles in mutation induction (Shen et al. 1998) and maintaining genomic integrity (Cleaver, Afzal, et al. 1999, Cleaver, Thompson, et al. 1999, Hinz et al. 2007), following radiation damage were examined. The role of repair genes associated with radiation-induced DNA damage in regulating the genomic stability of cells and the induction of radiation-related genomic instability is discussed in greater detail in chapter 5.

A major approach to understanding DNA repair has been to use model experimental systems, human families, populations, and cell types deficient in DNA repair to determine the genes and mutations involved in these deficiencies. Several major genetic diseases proved to be very useful in defining the genes involved in DNA repair and the role of these genes in the disease (Thompson and Schild 2002). The cells and tissues from individuals with genetic disease were exposed to radiation. The influences of radiation on many biological endpoints were determined in these deficient cells and populations, then were characterized and compared to normal responses.

One disease recognized early on as resulting from a DNA repair deficiency was Xeroderma pigmentosum (Cleaver, Afzal, et al. 1999, Cleaver, Thompson, et al. 1999, Cappelli et al. 2000). Genes from this disease (a rare genetic disorder in which the ability to repair damage caused by ultraviolet (UV) light is deficient), model systems, and other diseases have been characterized and their role in DNA repair mapped and evaluated. Other syndromes of importance in defining DNA repair genes include Cockayne syndrome and trichotriodystrophy (Cleaver, Thompson, et al. 1999), Fanconi anemia (Yamamoto et al. 2003, Yamamoto et al. 2005, Thompson et al. 2005, Tebbs et al. 2005, Hinz et al. 2006, Hinz et al. 2007), Nijmegen breakage syndrome (Pluth et al. 2008, Williams et al. 2002), and more recently, genes that influence the production of diseases such as breast cancer (Easton et al. 2007).

There were many publications in these same areas of research outside the program that are not included in this book. These publications and those from the program constructed a firm understanding of the role of radiation-induced DNA damage and repair and the consequence of that damage.

2.5 Cell Killing and Apoptosis
2.5.1 Cell Killing

A major focus in radiation biology is cancer therapy research; thus, cell killing was a hallmark of much of the early research. As it became possible to culture mammalian cells, methods were developed to determine the ability of the cells to divide and form colonies following radiation exposure. Because radiation did not easily kill many cell types in interphase but would prevent them from dividing and forming viable colonies, the colony formation method became the standard for evaluating radiation-induced cell killing.

To conduct this technique, a known number of cells were seeded in a dish, irradiated, and allowed to divide and form colonies. The colonies were counted, and the number in the exposed dishes compared to those in the control dish to estimate the radiation-induced cell killing. Cell survival curves were characterized for many different types of radiation exposure, different dose-rates, fractionation schedules, cell types, and tissue types. Normal and genetically altered cells were studied to provide information on the role of many genes in the induction and repair of radiation-induced damage. The methods to conduct cell survival studies and the shape and slope of these survival curves have been carefully reviewed in the light of modern biology (Hall and Giaccia 2006). The use of survival curves was critical in the development of the hit theory, which needs to be revised in light of more recent data on cell survival.

It has been well established that exposure of cells in culture to high-LET radiation (alpha particles and neutrons) produced a linear decrease in survival as a function of dose when plotted on semi-log paper. It was also established that after exposure to high-LET radiation, cell survival had only minor dependence on dose-rate or dose fractionation, so the radiation response could be easily described and quantified. From such data, it was assumed that single hits from an alpha particle were responsible for cell death, and there was little repair following this type of radiation.

However, following exposure to low-LET radiation (X-rays, gamma rays, and beta particles) there seemed to be a threshold dose range over which cell killing did not change from that observed in the controls. Following exposure to low-LET radiation, the response was decreased by reducing the dose-rate and allowing time for "recovery" by dose fractionation schedules. These data suggested that there was repair of the lesions and that multiple hits were required to kill cells. The width of this threshold or shoulder was a function of the cell type, radiation dose-rate, and radiation type. However, the colony formation technique was not sensitive enough to determine the fine structure of the dose-response relationship for low-LET radiation exposure in the low-dose region of importance to the program.

Studies conducted on the induction of damage from small doses *in vivo* in mouse skin (Joiner, Denekamp, and Maughan 1986) and in mouse renal tissue (Joiner and Johns 1988) suggested that the effectiveness of small doses of low-LET radiation was higher per unit of dose in producing cell killing than larger doses. Determining this low-dose sensitivity *in vitro* became possible after development of a dynamic microscopic image processing scanner (DMIPS) cell analyzer, which made it possible to locate each cell on a dish and, after exposure to radiation, directly measure the number of cells that survived and formed colonies (Marples and Joiner 1993). This technique made it possible to more accurately measure cell survival in the low-dose region and added an important tool for use in study of low-dose radiation effects. It also formed the basis for studies that determined that the apparent plateau in the low-dose region was really an area where the cells were more sensitive per unit dose and then became resistant as the dose increased. Thus, there was structure in the low-dose region that was not appreciated in the past.

With the development of the flow cytometer, additional sensitive techniques for the detection of cell killing were developed (Bogen et al. 2001, Short et al. 1999). These techniques were faster and also provided information on the stage of the cell cycle during radiation exposure (Short et al. 2003) that helped define the mechanisms of action for this observed fine structure in the dose-response relationship.

It is important to provide a brief description of the fine structure in the dose-response relationships in the low-dose region. At very low doses, there was a steep curve for cell killing called hyper-radiosensitivity (HRS), followed by a upswing in the survival over a narrow dose range, called increased radioresistance (IRR), and finally, as the dose continued to increase, the final slope of the dose response as detected by other less-sensitive techniques was evident (Marples and Joiner 1993). Thus, the plateau over the low-dose region was not a true plateau, but an area of low-dose hypersensitivity (HRS) followed by radiation-induced resistance. This observation was made in many cell systems and seems to be a biological generality. Some of the cell types where HRS and IRR were measured and observed included human cells, cancer cells, and immortalized cell lines from humans and animals (Short et al. 1999, Mitchell, Folkard, and Joiner 2002, Chalmers et al. 2004, Harney, Shah, et al. 2004, Harney, Short, et al. 2004).

2.5.2 Apoptosis

Another very useful area of rapid development in cell and molecular biology in the program was the increased understanding of programmed cell death (apoptosis). Programmed cell death has been well understood during fetal development as cells divide, differentiate and die in a well programmed pattern. It was not appreciated that following exposure to ionizing radiation some cell types undergo this programmed cell death and that this can be related not only to the cell type but also to the status of the cells in their progression from normal cells to cancer cells (Boreham et al. 2000, Kagawa et al. 2001, Bassi et al. 2003, Bauer 2007a, Bauer 2007b).

This field has expanded rapidly and is becoming very important in cancer therapy. Detailed information on apoptosis and its role in cancer induction and therapy can be found in the research and will not be reviewed in detail here.

Several methods can detect the increase in apoptosis. The first is the TUNEL assay (Terminal deoxynucleotide transferase dUTP Nick End Labeling), with which the changes in nuclear morphology and staining characteristics were detected using a microscope. Another method combined TUNEL staining techniques with the flow cytometer to measure apoptotic cells. Again, the flow cytometer method made it possible to determine the stage of the cell cycle when the cells were undergoing programmed cell death. Finally, time-lapse photography has proven useful in evaluating the role of cell cycle, mitotic arrest, differentiation, mitotic catastrophe, mitotic death, and apoptosis in radiation-induced cell killing (Chu et al. 2002, Chu et al. 2004).

Using microbeams and other techniques, it was determined early in the program that radiation induces apoptosis not only in cells hit by radiation, but also in bystander cells (Belyakov et al. 2002, Lyng, Seymour, and Mothersill 2000, 2002b, a). Early research focused on the signals and the critical genes and proteins involved in the induction of apoptosis. It was determined that different forms of the stress-inducible polypeptides called clusterin played a key role in radiation-induced apoptosis (Kalka et al. 2000, Leskov et al. 2001, Leskov et al. 2003, Araki et al. 2005). Radiation-induced activation of critical genes and proteins including nuclear clusterin were found to play an important role in radiation-induced apoptosis (Klokov et al. 2004, Yang et al. 2000). It was also determined that clusterin played a key role in signaling and acted as a molecular sensor between DNA damage and cytoplasmic responses (Davis et al. 2001, Huang et al. 2000). Thus, transcription factors activated by low doses of radiation resulted in apoptosis and depended on the p53 status of the cells (Criswell, Klokov, et al. 2003, Criswell, Leskov, et al. 2003). It was suggested that this cell killing in bystander cells could be selective against cells with genomic instability and transformed cells. Such differential cell killing was postulated to result in antitumor activity and a protective effect from low doses of radiation exposure (Boreham et al. 2000, Kagawa et al. 2001, Bassi et al. 2003, Bauer 2007a, Bauer 2007b).

2.5.3 Teratogenic Effects

A biological major change observed in the A-bomb survivors in Japan was the development of birth defects in individuals exposed during fetal development. The relationship of these effects to low doses of radiation and the role of cell killing during the development of the embryo remains an important question.

The Low Dose Radiation Research Program has funded a number of studies using fish embryos to address, at the cellular and molecular levels, the impact of low doses of radiation on embryonic development. In these studies, zebrafish were exposed to low levels of radiation during embryogenesis and the effects monitored. With this experimental system it was possible to irradiate different parts of the developing embryo, quantify the induction of cell death in situ, and determine the impact of killing cells in these different regions on the development of birth defects (Bladen, Flowers, et al. 2007). Studies were also conducted at the molecular level to determine if there were biological responses that could protect these fish from exposure to low doses of radiation.

It was determined that increased expression of the subunit XRCC6 of the Ku70 proteins, which are involved in regulation of the cell cycle, protected the zebrafish against the development of birth defects (Bladen, Navarre, et al. 2007). With further development, these systems have the potential to provide important basic mechanistic information on the role of radiation in the development of birth defects if humans were exposed in utero.

All these new techniques and biological systems have made it possible to address important questions in low-dose radiation biology and to generate a large amount of data on the response of many biological systems to exposure to low doses of ionizing radiation. These observations are reviewed in chapter 5, and many have resulted in new paradigms in radiation biology. The data from the program research conducted using these techniques has also helped address many practical problems, such as developing new methods for biodosimetry, understanding low-dose-rate effects, and evaluating the potential usefulness of some of the factors used in radiation protection, such as the DDREF and radiation weighting factors (w_D).

3. Major Points: Application of New Technology

The Low Dose Radiation Research Program was essential for the development of microbeam and biological and molecular technology.

- Single-cell irradiation systems using alpha particles were developed at several locations.
- An electron gun was developed to expose small numbers of cells to beta particles.
- Equipment was developed to deliver focused X-rays to individual cells. These used a range of energies to represent the types of gamma rays present in the environment.
- Technology developed to differentially stain individual chromosomes was used by the program along with cell sorting to characterize cytogenetic damage as a function of dose, dose-rate, and radiation type.
- The use of cell and molecular techniques developed in the Human Genome Program made it possible to measure changes in radiation-induced gene expression in large numbers of genes as a function of radiation dose. Some genes were turned on at low doses and others at high doses.
- New DNA repair technology using γH2AX foci made it possible to determine the number and location of nuclear traversals from microbeam irradiation.
- Techniques were developed that made it possible to detect multiple damage sites in DNA.
- Using creative assays, DNA repair genes were identified and characterized.
- Methods to measure cell killing were improved to define the fine structure in the dose response relationships. This demonstrated non-linear responses in the low-dose region.
- The identification of apoptosis was improved using modern technology. It became possible to identify selective cell killing of transformed cells through apoptosis.

- Zebra fish were exposed with the microbeam to irradiate different parts of the developing embryo and relate cell killing to birth defects.
- Early technique developments in proteomics, secretomics, and metabolomics were important in detecting metabolic biological changes as a function of radiation dose.

CHAPTER 5

Paradigm Shifts in Low Dose Radiation Biology and Application of Data

With the development of the new tools and more sensitive techniques described in chapter 4, the major technological obstacles to the study of biological responses to low dose radiation had been knocked down. The program's researchers rapidly began to make important observations. Its major early accomplishment was the discovery and characterization of three unique biological responses: bystander effects, adaptive responses, and genomic instability. These discoveries presented broad challenges to the traditional thinking in radiation biology, triggering a still-ongoing paradigm shift on the response of biological systems to low doses of radiation. This chapter discusses the program's new scientific observations and how they led to new paradigms.

Bystander effects are the result of cells and tissues communicating with each other; when radiation insults one cell, it results in a response in the neighboring cells that have no radiation energy deposited in them. It has long been known that there is extensive cell-cell, cell-matrix, and cell-tissue communication, and the matrix and cell-cell interaction influence changes in gene expression (Bissell & Aggeler 1987; Bissell & Barcelloshoff 1987). The functional units for cancer induction were thus shown to be units larger than cells, and it was suggested that it takes a tissue to make a cancer, not simply changes in the individual cell (Barcellos-Hoff & Brooks 2001).

With the development of microbeams and other techniques, this became obvious to the field of radiation biology. Early studies made this observation for many different biological systems and were able to relate such a response back to previous research that suggested that the "hit theory" for describing radiation response needed to be modified because the targets for biological response were much larger than individual cells (Brooks 2005).

In the early part of the program, adaptive responses were described as "any responses to low doses of radiation that changed the magnitude and direction of the biological response to subsequent radiation exposure." This term was later expanded to include the observation that low doses of radiation could also reduce the background level of biological alterations for a wide range of different biological systems.

In many different biological systems, adaptive responses were shown to have a marked influence on the shape of dose-response relationships in the low-dose region. It decreased the magnitude of the response in the low-dose region below that predicted from a linear extrapolation from the high-dose region. Because of this decrease, the adaptive response has also been called a protective or a "hormetic" response to low doses of radiation. This phenomenon has been the center of many scientific discussions and arguments on the shape of the radiation-induced dose-response relationship for the induction of disease (Tubiana 2005; NRC 2006).

Radiation can alter the genomic stability of cells and tissues. Early research was reviewed and it demonstrated that radiation-induced genomic instability was observed after many cell divisions following radiation exposure. Genomic instability was manifested by an increase in the frequency and type of chromosome aberrations (Morgan et al. 1996). Soon after radiation, cells would divide and return to a "normal" state. After multiple cell divisions, genomic instability would develop, and many cells with chromosome abnormalities, most of which were not the result of clone formation, were observed in the population. Genomic instability was shown to be a frequent event per unit of radiation dose, so the target for its induction was much larger than a single gene. It was thus not related to a simple mutation in a single gene (Limoli et al. 1999; Ullrich 2003). However, it was difficult to demonstrate the changes in the frequency of genomic instability following low doses of radiation.

Kadhim et al. observed radiation-induced genomic instability in bone marrow cells from both humans and rodents (Kadhim et al. 1995). Ponnaiya et al. linked genomic instability to the sensitivity of different strains of mice to the induction of breast cancer, suggesting that it is an important step in radiation-induced cancer (Ponnaiya et al. 1997a). The observation of genomic instability has made it necessary to alter paradigms associated with the influence of single mutations on the induction of cancer and to take a more holistic view suggesting that tissue responses to genomic instability may be an important part of radiation-related cancer.

It was also demonstrated that the genetic background of the cells/tissues and organisms was very important in the magnitude and frequency of each of these new phenomena.

1. Bystander Effects: Cell-Cell and Cell-Tissue Communication

Before the program, cell-cell and cell-matrix communication and interactions were recognized as important in altering biological responses to many environmental insults. Such interactions play an important role on malignant phenotype during radiation-induced cancer (Bissell & Barcelloshoff 1987; Trosko et al. 1990; Park et al. 2000).

Nagasawa and Little (1992) published one of the earliest reports demonstrating that the target for the effects of ionizing radiation was larger than the cell nucleus. They observed that when CHO cells were exposed to a collimated alpha source at very low doses (0.31 mGy), 30% of the cells had an increased frequency of sister chromatid exchanges (SCE), even though fewer than 1% of the cells were calculated to have been traversed by an alpha particle.

Additional studies confirmed that many more cells were responding with an increased frequency of SCEs than had energy deposited in them (Lehnert & Goodwin 1997). This observation was in direct conflict with the current target theory and resulted in major discussions. The importance of bystander effects in the induction of SCE on cancer risk was questioned. There was concern that the observed increase in the frequency of SCEs as a bystander effect may not impact radiation risk (Bonassi et al. 2004). In addition, at this time it was impossible to tell which cells had energy deposited in them and which affected cells were neighbors, or bystanders. It was only known that fewer cells were hit than were responding.

1.1 Bystander effects demonstrated with Microbeam

The microbeam made it possible to place the alpha particles in a defined cell so that the cells that were hit and had energy deposited in them could be identified. This enabled biological changes to be studied in cells with and without energy deposited in them (Prise et al. 1998; Prise et al. 2002; Braby et al. 2006). After alpha particle microbeams were developed, additional equipment was developed that made it possible to use focused X-rays (Prise et al. 2003) and beta particles (Sowa et al. 2005; Persaud et al. 2007) to place energy in known cells and study the biological responses of both the hit cells and the bystanders. These studies suggested that bystander effects such as cell killing and micronuclei could also be seen following the exposure of individual cells to low-LET radiation (Prise et al. 2003; Schettino et al. 2003; Resat & Morgan 2004b; Persaud et al. 2007). All the cells could then be evaluated. It was determined that many cells were responding with biological changes without the deposition of energy.

The use of these tools and other techniques on both hit cells and non-hit cells in the same culture dish, separated with medium such that there could be no energy deposited in one set of cells while the others were being hit (Geard et al. 2002), or using the media from hit cells to initiate responses in cells that were not exposed to radiation, resulted in several publications. (Prise et al. 1998; Lyng et al. 2000; Azzam et al. 2002; Suzuki et al. 2004; Yang et al. 2005).

From these studies, it became evident that there are two basic types of bystander effects. First, there is direct cell-cell and cell-matrix communication that requires that the cells be in direct contact with each other (Azzam et al. 1998; Azzam et al. 2001; Nagasawa et al. 2002; Azzam et al. 2003a, b; Mitchell et al. 2004b) or with the matrix (Barcellos-Hoff & Ravani 2000; Park et al. 2000; Barcellos-Hoff & Brooks 2001). This contact depends on the presence of gap junctions between the cells and can be blocked by substances that inhabit gap junction function (Azzam et al. 2001,2003a, b).

The second type of bystander response demonstrates that cells with energy deposited in them release soluble factors, hormones, cytokines, or clastogenic factors into the media or the tissues. These produce alterations in other cells that do not have energy deposited directly in them (Lyng et al. 2000; Mothersill & Seymour 2001; Suzuki et al. 2004; Mothersill et al. 2005; Yang et al. 2005). These "media transfer" studies were conducted for a wide range of different systems where the media from irradiated cells (both following high- and low-LET exposure) were transferred to non-irradiated cells and biological effects were measured and demonstrated in the non-exposed cells.

It is important to recognize the wide range of biological endpoints that are modified by bystander effects. The earliest research on bystander effects using the microbeam was conducted in monolayer tissue cultures *in vitro*. Cell killing through apoptosis was an early endpoint that could be easily measured both in cells that had energy deposited in them and in their neighbors (Prise et al. 1998; Belyakov et al. 2001; Prise et al. 2002). Elimination of cells by apoptosis can potentially result in a decrease in damaged or transformed cells from tissues or organs and could result in a protective effect. These effects will be discussed in more detail in section two of this chapter, Adaptive Responses.

Early in the program, research was conducted on the ability of alpha particles to cause cell transformation. The tissue culture cells used in transformation studies were already altered and well on the way to the development of cancer. The transformation endpoint measured is an indication that radiation can move the cells through the final steps needed to move normal cells to cancer cells.

These early studies demonstrated that cell transformation *in vitro* could be induced by a single alpha particle (Miller et al. 1999). In tissue cultures where every cell had one alpha particle deposited in it, the transformation frequency was lower than in cultures where the alpha particles were randomly distributed with an average of one alpha particle per cell. It was postulated at the time that perhaps more than one alpha particle was necessary to induce cell transformation. As research has continued, it became obvious that many of the transformed cells may have been bystanders with no energy deposited in them. By conducting studies where only a small fraction of the cells in the population had alpha particles deposited in them, it was determined that in cells that did not have energy deposition in them, bystanders could also be transformed (Sawant et al. 2001b; Mitchell et al. 2004c). This transformation did not depend on the number of alpha particles deposited in the cells or the fraction of the cells exposed. Thus, these studies demonstrated that bystander cells that did not receive any energy deposition could be transformed by direct communication from exposed cells through those final stages from normal to cancer.

The kinetics of the initiation of cell transformation demonstrated that there was an early rise in transformation frequency with exposure to a small number of cells to single alpha particles, and that the frequency of transformation remained rather constant as a function of the number of cells exposed or the number of alpha particles that traversed the cells (Sawant

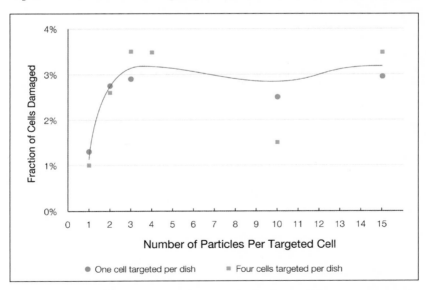

Figure 10. Bystander effect showing off-or-on type of dose response. Increasing the number of alpha particles per hit cell did not modify the frequency of micronuclei in non-hit cells (Bennett et al. 2007).

et al. 2001a). Figure 10 demonstrates these "on/off" or "all or none" nonlinear kinetics for the induction of cell transformation and micronuclei in bystander cells, which have been demonstrated in a number of cell systems and for a variety of endpoints including the induction of micronuclei (Belyakov et al. 2001; Azzam et al. 2002; Ponnaiya et al. 2004).

The slope of the hit-response relationship for the induction of micronuclei in cells that had alpha particles deposited in them was similar to that reported in other studies (Nelson et al. 1996; NRC 2005). However, in the bystander non-hit cells there was a non-dose-dependent increase in the number of micronuclei above that observed in control cells (about twofold). These types of binary behavior in the dose-response kinetics were also demonstrated using low-LET microbeams (Schettino et al. 2005). These dose-response relationships have been evaluated and many models developed to predict the impact of bystander cells on risk. These models will be further evaluated in chapter 8.

Because of the ease of scoring, the frequency of micronuclei has been used as an endpoint to evaluate the role of many experimental conditions on the induction of bystander effects. Research demonstrated that the target for induction of bystander effects was the nucleus. Cells were radiated with microbeams under conditions where only the nucleus and cytoplasm had energy deposited in them (Shao et al. 2004), and bystander effects were observed only when the nucleus was "hit." The influence of genetic background of the system studied had a marked influence on bystander effects (Zhou et al. 2005). It was determined that intercellular communication (Azzam et al. 1998; Azzam et al. 2001; Azzam et al. 2002; Shao et al. 2003a) was essential for the induction of bystander effects. The oxidative status of the cells altered the frequency of micronuclei and bystander effects (Azzam et al. 2002). It was determined that cells in different stages of the cell cycle had marked influence on the observation of bystander effects (Balajee et al. 2004), and the influence of time and distance between the cells with energy deposited in them and the bystander cells (Belyakov et al. 2002). Wu et al. also determined that as the energy across the Bragg peak changes there is little change in the frequency on micronuclei as a function of energy deposited in a localized area (2006). This again suggests that bystander effects are acting across this system to result in similar responses.

The types of cell and molecular-level damage observed in cells with direct passage of alpha particles were demonstrated to be different than those in bystander cells. In cells with direct energy deposition, scientists observed that the majority of the mutations were of the deletion and loss type, the same as was previously demonstrated for the induction of mutations from ionizing radiation (Jostes et al. 1994; Schwartz et al. 1994). However, most of the mutations induced in bystander cells were point mutations, base substitutions, and base changes that are more closely related to spontaneous mutations observed in control cells and mutations induced by chemicals (Jostes et al. 1994; Schwartz et al. 1994; Huo et al. 2001; Zhou et al. 2003). This important difference in mutation types could markedly affect the impact on radiation risk. However, the radiation risk associated with the induction of genetic disease has been evaluated, and it has been determined that the risk for genetic disease is 0.4 to 0.6% per gray of the very high baseline frequency (738,000 per million) (NAS/NRC 2006). This low value illustrates that the risk for genetic disease is much lower than the risk for induction of cancer, about 10% per Gy against a background frequency of cancer of 40%, and is not the primary concern when setting radiation standards (NAS/NRC 2006).

Studies on the induction of chromosome aberrations in bystander cells also demonstrated a marked difference in the type and frequency of chromosome aberrations induced in cells that were directly hit with ionizing radiation and those induced in bystander cells. It was again observed that the slope of the linear dose-response relationships for the induction of chromosome aberrations and micronuclei by alpha particles was similar to that previously reported (Brooks et al. 1994; Miller et al. 1996; Nelson et al. 1996; 1999; Geard et al. 2002).

1.2 Bystander Effects soluble factors

In other bystander studies, a technique was used in which cells that were exposed were separated by medium. The thickness of the media was such that the alpha particles would reach cells on one side of the flask resulting in energy being deposited in these cells, while cells grown on the other surface were out of the range of the alpha particles. Thus, no energy or direct alpha hits resulted in these cells (Geard et al. 2002; Suzuki et al. 2004). By quantifying the aberrations in cells with or without energy deposition, it was possible to identify damage produced in "hit cells" and damage from bystander effects in the "non-hit" cells. Cellular alterations were scored using the premature chromosome condensation (PCC) technique. It was discovered that cells with direct energy deposited in them contained chromosome-type aberrations (Suzuki et al. 2004). Again there was a non-linear dose-response relationship in the bystander cells, with a marked increase in bystander cell aberrations, above that seen in the control cells. The other important observation in bystander cells was that most of the aberrations were of the chromatid type. These types of aberrations are not normally produced by radiation of cells that are in the G_1 stage of the cell cycle. This indicates that the chromatin must have been damaged in the bystander cells to produce these aberrations. Chromatid-type aberrations are also a hallmark of cells that are becoming genomically unstable (Kadhim et al. 1995).

As studies have expanded, it has become increasingly evident that many biological changes are induced in bystander cells. An important observation has been that cells that do not have energy deposited in them have a change in gene expression following irradiation of cells that communicate with them (Azzam et al. 1998). Changes in gene expression have been very useful in understanding both bystander effects and other low-dose radiation effects. In addition to the formation of γH2AX in cells directly traversed by alpha particles (Prise et al. 2002), it has also been demonstrated that bystander cells have increased frequency of γH2AX and DNA DSBs (Sokolov et al. 2005; Smilenov et al. 2006). During repair (Little 2003) these DSBs can result in the formation of chromosome aberrations and γH2AX focus. This is expected because bystander cells also induce chromosome damage resulting in both chromatid aberrations and micronuclei.

The biology of bystander cells is modified in many ways. Studies have demonstrated that bystander cells have modifications in the cell cycle (Balajee et al. 2004), and there is evidence for chromatin damage in bystander cells that results in the induction of chromatid-type aberrations (Suzuki et al. 2004). In organized tissues, bystander cells seem to be forced to differentiate in non-standard ways (Belyakov et al. 2002, 2006).

It is important to understand the physical and biological variables associated with the bystander response. A number of important scientific questions relative to the bystander effects are posed, and the early research addressing these questions will be examined. To fully understand the significance of bystander effects to cancer risk, the following questions need to be answered:

- What is the cellular target for the initiation of the bystander response?
- Which cell types can communicate with each other?
- Can all the cells in a tissue respond to bystander signals?
- What molecules and structures are involved in communication of the bystander effects?
- What is the time required for the communication?
- Over what distance can cell-cell communication be observed for the cell-cell contact type of bystander effect?
- Do the bystander effects occur in whole organisms within tissues and between different tissues?

Many studies using microbeams have demonstrated that the prime target for the induction of bystander cells is from deposition of energy in the nucleus of the cell using either high- or low- LET radiation (Prise et al. 2002; Morgan 2003b; Schettino et al. 2003; Zhou et al. 2003; Hall 2006). Exposure of only the cytoplasm has also been shown to initiate bystander effects (Shao et al. 2004), but the response is not as robust as deposition of energy in the nucleus.

It has been noted that the communication between cells is almost universal when the cells are the same type. Thus, studies demonstrated that fibroblasts, epithelial cells, and other cells can communicate with each cell type. However, it has also been demonstrated that only a limited number of individual cells within any tissue can respond to a bystander signal, and that material from one cell type can stimulate a bystander response in other cell types. These studies have been carefully reviewed (Lorimore et al. 2003; Morgan & Sowa 2005; Hall 2006). Nevertheless, it has been demonstrated that communication between some different cell types can be limited. For example, fibroblasts and glioma cells communicate within each cell type, but communication between the different cell types appears to be unidirectional (Shao et al. 2004). The fibroblasts can communicate and produce responses in glioma cells, but exposure of the glioma cells does not seem to be able to produce changes in the co-cultured fibroblasts.

Extensive research has been conducted to determine how the signals are transmitted from one cell to the next during the initiation of the bystander response. It has been well established that the membranes between the cells play an active role in the communication of the messages between cells (Azzam et al. 2001; Nagasawa et al. 2002) and that the extracellular matrix also plays an important role in signaling and controlling the fate of cells (Barcellos-Hoff & Brooks 2001).

The physiological state of cells plays an important role in bystander communication. For example, it has been established that inflammatory responses play an important role in cell-cell communication (Lorimore et al. 2003). The redox status of the cells (Spitz et al. 2004; Hu et al. 2006), the energy and oxidative metabolism (Mothersill et al. 2000; Azzam et al. 2002), and the oxidative stress-related pathways responses and the molecules associated with them are all involved in bystander effects (Azzam et al. 2002; Spitz et al. 2004; Hu et al. 2006). The nutritional status of the cells and cell-cell contact are important in bystander responses and in the production of radiation-induced hypersensitivity (Chandna et al. 2002). Studies continue on the molecules involved. The molecules and mechanisms involved in the communication are described in chapter 4.

The length of time required for an exposed cell to produce a response in a bystander cell has also been the subject of research and has been reviewed (Morgan 2003b). For direct cell-cell communication, the length of time required for transmission of signals is very short (<1 minute). Since the time required for communication in many systems was much longer, direct cell-cell communication is not involved (Little et al. 2002a; Banaz-Yasar et al. 2006). After stimulus of cells by radiation, substances responsible for bystander effects are secreted into the medium very quickly.

The distance over which bystander effects can be seen has been an important area of research. Using culture systems, it was evident that the bystander cells can be detected throughout the whole culture dish (Azzam et al. 1998; Prise et al. 1998; Azzam et al. 2001; Prise et al. 2002; Belyakov et al. 2001; Sawant et al. 2001b; Ponnaiya et al. 2004) and that the distribution of the cells displaying bystander effects seems to be randomly distributed over the dish (Azzam et al. 2001; Prise et al. 2002). These data have been reviewed (Morgan 2003b) and suggest that all the cells on the dish communicate with each other.

When more specialized three-dimensional cell systems were devised *in vitro*, a range of different results were observed. In a system where a strip of cells was exposed using a specialized microbeam, the communication was limited to a few cells in close contact with the exposed cells (Belyakov et al. 2006). In this system the cell-cell communication was only a matter of a few cell diameters. More complex tissue systems were developed using both human and porcine urothelial explants where it

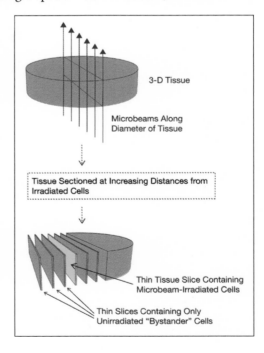

Figure 11. A plane of cells was irradiated using a microbeam. Bystander response was measured in cells at different distances and times from the hit cells, making it possible to determine how fast and far from the hit cells the bystander response could be detected (Belyakov et al. 2005).

was possible to detect a wide range of different cell responses under more realistic physiological conditions (Belyakov et al. 2002, 2003). This system was exploited to detect bystander changes as a function of distance from the exposure and is illustrated in Figure 11.

The tissue was exposed at a known location, and sections were taken as a function of distance from the exposure. The frequency of apoptotic cells (Belyakov et al. 2002) and cells undergoing premature differentiation (Belyakov et al. 2006) was measured in each section. It was determined that a constant frequency of altered bystander cells could be detected as a function of distance from the target. These changes were detected over a range of up to 1 millimeter from the radiation site (Belyakov et al. 2005). This system was also very useful for measuring bystander-induced cell proliferation (Belyakov et al. 2003).

1.3 Bystander effects in whole animals

Such measurements suggested that bystander effects could be detected in whole human or animal tissues. The existence of clastogenic factors that cause chromosome damage when cells are treated with serum from radiation-exposed animals has long been recognized and has been reviewed (Morgan 2003a). When one part of an organ, the lung, was exposed to high doses of radiation, responses (micronuclei formation) were seen to be greatly elevated in the remainder of the lung. It seems that certain lobes of the lung communicate the existence of damage in a directional fashion, from the bottom to the top. These responses were limited to the lung (Khan et al. 1998).

Other evidence for bystander effects were observed in the liver *in vivo*. These data were reviewed and include both the formation of chromosome aberrations and cancer (Brooks 2004). In these studies, the frequency of both chromosome aberrations and liver cancer was the same whether 1% or 100% of the liver cells had energy deposited in them from internally deposited ^{239}PuO$_2$ particles. The frequency of change in both endpoints depended on the total dose to the organ and not on the number of cells hit or the dose to individual cells. However, when animals were exposed to internally deposited radioactive materials, which concentrated in target organs, it was noted that clastogenic factors and effects in organs not exposed effects were not evident. Only the organs with the isotopes deposited in them formed cancer (Stannard and Baalman 1988).

It has been postulated that a small amount of radiation over time, such as the low dose-rates from these exposures, results in a low level of clastogenic or any other soluble factors. It has been noted that following high dose-rate exposure to high total doses, soluble compounds seem to result in either the cause or cure of cancers at sites distant from the radiation. The role of ascopal effects, bystander effects, and clastogenic factors in the induction of genomic instability and cancer was reviewed (Morgan 2003a).

A number of reviews have been written on bystander effects and provide extensive additional information on these effects (Hall 2000b; Mothersill & Seymour 2001; Hall 2003; Morgan 2003b, a; Azzam & Little 2004; Hall 2006). These reviews presented conclusive data demonstrating that the cells did not need to have energy deposited in them to elicit a wide variety of biological responses. It seems that many small molecules including calcium (Lyng et al. 2006) and nitric oxide molecules may be involved (Shao et al. 2003b) in transmission of the bystander responses between cells. Additional discussion on the pathways, such as the MAP kinase signaling pathway (Lyng et al. 2006) and other mechanistic studies, is found in chapter 7.

The observation of bystander effects in so many molecular, cellular, and whole-animal studies impacts the use of hit theory to understand the relationships that exist between radiation dose and biological responses. The use of hit theory must be modified in light of the bystander effects to show that the target size for the biological response is much larger than the cell nucleus.

Cell transformation and other biological changes in the DNA are important endpoints that suggest that bystander effects could increase the risk at low radiation doses (Hall 2000a). The importance of these responses in terms of risk assessment remains an open question and has been reviewed (Morgan & Sowa 2009). However, a recent publication suggested that researchers were not able to detect bystander effects following exposure to high-LET radiation (Groesser et al. 2008). These studies measured several cytogenetic endpoints including the induction of micronuclei, γH2AX, and cell killing. Additional studies need to be conducted to help resolve this observation with the literature on the induction of bystander effects by high-LET radiation using several experimental systems.

Several studies have been published on the impact of bystander effects on genomic instability and the risk for the induction of cancer (Brenner & Elliston 2001; Brenner & Sachs 2002a, 2003; Morgan & Sowa 2009). From these studies it was concluded that the bystander effects of alpha particles may influence the shape of the dose-response curve, but that the risk currently used to estimate radiation risk, for example for radon, may not be markedly influenced by the bystander effects (Brenner & Sachs

2003). Additional discussion of the risks associated with bystander effects is covered in chapter 9.

1.4 Major Points: Bystander Effects
- Cells that have energy deposited in them communicate with neighboring cells, which do not have energy deposition. This result in responses that may potentially be protective or detrimental. Bystander effects exist both *in vitro* and *in vivo*.
- There are two different types of bystander effects; those that require direct cell-cell and cell/matrix contact and those that result from release of substances into the media or blood.
- The bystander effect results in changes in several different biological endpoints and depends on the physiological and oxidative status of the cells and tissues.
- The dose-response relationships for the induction of bystander effects are non-linear, with a low-dose resulting in the maximum response followed by a plateau as the dose increases.
- The type of damage in bystander cells differs from the type of damage induced in cells with energy deposited in them.
- Because of bystander effects, tissues respond as a whole to ionizing radiation and not as single cells. These tissue responses are non-linear.
- There is evidence that bystander cells may either increase or decrease the radiation-related cancer risk.

2. Adaptive Responses

2.1 Background
Many physical and chemical agents are toxic when given at high doses (e.g., vitamins, aspirin, many toxic agents, stress) but have protective and beneficial effects when given at low doses (Luckey 1991; Calabrese 2004). For example, vitamins are very toxic at high doses but are essential for life at low doses. Exercise also has beneficial health effects even though it generates many free radicals that are known to be damaging and increase cancer risk when levels are too high. These protective non-linear relationships between exposure, dose, and response have been extensively reviewed and are collectively termed hormesis (Calabrese & Baldwin 2003). Hormesis is the production of a beneficial effect caused by a low dose of an insult.

For radiation, it has long been assumed that each ionization has the potential to produce DNA damage, and that DNA damage is linearly linked to the formation of adverse health effects, including cancer and genetic effects. Thus, models have been developed that predict that damage and risk from radiation exposures increase linearly with radiation dose. Data support this idea for the induction of DNA damage. However, recent research, much of which was funded by the Low Dose Radiation Research Program, has demonstrated that the processing of radiation-induced damage and the total response to radiation in the low-dose region is non-linear. This non-linear processing can result in "protective adaptive responses" in the low-dose region. Many biological responses to radiation are very different in the low-dose region from those seen in the high-dose region. Such research predicts that the mechanisms of action for biological responses change as a function of radiation dose.

The prime argument against a protective adaptive response for radiation has involved the way that radiation interacts with and alters cells. Radiation deposits its energy in discrete sites in cells and molecules. The biological action on these sites is randomly distributed in the tissue and is unique. Because of these facts, it was suggested that radiation-induced damage cannot be compared to damage from chemicals or stress, in which all molecules and cells in a tissue organ or organism will receive the exposure uniformly and have the potential for being affected. However, extensive research reviewed in the last section of this chapter illustrates the role of bystander effects following radiation exposure. These effects demonstrate that even though the energy is deposited in random defined sites and the initial DNA damage increases as a linear function of exposure, radiation effects are not limited to the individual cells where the energy is deposited. The whole biological system responds to the insult in the same way as seen for chemicals, and the processing of the radiation damage is non-linear. This non-linear processing of radiation-induced change supports the existence of protective adaptive responses.

Because of these observations and the biological complexity associated with cancer induction, a systems approach rather than the use of the hit theory to predict radiation effects is required and must be the focus of future research. Thus, both scientific opinion (Jenkins-Smith et al. 2009) and research reviewed here currently support the existence of non-linear dose-response relationships in the low-dose region.

The field of radiation-induced hormesis has been carefully reviewed and several literature citations gathered to support the concept that low doses of radiation can have a protective effect for many different endpoints (Calabrese & Baldwin 2003). From this research it is suggested that in the low-dose region hormesis predicts the response to many agents better than other models, including the LNT and threshold models (Calabrese et al. 2007). The hormetic, or potential protective effect, from low doses of ionizing radiation continues to be debated, and the impact of low doses on radiation risk is a major subject of controversy (Brenner et al. 2003; Feinendegen 2005; NRC 2006; Averbeck 2009; Brenner 2009).

2.2 Adaptive response, priming dose and challenge dose

The diminished response of a biological system to low doses of radiation has been termed "adaptive response." Adaptive response was first demonstrated in studies where cells were treated with tritiated thymidine followed by exposure to large doses of X-rays (Olivieri et al. 1984). Surprisingly, the frequency of chromosome aberrations with the tritiated thymidine followed by the high dose of X-rays was lower than when the X-rays were given alone. These studies were followed up by exposing cells to a priming dose, very low doses of X-rays (10-50 mGy), followed soon after by a larger (1.0–2.0 Gy) challenge dose. The frequency of chromosome aberrations induced by the challenge dose when the cells had received a prior small or "tickle" dose was reduced relative to that observed when the challenge dose was given alone. This observation is illustrated in Figure 12.

Many publications resulted on the reduction in the frequency of chromosome aberrations as an indication of the induction of adaptive responses. The results of these studies were carefully reviewed by Wolff (1998), who noted that the genetic background of the biological materials used in the test plays an important role in the adaptive response. For the people tested for adaptive responses, many individuals were "responders," and others were not. The importance of genetic background on the induction

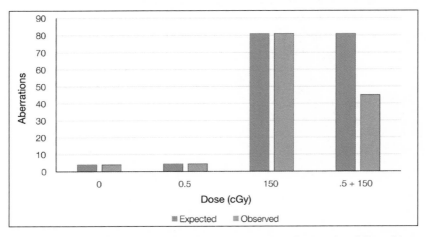

Figure 12. Expected and observed results from a small radiation dose followed by a larger dose showing that a small dose decreases the response, indicating the presence of an adaptive response (Shadley et al. 1987).

of bystander effects, adaptive responses, and genomic stability will be discussed further as it applies to each of these biological observations following low doses of radiation. The classic "adaptive response" has been demonstrated for several different biological endpoints including the induction of apoptosis (Boothman et al. 1998), cell killing (Sahijdak et al. 1994; Mitchell & Joiner 2002), micronuclei (Shankar et al. 2006), cell cycle changes (Boothman et al. 1996), gene expression (Coleman et al. 2005), mutations (Zhou et al. 2003; Sykes et al. 2006a; Tsai et al. 2006), and cell transformation (Redpath et al. 1987; Azzam et al. 1994).

An important system for measuring the adaptive response *in vivo* was to measure the frequency of recombinational events using the pKZ1 recombinational mutation assay in mice (Sykes et al. 2006a). In this system it was possible to measure responses to very low doses (0.005 and 0.001 mGy). These very low doses were responsible for depressing the response induced by a high dose of radiation. In many of the research projects conducted, the level of mutation found was depressed below the spontaneous mutation frequency. With this experimental approach, there was a very complicated dose-response relationship between the exposure to very low doses as a priming dose and the frequency of mutations induced by the challenge dose. Following exposure to high doses (1 Gy) given alone, the frequency of mutations showed a marked increase above the spontaneous level. Low doses given before the high dose resulted in a decrease in the mutation frequency below that observed in the control animals. As the tickle doses continued to decrease there was an increase in the mutation frequency.

There was some concern that this system was unique to cells involved in the immune system. The argument was that the cells used in the assay were a part of the immune system and that the observed adaptive response was a reflection of the spontaneous rearrangements known to be made as adaptive responses of these cells to antigens. Thus, it was suggested this adaptive response might not be related to radiation-induced cancer.

To evaluate this potential concern, other tissues in the mice, prostate (Hooker et al. 2004a) and spleen (Day et al. 2007a), were measured. It was determined that low doses to other somatic tissues showed the same decrease in the frequency of mutations, below that observed in the tissues of non-exposed animals, as was seen in the cells associated with the immune system. These studies resulted in very interesting dose-response relationships that require additional study. They demonstrated that genetic background of the mice had a marked role on the induction of this adaptive response (Hooker et al. 2004b).

Additionally, this system is the only one that has been able to demonstrate that the order of delivery of the doses, large challenge and small tickle, is not critical and that adaptive responses can be generated with both exposure schedules (Day et al. 2007a; Day et al. 2007b). The scientific concern about the importance of the classic adaptive response—that induced by a small priming dose of radiation followed by a large challenge dose—in terms of radiation-induced cancer risk is that this type of "adaptive response" induced by the priming dose was only active for a short period. Thus, it was thought that the classic adaptive responses might have little impact on radiation risk.

2.3 Adaptive response: decrease in response below background

In addition to the adaptive responses induced by a low priming dose followed by a challenge dose, adaptive response studies have been conducted on the influence of low doses of ionizing radiation on the background frequency of biological changes. This type of adaptive response may be more important in terms of the impact of low doses of radiation on cancer risk because it suggests low doses can be protective against many biological changes induced by other types of exposure as well as from the genetic background that may be involved in cancer induction. To conduct these types of studies, it was necessary for the endpoint of interest to have a rather high background rate, such as is seen for cell transformation and cancer.

A number of systems were developed to measure the influence of low doses of radiation on the spontaneous frequency of biological alterations and changes related to cancer induction. The most widely used of these systems measured the frequency of cell transformation as an endpoint. The primary cell transformation systems used were either a human hybrid cell system that has a high spontaneous frequency of cell transformation (Redpath et al. 1987) or the mouse embryo C3H 10T1/2 cell system (Azzam et al. 1994; Mitchel et al. 1997). With these tools, it was possible to expose the cells to low doses of radiation and determine the change in the frequency of cell transformation as a function of different dose parameters.

Many studies were conducted measuring cell transformation that showed low doses of ionizing radiation decreased the spontaneous frequency of cell transformation below that observed in control cells receiving no radiation exposure (Azzam et al. 1994; Azzam et al. 1996; Redpath 2004, 2006a, b). An example of the type of results demonstrated in many of these studies is shown in Figure 13.

Extensive studies have been conducted to determine the role of exposure variables on the induction of adaptive responses that decrease the spontaneous frequency of cell transformation. It was important to determine the role of total dose (**Redpath 2006a, b**) and dose-rate (**Elmore et al. 2006; Elmore et al. 2008**) on the induction of adaptive responses. Figure 13 shows the results of low-dose exposures (10–100 mGy) in decreasing the background rate of cell transformation. In the dose region below 100 mGy the frequency of cell transformation was decreased below that observed in the control non-exposed cells (**Redpath 2006b**). Thus, low total dose can reduce the frequency of transformed cells and may be protective.

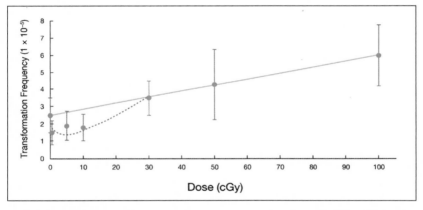

Figure 13. Sub-linear dose response demonstrating protective adaption where the low doses result in less damage than seen in the background response (Redpath et al. 2001).

When the dose-rate was decreased, the response in the exposed cells depended on both the total dose and the dose-rate. If the dose-rate was delivered at 1.9 mGy/min, there was a positive dose-response relationship for all dose groups up to as high as 1000 mGy. However, as the dose-rate was decreased to 0.47 mGy/min and below, the response in the control cultures was higher than that observed in the groups exposed to ionizing radiation at all the doses evaluated, up to 1000 mGy. Thus, low dose-rate exposure may have a protective effect over a much broader total dose range than observed for single acute exposure.

The dose-response relationship for the induction of micronuclei in cultured cells showed different responses following exposure to high or low dose-rates. Studies *in vitro* using normal human fibroblasts demonstrated that following a high dose rate exposure to a low total dose of 0.1

Gy (10 cGy), the frequency of micronuclei per cell was higher than the control values. However, when the same dose was given at a low dose rate and protracted over 48 hours, the level of micronuclei observed in the irradiated cells was lower than observed in control cells (deToledo et al. 2006). These data are shown in Figure 14 where frequency of micronuclei per 100 cells induced by a low dose (0.1 Gy) is related to radiation exposure time and thus, dose rate.

2.4 Adaptive response: dose rate effects

A marked dose rate effect was also observed in the lack of increase in the frequency of micronuclei in bank voles living in the zones of high radiation following Chernobyl. In spite of having calculated doses that were greater than 1.0 Gy/year, delivered at a low dose rate, there was no detectable increase in the frequency of micronuclei (Rogers et al. 2000). Such data on micronuclei support the concept that radiation adaptive protection for the induction of chromosome aberrations exists in the low dose and low dose-rate region (deToledo et al. 2006; Dauer et al. 2010; Feinendegen et al. 2011) and suggest the potential induction of a protective response that may require a negative term in modeling risk.

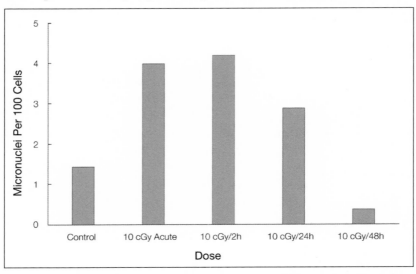

Figure 14. Influence of Dose Rate on Chromosome Damage. The major point of interest is that when the radiation exposure was protracted over a forty-eight hour time period the frequency of micronuclei was less than observed in the control cells. The role of cell cycle and the elimination of damaged cells through cell division could be a competing factor.

It was also determined that cells exposed to chronic low dose-rate exposure had a higher survival rate when challenged with a subsequent acute radiation exposure than cells that had the challenge dose only (Mitchell & Joiner 2002). Thus, low dose-rate exposure seemed to be protective for cell survival as well as cell transformation. When radiation dose was fractionated there was also a decrease in cell transformation (Bennett et al. 2007). It was important to determine if the types of radiation used in diagnostic procedures would induce such a response, since it has been predicted that the use of diagnostic radiation procedures, especially CT scans, could potentially result in a marked increase in total cancers in the population (Brenner & Elliston 2004). Research was conducted on the influence of diagnostic energy X-rays, (Redpath et al. 2003), and on mammographic energy X-rays (Ko et al. 2004). Both of these types of exposures resulted in a decrease in the frequency of cell transformation in the dose range used for diagnostic evaluations. The results of these types of studies were reviewed, and it was concluded that the low photon energies used in medical imaging all produced a reduction in the frequency of cell transformation (Redpath 2006a).

Additional studies on adaptive responses were conducted to simulate exposures found in the space environment. In these studies, low doses of proton exposure were given prior to exposure to HZE particles. This resulted in an adaptive response in a cell system using anchorage-independent growth in primary human fibroblasts as an endpoint, another measure of cell transformation (Zhou et al. 2006). The role of radiation type and changing LET on the induction of adaptive responses and bystander effects has been carefully reviewed (de Toledo et al. 2006). In this review, an adaptive response was observed over a wide range of LETs. Such a manuscript provides a useful reference for further study of these exposure variables on biological responses.

2.5 Adaptive Response: Mechanisms of action

The next task at hand was to link the changes in gene expression to a measurable biological response. This was done by developing cell lines that were unique in their ability to mount an adaptive response. Cell lines were classified as non-adaptive or adaptive cells. The cell lines that were able to initiate an adaptive response for radiation-induced chromosome aberrations had a different gene expression profile than cells with the genetic make-up that made them non-adaptive for the same endpoint (Coleman et al. 2005). All of the cells were exposed to 50 mGy and their genome evaluated for changes in gene expression. Of a total of 12,000 genes evaluated, the number of genes that had their gene expression significantly either up- or down-regulated in non-adaptive cells was 57. The number of unique changes in gene expression in adaptive cells was 45, and genes that changed gene expression in both cell lines after an exposure to 50 mGy totaled 47. These changes in gene expression are illustrated in Figure 15.

Extensive advances in biological and physical sciences over the past twenty years have made it possible to rapidly measure radiation-induced biological changes in gene expression in many genes at one time. These advances have been spurred on by the sequencing of the genome and the development and use of gene expression arrays and other more informative methods such as RNA sequencing.

Yin et al. (2003) used gene expression arrays to measure radiation-induced changes in gene expression in thousands of genes at the same time. Following high dose-rate exposure over a wide range of total doses, the pattern of gene expression changed markedly as a function of total dose. Research determined that changes in gene expression include, "… genes that respond only to low-dose exposures, genes that are unique to high-dose exposures, and genes that are modulated in their expression at both high and low doses" (Coleman and Wyrobek 2006). The changes have been measured in human and mouse cells. A brief summary of these data are provided in Table 1.

Table 1. High and Low Dose Reponses

Response	High Dose	Low Dose
Cell killing	High	Low
DNA damage	High	Low or not detected
Gene expression	Damage	None or protective
Epigenetic effects	?	Protective
Free radicals	Increased	Decreased[a]
Apoptosis	Increased	Selective[b]
Mutation frequency	Increased	Decreased
Cell transformation	Increased	Decreased
Immune response	Decreased	Increased
Carcinogenesis	5%/Sv	?

(a) Stimulates production of the antioxidants MnSOD and glutathione.
(b) Preferentially eliminates damaged or transformed cells.

Table 1. Radiation induced changes in gene expression in critical biological pathways. These data demonstrate that genes in many pathways respond to either high or low total doses suggesting unique mechanisms (Coleman and Wyrobek 2006).

The table illustrates that some pathways that change gene response in human cells *in vitro* differ from those in mouse *in vivo*. Also of relevance to the present topic are the unique responses in the low dose region for genes responsible for heat shock proteins, immune response and protein synthesis. The data suggest that cell cycle genes and those involved in cell signaling are modified by both low and high doses in each of the systems tested. Such results make it difficult to use these data for human risk assessment. Extensive research on gene expression levels has been conducted and for the most part support the data presented in Table 1 (Fornace et al. 1999; Amundson et al. 2003a, 2003b; Ding et al. 2005). Gene expression patterns change as a function of dose. Some of the genes involved in these radiation induced changes are very important in the critical pathways to cancer. A review of the changes in gene expression as a function of dose suggest that they reflect a change in the mechanisms of action (Dauer et al. 2010). Such studies demonstrate that the responses to this key event (gene expression) in the critical cancer pathways change as a function of radiation dose. High doses trigger pathways that have been postulated to be damaging while the molecular pathways up-regulated by low doses have been considered to be protective.

Such information provides the basis for postulating that the cellular responses to radiation-induced damage are different following low doses than after high doses. These mechanistic studies need to be expanded to high and low dose-rate radiation responses that can be used to predict cancer outcome based on mechanistic understanding.

Radiation alters gene expression patterns and induces signaling processes. Such signaling processes initiated by DNA damage are an important factor in determining radiation responses. These responses are involved in activation of many genes associated with stress following radiation exposure (Amundson et al. 2003a). Many of these stress genes are associated with transcription factor p53, "one of the key elements in cellular response, which can regulate nearly 100 genes that have already been identified" (Fornace et al. 1999).

The functions of the unique genes in each of these cell lines were evaluated and placed into four categories: (1) genes that were up-regulated in all the cells, (2) genes that were down-regulated in all the cells, (3) genes that were up-regulated in adaptive cells and down-regulated in non-adaptive cells, and (4) genes that were down-regulated in adaptive cells and up-regulated in non-adaptive cells. It was determined that the third group included the genes involved in DNA repair and cellular responses to stress. The group four genes were associated with the induction of apoptosis and the regulation of cell cycles.

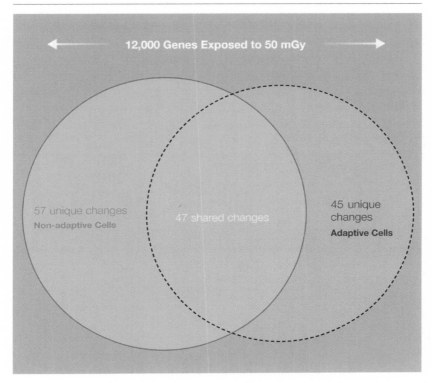

Figure 15. Radiation-induced changes in gene expression in adaptive and non-adaptive cells. Cells were given an adaptive dose of 50 mGy, and 12,000 genes were evaluated. A difference in gene expression was shown between adaptive and non-adaptive (Coleman et al. 2005).

It has been well established in human skin that gene expression changes as a function of both dose and time after exposure (Goldberg et al. 2004). Several critical genes were evaluated in human skin biopsy as a function of distance from the treatment area for prostate cancer as a function of both dose and time after exposure (Figure 16a and b).

These studies demonstrate that critical genes are both up- and down-regulate as a function of time after exposure and dose. These sets of genes were all down-regulated at low dose (1 cGy) when sampled at 24 hours. However, the same set of genes were either not changed or up-regulated when sampled at 1 hour after exposure to a 10-cGy radiation dose.

Such studies help identify the potential mechanisms of action associated with changes in adaptive responses. Other human studies have indicated that there are transient genome-wide changes in transcriptional

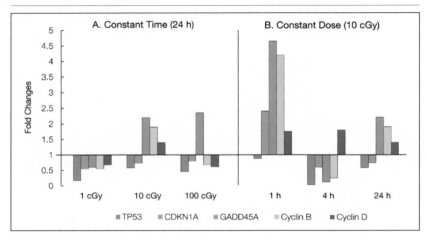

Figure 16. Gene expression changes as a function of both (A) radiation dose and (B) time after exposure (Goldberg et al. 2004).

responses following exposure to ionizing radiation (Berglund et al. 2008). Many of the changes in gene response are directly related to stress responses (Sheikh & Fornace 1999). Thus, even following low doses of ionizing radiation the cells recognize exposure to ionizing radiation as a form of stress. Such stress responses have also been implicated in many processes related to aging. Radiation exposure of the brain demonstrated that many of the changes in gene expression could be related to changes in cognitive functions, Alzheimer's disease, and other changes related to advanced aging (Lowe et al. 2009). However, there has been little evidence to document that radiation causes changes in the aging processes. This may be an area of fruitful future research.

It is important to remember that gene expression and protein production and the lifetime of the proteins may not be linearly related. Relating gene expression to protein expression following radiation is an important research area (**Coleman et al. 2000**). The next important step is to relate changes in proteins to the development of cancer. This was done using a proteomics approach to study leukemia induction in mice (**Rithidech et al. 2007a; Rithidech et al. 2007b**). Various *in vitro* systems have been used to study the role of low doses in altering the spontaneous frequency of micronuclei, cell killing, and apoptosis (**Sahijdak et al. 1994; Kagawa et al. 2001**). It has been postulated that changes in apoptosis may be responsible for many of the adaptive responses seen following low doses of radiation and could differentially remove transformed cells from a cell population, reducing the background frequency of cell transformation and potentially reducing the risk for cancer in the low-dose region below the spontaneous level (Bauer 2007a).

Although the field of radiation-induced apoptosis has expanded very rapidly at the time of the Low Dose Radiation Research Program, it will not be reviewed in detail here. The research conducted on radiation-induced apoptosis under the program and its implications will be discussed. Apoptosis has been postulated to play an important role in adaptive responses for both the formation of cancer (Bauer 2007a) and mutation induction (Sykes et al. 2006a). Building such relationships between the different cellular endpoints and the induction of the adaptive response provides critical information required to link molecular changes, cellular changes, changes induced in whole animals to the cancer induction, and risk in humans. The mechanisms for the induction of apoptosis show that this radiation-induced process plays an important role in cancer risks at low doses and will be further evaluated in chapter 7.

Research has demonstrated that low doses of ionizing radiation can also have effects in whole-animal *in vivo* studies. Individual chemical changes induced by low doses of radiation show protective effects. This was demonstrated as radiation-induced increases in apigenin reduced the frequency of chromosome damage in human lymphocytes (Rithidech et al. 2005). These studies demonstrated that low doses of radiation can depress the frequency of birth defects during embryogenesis (Wang et al. 2000). Low doses can also decrease the frequency and increase the time of onset of cancer in mice (Mitchel 2006). It has also been demonstrated that low dose-rates from low-LET radiation delivered to mice can result in a decrease in the frequency of cancer (Sakai et al. 2003). Low doses of low-LET radiation have been suggested to decrease the frequency of plutonium- or smoking-induced lung cancer (Sanders & Scott 2008). Exposure of dogs to internally deposited radioactive materials that emit low-LET radiation resulted in no change in the frequency of lung cancer at total cumulative lifetime doses of less than 20 Gy (Brooks et al. 2009).

Radiation-induced adaptive responses are thus well established in molecular, cellular, and whole-animal studies. However, to have an impact on standard setting it is important to demonstrate that such changes may exist in human populations exposed to low doses of radiation. The epidemiological studies of human populations exposed to low doses and dose-rates have been carefully reviewed (NRC 2006). The ability to determine the shape of the dose-response in the low-dose region is very difficult using epidemiological methods (Shore 2009). Many studies of cancer incidence in populations exposed to elevated low dose-rates from natural background show no response, less-than-predicted responses, or small positive responses (NRC 2006). Other studies of nuclear workers exposed to low doses of radiation showed a small but insignificant increase above the background level. Thus, human studies of populations exposed to low dose or dose-rates from radiation suggest that there is an increase in risk from low-dose exposures but fail to demonstrate either a significant positive or negative dose-response relationship.

However, "the risk from radiation exposure is highly quantifiable in terms of modifying factors such as age and sex, exposure to other carcinogens such as tobacco smoke and the measurable effects of other factors, usually unknown, that influence variations in baseline cancer rates by populations" (Land 2009). The difficulty is in part related to the variables in the baseline to cancer lifetime radiation-induced risk. The risk of cancer diagnosis in humans is about 45%, and lifetime risk of cancer mortality is about 25% (NRC 2006). Even in human populations that received a substantial radiation dose delivered at a low dose rate, controversy remains. Papers have been published stating that ionizing radiation delivered at a low dose rate can reduce the cancer frequency in humans (Chen et al. 2004a).

Studies have also been conducted that produced mixed findings on the usefulness of the adaptive responses or low dose-rate exposure as a way to protect normal tissue during radiation therapy for cancer (Chen et al. 2004a; Redpath 2007). However, because most of these clinical studies were not conducted as part of the DOE Program, they are not reviewed here.

The major points associated with radiation-induced adaptive response are summarized below and demonstrate that the adaptive response is a real and important biological phenomena that must be considered when evaluating radiation standards. If nothing else, the data demonstrate that the LNT hypothesis currently recommended for setting standards is conservative in the low-dose region and that there is extensive well-documented scientific data on the responses in this region that support this statement.

2.6 Major Points: Adaptive Response

There is a long and well-documented history of hormesis research that demonstrates many chemical and physical agents that produce damage at high doses, but elicit protective responses at low doses. Much research supports protective adaptive responses for low doses of radiation.

There are two major types of adaptive responses: (1) when a small tickle dose of radiation is given prior to or shortly after a large radiation dose the response is less than if the large dose is given alone, and (2) low

doses of ionizing radiation produce a reduction in the background frequency of many biological responses.
- The cellular and molecular responses following exposure to low doses of radiation are different from those induced by high doses, suggesting different mechanisms of action for high and low doses.
- The radiation-induced adaptive response is a very general biological phenomenon and has been carefully documented for many important biological endpoints including the induction of DNA damage, mutations, micronuclei, chromosome aberrations, cell killing, apoptosis, genomic instability, and cell transformation.
- The adaptive response has been demonstrated both *in vitro* and *in vivo*.
- Adaptive responses in many cellular systems result in responses that are lower than observed in the unexposed control cells.
- The adaptive response suggests that there is a need for a change in the current paradigms associated with the LNT biophysical models used to estimate risk.
- The extensive data generated from research on the adaptive response suggests that following exposure to low doses of ionizing radiation, the LNT assumption is conservative.

3. Genomic Instability
3.1 Background
When cells are exposed to ionizing radiation, immediate changes that influence the genetic status of the cells are observed. Some of these include the induction of DNA breakage, changes in DNA bases, mutations, chromosome aberrations, and cell killing. These changes are related directly to the exposure conditions including dose, dose-rate, and dose distribution. The frequency of these changes increases as a function of the radiation exposure and dose. The frequency of the radiation-induced changes also depends on the tissue where the changes are measured. After the initial response, the damage induced by the exposure decreases as a function of time after the exposure until the frequency of cells showing changes in the system returns to near normal levels of damage. Such phenomena have been investigated for many years in multiple systems and have been summarized nicely in thousands of publications. These genetic alterations are correlated with the induction of genetic disease and cancer.

It was noted that there are multiple genetic changes in many types of cancers that reflect the loss of genetic stability of the cells. This loss of genomic stability seems to be one of the hallmarks of the cancer process (Hanahan & Weinberg 2000) and is critical as the cells take on a cancer phenotype. Genomic instability is a marker of cancer and is widespread in many cancer types (Lengauer et al. 1998). However, it is not known if the genomic instability is actually induced by the agent (such as radiation) that "caused" the cancer or is simply a reflection of the cancer process where cells have escaped genetic control present in normal tissues.

Extensive research has been conducted that demonstrates the many ways that cells maintain their genetic stability. Research conducted in the program in this subject area identified many genes involved in control of the genomic stability of the cells. Specific genes have been identified that control against the induction of DNA cross-linking (Liu et al. 1998). Other important genes that maintain genomic stability, including chromosomal stability, have also been characterized (Liu et al. 1998; Honma et al. 2000; Fujimori et al. 2001). For example, it has been shown that specific genes control the processes in DNA repair that are essential in maintaining normal genomic stability (Thompson & West 2000). Studies were conducted using RNA interference to study DNA processing, genomic instability, mutations, and cancer. All these processes seem to be linked to common pathways, with the repair of DNA damage being one of the critical pathway elements that leads to genetic damage (Bedford & Liber 2003). Genes involved in two of the major DNA repair pathways involved in genomic stability were identified. These include genes important in DNA excision repair (Amundson et al. 2002) as well as homologous recombinational DNA repair (Thompson & Schild 1999). These genes have all been shown to be essential for maintaining genomic stability. Other processes important in genomic stability involve the balance of the reactive oxygen species (ROS) status of the cells. Radiation-induced genes have been identified that control the ROS status of the tissues by increasing the free radical scavengers in cells (Limoli et al. 2001a).

Following radiation exposure, cells can make multiple, apparently normal cell divisions, then a fraction of the irradiated cells can lose control of their genome. In the field of radiation biology, this was unappreciated in the past. This "genomic instability" or loss of genetic control, results in multiple genetic changes in the cells. Genomic instability has been defined as the increased rate of acquisition of genetic alterations in the progeny of an irradiated cell (Morgan et al. 1996). These changes are similar to those observed a short time after exposure.

One of the earliest reports of genomic instability *in vitro* was related to the induction of DNA damage and its role on the induction of chromosome instability (Marder & Morgan 1993). In this system, CHO cells containing a human chromosome #4 (CM10114 cells) were exposed to radiation, clonally expanded, and the progeny examined for the induction and change in type and frequency of chromosome aberrations. Radiation-induced, late-occurring multiple genetic changes were first observed using an *in vivo/in vitro* method in the bone marrow of mice that were exposed to ^{239}Pu (Kadhim et al. 1992). The mice were exposed to ^{239}Pu and the bone marrow examined *in vitro* after several cell divisions in tissue culture for the presence of abnormal chromosomes. The frequency of chromatid aberrations was increased, showing that the aberrations were not being produced by direct exposure to the alpha particles from the internal emitter. Similar changes were identified in primary cultures of irradiated human bone marrow cells. In these studies, the genomic instability was measured as both an increase in apoptosis and as chromatid type aberrations (Kadhim et al. 1995).

After the initial observations of radiation-induced genomic instability both *in vivo* and *in vitro*, many systems were developed where the genomic instability could be quantified and carefully followed in individual cells. Genomic instability has been demonstrated in a range of different cell systems and cell types, suggesting that it is an important biological endpoint in the development of cancer. Genomic instability has been induced by radiation exposure in CHO cells that contained a copy of human chromosome number 4 (CM10115 cells) (Morgan et al. 1996; Morgan 2003c; Morgan 2003b), in mammary cells (Ponnaiya et al. 1997b), in mouse bone marrow (Kadhim et al. 1992; Bowler et al. 2006), in human lymphocytes (Lou et al. 2007), and in human bone marrow cells (Kadhim et al. 1995).

A wide range of different endpoints were used as a measure of the induction of genomic instability. One of the major endpoints identified was the change in the frequency and type of chromosome aberrations as a function of time after the radiation exposure in a way that was inconsistent with the development of clones of abnormal cells. Late-occurring chromosome aberrations thus became one of the standards for measuring genomic instability (Marder & Morgan 1993; Morgan et al. 1996; Ponnaiya et al. 1997b; Schwartz et al. 2001). This endpoint was then related to many other changes that were indicative of genomic instability in the cells (Romney et al. 2001). The induction of chromosome instability (Schwartz et al. 2001) was related to changes in DNA repair including defective recombinational repair (Takata et al. 2001), DNA cross links (Donoho et al. 2003), and the induction of micronuclei and DNA damage measured by the comet assay (Lou et al. 2007). In addition to measuring chromosome instability, other endpoints of genetic instability were developed including delayed hyper-recombination (Huang et al. 2004), induction of lethal mutations (Mothersill & Seymour 1997), and changes in DNA copy number (Kimmel et al. 2008).

However, studies have been conducted suggesting that genomic instability cannot be induced in stable normal cells. This inability of radiation to cause genomic instability has been demonstrated in human cells (RKO cells) in cultures (Huang et al. 2007), as well as in other normal human and animal cell lines (Dugan & Bedford 2003). The failure to demonstrate genomic instability in normal cells suggests that it may be part of the process of the cancer development and not induced by the radiation insult. If this is the case, genomic instability would not be detected in A-bomb survivors that received the radiation exposure but had not developed cancer. There has been no genomic instability demonstrated in follow-up of the A-bomb survivors.

Studies have been published to try to resolve the differences in the induction of genomic instability seen in experimental systems and the failure to demonstrate it in normal cell populations and in human populations (Morgan & Sowa 2007). One study demonstrated that when normal human cells (RKO) were incubated with growth medium conditioned by cells that were genomically unstable, high doses of radiation (5 Gy) could induce genomic instability. This demonstrated that in addition to the radiation, the growth medium was critical for the induction of genomic instability. Using this cell system, it was also determined that the frequency of radiation-induced genomic instability could be decreased by previous exposure to a low dose of radiation, a phenomenon indicative of an adaptive response (Huang et al. 2007). Thus, genomic instability and adaptive response appear to be closely related.

Studies were conducted to determine the cellular target for the induction of genomic instability. DNA damage (Morgan et al. 1996) and the cell nucleus (Kaplan & Morgan 1998) were implicated as the target for the induction of genomic instability. Studies on the induction of chromosome instability using different dose and dose-rates also defined the nucleus as the target for genomic instability (Limoli et al. 1999). More recent research has extended these studies and demonstrated that genomic instability can be induced by exposing a single chromosome to ionizing radiation (Mukaida et al. 2007). Thus, the target for the induction of genomic instability has been demonstrated to be the nucleus of the cell.

3.2 Physical Variables That Influence the Induction of Genomic Instability

One of the major concerns associated with the induction of genomic instability was the role of LET on this endpoint (Evans et al. 2003; Smith et al. 2003). This concern was related to the radiation environment in space where a range of different high Z particles, such as iron-56 and carbon, exist. These high Z particles contribute only a small fraction of the total energy deposited, but may be very important in evaluating the radiation risk related to space travel. It was of interest to NASA to determine the risk for exposure to these high Z particle in the space environment. Thus, a number of studies were jointly funded by the DOE Low Dose Radiation Research Program and NASA to extend the research from an interest in low-LET radiation to cover a range of different types of radiation.

Many of these studies of high Z particles were conducted at Brookhaven National Laboratory where high Z particles can be delivered over a range of well-defined doses and dose-rates to both cells and whole animals. Research demonstrated that genomic instability was induced by many different types of radiation exposure including UV light (Durant et al. 2006), gamma rays and neutrons (Ponnaiya et al. 1997b), Carbon ions and X-rays (Hofman-Huther et al. 2006), and heavy ions like iron-56 (Limoli et al. 2000b; Evans et al. 2003). Studies were also conducted to determine the role of dose-rate on the induction of genomic instability. Over a limited dose-rate range, dose-rate had little influence on the frequency of genomic instability induced by high Z particles (Limoli et al. 1999).

Studies were conducted to define the dose-response relationships and the role of LET on the induction of genomic instability. The frequency of radiation-induced genomic instability was found to be high, about 3-4% increase in genomic instability per Sv, and similar following exposure to low-LET radiation, alpha particles, or HZE particles (Limoli et al. 2000b; Limoli et al. 2000a; Evans et al. 2003; Kadhim et al. 2006). This very high frequency of genomic instability induced per unit of exposure suggests that the target for induction of genomic instability is much larger than a single gene and is not a simple mutation or combination of mutations.

The fact that a single dose of radiation can produce genomic instability suggests that genomic instability may be a potential mechanism to induce the multiple cellular changes required to change a normal cell into a cancer cell. Such information provides evidence that radiation-induced genomic instability may be a major pathway for a radiation-induced cancer. Over the dose range where significant increases in the frequency of genomic instability were observed there was a linear increase as a function of dose. Thus, radiation-induced genomic instability could provide a mechanism that would result in an LNT dose-response relationship for the induction of cancer.

However, because of the high signal-to-noise ratio, it was very difficult to determine the induction of genomic instablility in the low-dose region. It was noted that the control level of genomic instability was higher than the level of genomic instability observed following low doses of radiation (<0.5 Gy, Figure 17) (Limoli et al. 2000b; Limoli et al. 2000a).

In these studies, linear dose-response models were used to estimate the frequency of induction of genomic instability and to extrapolate the results into the low-dose region where significant responses were not seen. Thus, the data do not preclude the existence of non-linear dose-responses

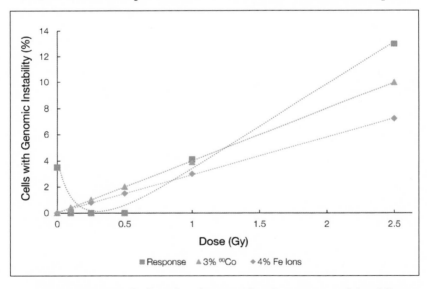

Figure 17. Percentage of cells with radiation-induced genomic instability following exposure to either low-LET ^{60}Co gamma rays or high-energy iron nuclei. A high frequency of genomic instability was induced by either treatment. Control level of genomic instability was higher than the level of genomic instability observed following low doses of radiation (<0.5 Gy) (Limoli et al. 2000b).

and the potential for an adaptive response in the low-dose region for the induction of genomic instability.

As the data on the role of LET on the induction of genomic instability in bystander cells were reviewed and evaluated, it was found that radiation quality has a minimal effect on the frequency of radiation-induced genomic instability in bystander cells (Smith et al. 2003; Kadhim et al. 2006). However, all doses of high-LET radiation produced genomic instability in their cell systems while 3.0 Gy of low-LET radiation did not produce genomic instability.

3.3 Biological Processes That Influence Genomic Instability

3.3.1 ROS Status of the Cells

Early in the research on genomic instability, it was recognized that oxidative stress played an important role in producing cells with chromosome instability. In genomically unstable cells, persistent increased oxidative stress was one of the physiological alterations noted (Limoli et al. 2003). In addition to oxidative stress, other cellular changes were observed such as apoptosis, cell cycle checkpoint modifications, and reproductive failure during cell division. Thus, changes in oxidative stress modified many cellular functions that ultimately resulted in chromosome instability (Limoli et al. 1998; Limoli et al. 2004).

It was determined that stress from non-DNA-damaging agents could also result in genomic instability (Li et al. 2001). Because the mitochondria are the major organelles involved in production and maintenance of the proper level of reactive oxygen in the cell, studies were conducted to evaluate their function in cells that demonstrated genomic instability (Kim et al. 2000). These studies showed that mitochondrial dysfunction was a landmark of genomically unstable cells. This suggested that in addition to modifications in DNA, modification of the mitochondria also plays an important role in genomic instability. The modification of radiation-induced oxidative stress by free radical scavengers and cell proliferation was shown to prevent the induction of genomic instability (Limoli et al. 2000b; Limoli et al. 2001b). These observations and others associated with the mechanisms involved in induction and maintenance of genomic instability are discussed in chapter 7 and suggest that the oxidative status of the cells is one of the most important factors involved in maintaining the genomic integrity of cells. This suggests that there are important non-DNA mechanisms involved in the induction and modification of genomic instability. Research suggests that reactive oxygen and oxidative stress in cells represent one of the major mechanisms involved in many of the radiation-induced changes associated with low dose exposures.

3.3.2 Apoptosis

Extensive research has been conducted in the past few years on radiation-induced apoptosis. This is an important mechanism involved in maintaining the genomic stability of cell populations by eliminating damaged cells. In fact, variation in the frequency of apoptosis was determined to be another marker of genomic instability (Nagar & Morgan 2005). Both delayed apoptosis and the induction of genomic instability were present following exposure to carbon ions (Hofman-Huther et al. 2006).

3.3.3 Death-Inducing Effects (DIE)

During the study of genomic instability in cell systems, it was discovered that soluble factors were secreted into the medium by genomically unstable cell lines that had a marked effect on the survival of normal cells (Nagar et al. 2003b). These factors were called death-inducing effects (DIE). Clones of cells with radiation-induced genomic instability released unique substances into the media. Very small concentrations of the media from the genomically unstable cell lines would result in 100% lethality in the normal cells. The role of these factors on the induction of genomic instability and the potential impact of DIE factors on radiation-induced cancer have been reviewed (Nagar & Morgan 2005). Because these factors did not seem to be produced by either normal or tumor cells, limited research or publications exist on the role of DIE on genomic instability or during the induction of cancer.

3.3.4 Genetic background, Cancer, and Genomic Instability

To link genomic instability to cancer, several important studies were conducted demonstrating that animals displaying an increase in genomic instability were also sensitive to cancer induction. From the early discovery of radiation-induced genomic instability, it was evident that the genetic background of the system being studied had a marked influence on the outcome of the studies. One of the earliest and best studies linking genetic background to the induction of genomic instability and cancer was conducted by exposing different strains of mice with different sensitivity to radiation-induced mammary cancer to radiation and measuring both the induction of genomic instability and mammary cancer in these mice. BALB/c mice with a high frequency of spontaneous mammary cancer were used along with C57BL/6 mice in which mammary cancer was not seen in the controls. In these studies, the frequency of radiation-induced mammary cancer was very high in BALB/c mice and low or non-existent in the C57BL/6 mice. The frequency of radiation-induced genomic instability was also high in the BALB/c mice and low in the C57BL/6 mice (Ponnaiya et al. 1997a, b).

Figure 18 demonstrates that after a number of cell population doublings, the cells of the BALB/c mice demonstrated a marked increase in chromosome aberrations as an indication of the induction of genomic instability while the C57BL/6 mice had no increase in chromosome damage. These observed strain differences made it possible to link genomic instability to the genetic background of the animals and the induction of cancer.

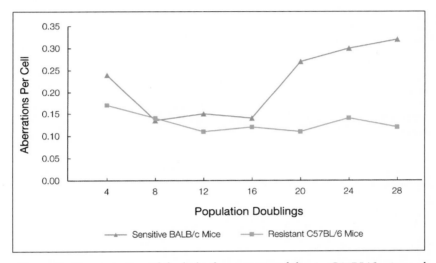

Figure 18. Demonstration of the lack of genomic instability in C57BL/6 mice and the induction of genomic instability in radiation-sensitive BALB/c mice. The genomic instability was seen after ~16 cell population doublings (Ponnaiya et al. 1997a).

Additional studies using the mammary epithelial cell system confirmed that p53 null mice could be induced to have chromosomal instability related to changes in hormonal status. This genotype seemed to be in part responsible for the difference in the biological responses observed between the sensitive (BALB/c) and resistant (C57BL/6) mouse strains to radiation-induced cancer and genomic instability (Pati et al. 2004). The importance of the TP53 gene in genomic instability was also demonstrated using other cell lines, including human lymphoblastoid cells (Schwartz et al. 2003). The genomic instability demonstrated in lymphoid cells seemed to play a role in the induction of mouse lymphoma in p53 heterozygous mice (Mao et al. 2005).

Radiation induced genomic instability is found when cells lose genetic control and many mutations and chromosome aberrations are formed in the cells. This is a concern since loss of genetic control and genomic instability are present in many solid cancers. It was demonstrated the induction of genomic instability by radiation was influenced by the genetic background (genotype) of the individuals or cells being studied. Geneotypes with modification important in DNA damage repair pathways were identified as being prone to radiation induced genomic instability. Extensive studies were conducted and reviewed to evaluate the role of DNA-repair-deficient mutants in Chinese Hamster Ovary (CHO) cells in radiation induced genomic instability (Somodi et al. 2005). It was determined that five cell lines that were defective in recombinational repair in (Rad 51) paralogs were involved in genomic instability (Takata et al. 2001). Genomic instability was also demonstrated in cells from animals with the gene Gadd45, a deficient mouse strain (Hollander et al. 1999). Rad51C gene deficiency was shown to destabilize the gene XRCC3, impair recombination, and increase radiosensitivity in cells in the S/G_2 stages of the cell cycle. Major changes in DNA crosslinking, chromosome instability, and mice lifespan were associated with alterations in Braca2 gene with a change in exon 27 (Honma et al. 2000). This research all points toward DNA damage and repair in the nucleus as the target for radiation-induced genomic instability. Finally, cells and tissues from people with genetic diseases such as Xeroderma pigmentosum, which is influenced by the p53 gene (Cleaver et al. 1999a), and retinoblastoma (Zheng et al. 2002) were found to display genomic instability. The importance of genetic background on radiation induced genomic instability has been carefully reviewed (Kadhim 2003).

Genomic instability has been related to other types of radiation-induced cancer in experimental animals. Exposure of mice to either radiation or benzene produced genomic instability and acute leukemia (Rithidech et al. 1999). Studies demonstrated that radiation could also produce genomic instability in the lymphocytes and lymphoma in p53 heterozygous mice (Mao et al. 2005; Bowler et al. 2006). These studies have established an association between cancer and genomic instability and have strengthened the link between genetic background and genomic instability. Such research also suggests that genomic instability is an important part of the radiation-related process as cells and tissue progress from normal to cancer.

Review papers on radiation-induced genomic instability (Little 1999, 2003; Morgan 2003b, a) provide additional information and will help complete this literature review.

A number of papers have used the data to link genomic instability to radiation-induced cancer in humans (Goldberg & Lehnert 2002;

Huang et al. 2003). These papers make a convincing argument that radiation-induced genomic instability could represent an early stage of radiation-induced cancer and may be important in the cancer process for many forms of cancer displaying genomic instability. Finally, it will be of interest to determine if the induction of genomic instability can be modified by low doses of radiation and other environmental insults and what, if any, the clinical implications of this process are (Goldberg 2003).

3.3.5 Epigenetics and Genomic Instability

Recently the role of epigenetic factors, or those not associated with direct changes in the DNA or genetic material, has been investigated. Early research demonstrated that changes in diet of special strains of mice can change the coat color of the offspring. These observations have been expanded to other systems to help understand the risk from exposure to different types of environmental insults such as diet and environmental pollutants (Jirtle & Skinner 2007). A mother's diet can modify the environment of a fetus during development, changing gene expression and modifying phenotype. It has been suggested that epigenetic influences during development may play an important role in the development of cancer and other genetic diseases in adults (Dolinoy et al. 2007b; Dolinoy et al. 2007a). Environmental exposures to a number of different insults can change the offspring in a way that suggests epigenetic mechanisms could play an important role in human health and disease (Dolinoy & Jirtle 2008).

The role that radiation plays in induction of epigenetic changes is a relatively new field and requires additional research. However, two research areas suggest that in addition to causing direct alterations to the genetic material, radiation also plays an epigenetic role in causing phenotype modifications and cancer risk without direct interaction with the DNA or chromosomes.

First, it has been suggested that the ability of radiation to modify non-genetic targets such as the extracellular matrix and influence the outcome of diseases plays a role in the induction of breast cancer (Asch & Barcellos-Hoff 2001). It has been demonstrated that radiation exposure of the stroma, which is connective tissue and contains no genetic information, has been linked to an increase in the frequency of breast cancer (Barcellos-Hoff & Ravani 2000). Such studies suggest that a tissue is responding as a whole to the exposure to radiation and not as a set of individual cells.

Second, it has been suggested that radiation can induce genomic instability through epigenetic mechanisms that could be passed from one generation to the next. Studies were conducted using a mouse model system where chimeric male mice were exposed to radiation and their radiation history was traced to the F(2) generation. Changes were shown in gene expression and enzyme activity in the many kinases in organs of offspring where the parents had been exposed to radiation (Baulch et al. 2002; Vance et al. 2002). These changes are indicative of the induction of genomic instability in the offspring and suggest the potential for increased cancer risk in the radiated mice.

Other studies with the same system indicated radiation-induced cellular reprogramming resulting in changes in gene and protein expression. These changes were followed in the offspring through the F(3) generation (Vance et al. 2002). These studies suggested that genomic instability had been induced in the offspring of the irradiated mice. The frequency of the changes in the offspring were so high that they could not be explained by known genetic transmission, but rather only by epigenetic mechanisms.

Epigenetic effects of radiation have been suggested in human and mouse populations exposed to either external radiation or from internally deposited radioactive materials. The role of radiation during the induction of epigenetic changes was reviewed early in the DOE Program (Nagar et al. 2003a) with little evidence of a marked effect on genetic or carcinogenic risks from these mechanisms (Nagar et al. 2003b). However, transgenerational epigenetic changes have been detected using changes in gene sequences in tandem repeated DNA loci (TRDLs). The high frequency of these changes suggests that they are induced by a mechanism different from that seen for direct radiation effects. These are measured in mini-satellites in human and expanded simple tandem repeats in mice.

A review of the literature involving the TRDLs noted important differences between the structure of mouse and human TRDLs and suggested that transgenerational effects associated with TRDLs are present in mice but may not be present in humans (Bouffler et al. 2006). Again, there is little evidence that these radiation-induced changes in TRDLs will play an important role in the risk for cancer or mutations in humans. The role of radiation-induced epigenetic effects is important and requires additional research because it could be one of the few areas in which the risk from radiation-induced damage could have been underestimated.

3.3.6 Genes and Proteins Involved in Genomic Instability

In studies of adaptive response, extensive research was conducted on the genes and proteins involved in the induction of these protective effects. These studies were important for understanding the mechanisms involved. However, only limited research has been done on the changes in gene expression and protein expression in cells demonstrating genomic instability. Research in this area could provide critical information on the role of genomic instability, especially the role of epigenetic effects, on cancer risk. Limited studies were undertaken to determine if changes in gene expression would provide clues for understanding the process of genomic instability (Snyder & Morgan 2004a, b). As the research continued, there was a lack of consensus of gene expression changes associated with radiation-induced chromosomal instability, and it became obvious that many processes may be involved during the induction of genomic instability.

With the development of new systems for rapidly detecting genomic instability using a stable, transfected, plasmid-based, green fluorescent protein assay that detects homologous recombination and delayed mutation/deletion events (Huang et al. 2004), it became possible to measure radiation-induced genomic instability after low doses of radiation. Research on the mechanisms involved in the induction of genomic instability will be covered in more detail in chapter 7.

3.4 Major Points: Genomic Instability

- Many cancers display genomic instability.
- Radiation produces genomic instability in a number of *in vitro* and *in vivo* experimental systems.
- Radiation-induced genomic instability is a frequent event. This suggests that it is not a process involving a single gene or small numbers of genes, but requires a larger target for its induction.
- The nucleus of the cells seems to be the target for the induction of genomic instability.
- The ROS status and mitochondrial metabolism play critical roles in the loss of genomic stability.
- Some genomically unstable cell lines secrete a substance (DIE factors) that is lethal to 100% of normal cells. To date, this factor or substance has not been characterized.
- The genes and proteins involved in the induction of genomic instability are not well defined at present.
- The genetic backgrounds of cells, animals, or humans are important in radiation-induced genomic instability.
- Animals that are sensitive to radiation-induced cancer are also sensitive to the induction of genomic instability.
- Evidence exists that small doses of radiation can produce an adaptive response for genomic instability and reduce genomic instability induced by large radiation doses.

- The presence of adaptive responses indicates a non-linear dose-response relationship for the induction of genomic instability. Nevertheless, genomic instability is a process triggered by a number of environmental factors and could potentially support the linear-no-threshold theory.
- Research is needed in the low-dose region to further define the mechanisms of action and the shape of the dose-response relationship for induction of this important process.

4. The Relationships of New Paradigms

The interrelationships between each of these new phenomena, bystander effects, adaptive response, and genomic instability, has been carefully documented (Sawant et al. 2001b; Lorimore et al. 2003; Zhou et al. 2003; Mitchell et al. 2004c; Lavin et al. 2005; Zhu et al. 2005; Hall 2006; Prise et al. 2006b), and are illustrated in Figure 19.

Each phenomenon is influenced by the genetic background and the genetic sensitivity of the system used to make the measurements. This is illustrated by the large circle over the whole figure. The close relationships that have been demonstrated between these seem to reflect manifestations of similar basic biological changes, and as studies have been expanded, adaptive responses, genomic instability, and bystander effects have been tightly linked.

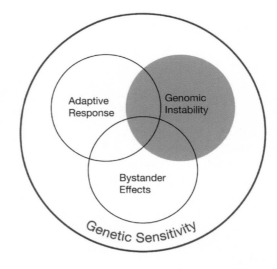

Figure 19. The overlap of biological responses induced by low doses of radiation. All these responses depend on the genetic background of the test system.

4.1 Bystander Effect and Genomic Instability

Early in the Low Dose Radiation Research Program, it was recognized that there were many biological changes present in bystander effects and genomic instability, that many of these changes could be related to the induction of cancer, and that the critical molecules and pathways were influenced in measurements of each of them (Barcellos-Hoff & Brooks 2001). The research on the relationships that exist between bystander effects and the induction of genomic instability has been reviewed (Maxwell et al. 2008). Much of this literature has focused on how this interaction could potentially increase the radiation risk by causing genomic instability in cells with no energy deposited in them (bystander cells) (Morgan 2002; Morgan et al. 2002; Hall & Hei 2003; Lorimore et al. 2003; Huang et al. 2007). This research suggested that the target for the induction of genomic instability was much larger than the number of cells traversed by radiation.

The response of bystander cells to produce genomic instability has been related to many physiological changes such as the induction of inflammatory-type responses induced by radiation (Lorimore & Wright 2003). It was suggested that secreted factors were important and research conducted to try to identify the important factors (Sowa Resat & Morgan 2004). The interactions between genomic instability and bystander cells have been the subject of a number of review articles (Lorimore et al. 2003; Maxwell et al. 2008).

4.2 Bystander Effects and Adaptive Responses

Early in the program, much of the research was conducted *in vitro* using a wide range of different cell types. As bystander effects, the importance of cell-cell, cell-matrix, and cell-tissue interactions, and total tissue responses became better understood (Bissell & Barcelloshoff 1987), the calls for proposals and the research direction in the program shifted to the study of organized cellular systems and tissues. Many biological endpoints were measured that demonstrated the bystander effects and suggested that this response represents cell-cell and cell/matrix communication capable of altering biological outcomes (Barcellos-Hoff 2005a; Andarawewa et al. 2007b).

This research demonstrated that many of the responses observed in a monolayer tissue culture were not observed in complex tissue. It also suggested that in complex tissues and whole animals other responses seemed to modify the observed biological responses on the single-cell level. It was demonstrated that adaptive responses are observed in cells that do not have energy directly deposited in them and that this could reduce the biological response (Schettino et al. 2005).

The control and modification of these single-cell processes by the environment had a major impact on the responses in the low-dose region and helped to demonstrate that interactions between different cell types and modification and processing of the initial damage controlled the induction of radiation-induced disease. These processes need to be carefully considered as models of radiation risk are developed.

As the result of this research, it has been suggested that adaptive processes might be in play during radiation therapy for cancer (Sgouros et al. 2007). With further research, clinical applications of adaptive responses and bystander effects may provide mechanistic understanding of the responses to low doses of radiation. The use of low-dose protocols and the induction of adaptive responses and bystander effects may be useful to modify cancer risk of secondary cancers and adverse acute effects produced during cancer therapy.

4.3 Genomic Instability and Adaptive Response

It has been suggested that genomic instability may increase the radiation-related cancer risk above the level predicted by the LNT, while adaptive responses may decrease the calculated risk in the low-dose region. Research has been conducted to determine the relationships between these two observed phenomena. In cells that display genomic instability, interleukin-8 was shown to produce a pro-mitogenic and pro-survival effect to modify radiation-induced genomically unstable cell lines (Laiakis et al. 2008). Dziegielewki et al. (2008) also demonstrated that chemicals that serve as radioprotectants such as WR-1065 act through their active metabolite, amifostine, to mitigate radiation-induced genomic instability.

These findings demonstrate potential protection against the induction of genomic instability and may also result in decreased incidence of radiation-related cancer. Research was conducted by Huang et al. (2007) to determine if a small radiation dose prior to a large challenge dose used to demonstrate adaptive responses could decrease the frequency of radiation-induced genomic instability. They determined that previous low doses of radiation resulted in an adaptive response that decreased the response to subsequent high doses of radiation by the induction of genomic instability. Such research provides strong associations between these two biological phenomena and demonstrates the potential for adaptive responses to modify genomic instability as well as the risk for cancer.

Figure 18 suggests that all three of these observations are closely linked. Published studies suggest reasons for these tight links and associations, one important one relating to the fact that there is biological variability in all biological systems and that what was being measured was only an indication of this variability (Schwartz 2007). Many changes in gene expression, metabolic pathways, biological processes, and individual chemicals altered by low doses of radiation were suggested to be common in all these observations (Morgan 2003b, a). This suggests common mechanisms are involved in each process.

As research has progressed, a need has arisen for a much more mechanistic approach (Kadhim et al. 2004). Many of these common mechanisms will be reviewed in chapter 7 by discussing the different mechanisms involved and how they impact the biological responses.

As the data have matured, it is obvious that bystander effects, genomic instability, and adaptive responses are seen in many biological systems and represent the body's attempts to deal with low doses of ionizing radiation. These discoveries have impacted current knowledge in the field of radiation biology and must be considered when discussing radiation effects and radiation protection, and when predicting radiation risk in the low-dose region.

This chapter has described these observations in many different systems to demonstrate that they are general biological phenomena and represent well-described biological responses to low doses of radiation. Chapter 6 will describe how these discoveries came to influence the rapidly growing field of biomarker research.

CHAPTER 6
Biomarkers of Radiation Exposure and Dose

Physical dosimetry is essential to estimate radiation exposure and dose accurately, both from external radiation exposure and internally deposited radioactive material. However, there are cases where this is impossible. When the dose is from internally deposited radioactive material, like radon gas, and no direct measurement of dose is possible, the measurement of biological changes can often be used to generate data for models to estimate radiation dose (NRC 2006). In addition, there have been a number of radiation accidents where no physical dosimetry was available. In such cases it was necessary to use biological changes in the exposed individuals to estimate radiation exposure or dose. Using biological changes to estimate dose after exposure is the field of biodosimetry. Biomarkers of radiation exposure, dose, and sensitivity have been successfully applied to a wide number of situations including radiation accidents.

When the dose is high, symptomatic markers, such as onset of nausea and vomiting, provide a first guess as to the dose received. At lower doses, other changes must be measured, such as the kinetics of lymphocyte depletion, where it is possible to relate the decline in lymphocyte numbers to the radiation dose and dose-rate (Goans et al. 2001; Dainiak et al. 2007). At still-lower doses, the gold standard for biodosimetry is the induction of chromosome damage. The use of chromosome damage as a biomarker of radiation dose for internally deposited radioactive materials (Brooks et al. 1997) and external acute radiation exposure (Bender et al. 1988) was a subject of extensive research before the program was initiated.

The program was not focused on biodosimetry or the measurements of biomarkers of radiation-induced changes because their major use had been to estimate dose following radiation accidents, terrorist activities, or the potential for future nuclear war. All of these events focus on saving lives following very high radiation doses.

Even though the focus of the program was not on biomarkers, its research has resulted in major scientific advances involving development of new biomarkers that could be used to provide rapid estimates of radiation dose. This research provided a foundation for additional studies using cell and molecular changes as radiation exposure and dose biomarkers.

It was recognized early in the program that biomarkers could be used to help define genetic background related to the induction of disease, as well as radiation dose (Albertini 1999). It became apparent that biomarkers that change as a function of radiation can measure different processes. So, in some cases, the biomarker is a measure of radiation dose. In other cases a marker provides an indication of radiation sensitivity, and finally, the marker may indicate the presence of a disease (Brooks 1999).

Chromosome aberrations are a very reliable biomarker of exposure and dose. However, they vary markedly between tissues (Bao et al. 1997), and the frequency of chromosome aberrations in a tissue is not necessarily related to cancer frequency in these tissues. In the past, chromosome aberrations and other biomarkers of radiation exposure have also provided a quick *in vitro* and *in vivo* method to determine the relative biological effectiveness (RBE) of many radiation types (Brooks et al. 1997; Groesser et al. 2007). These values were then used in radiation protection standards to estimate cancer risk. However, research from the program has shown that although chromosome aberration frequency reflects dose, in many cases it may not be a direct reflection of cancer risk (Brooks 2003).

As biomarker research has developed, its importance has been recognized by the scientific community. Other agencies have initiated extensive funding of biomarkers for rapid evaluation of exposure and dose in mass-exposure events. Centers have been developed for biomarker research and development of techniques for providing rapid and reliable measures of exposures of large populations to high radiation doses. For example, the "Center for High Throughput Minimally-Invasive Radiation Biodosimetry" was formed at Columbia University College of Physicians and Surgeons to apply modern radiation biological techniques to detect biological changes as a measure of radiation dose. In addition, special meetings have been held on biomarkers of radiation exposure that summarize the state of the art for biomarkers. These meeting were sponsored by organizations such as the National Council on Radiation Protection and Measurements (NCRP) (Brooks 2001), the military (Blakely et al. 2002), and NASA (Straume et al. 2008). Only the research conducted under the DOE Program that has had an impact on the field of biomarkers and biodosimetry will be briefly reviewed here.

1. DNA Damage

It has been well established that radiation can produce many changes in DNA. However, in the past this damage was detectable only after high radiation doses. Development of techniques that detect damage and repair of DNA at specific loci and the production of γH2AX foci made it possible to detect DNA damage in the low-dose region (Burma et al. 2001). These techniques have been improved (Nakamura et al. 2006) and automated using such techniques as flow cytometry (Kataoka et al. 2006), making it possible to use them as markers of radiation exposure. Such techniques are very useful for detecting acute radiation exposure; however, there is some question as to their usefulness following exposure to low dose rates of ionizing radiation. In a study where the radiation was delivered at a low dose rate, there was little change in the frequency of γH2AX, even following larger (up to 5 Gy) radiation doses (Ishizaki et al. 2004).

Development and use of clustered lesions in the DNA, described in chapter 4, were very useful in evaluating DNA damage over a wide range of radiation doses. These lesions have been suggested to be unique for radiation exposure, can be detected at relatively low exposure levels, and could be used as a biodosimeter to estimate radiation doses (Sutherland et al. 2001c).

2. Changes in Gene Expression

A major spinout from the program was the use of radiation-induced molecular changes to develop biomarkers of radiation exposure and dose. It was determined early in the program that changes in mRNA expression can be measured as a function of radiation dose and may be useful in identifying individuals or populations exposed to ionizing radiation (Amundson et al. 2000). Continued research on radiation-induced changes in gene expression, especially studies following low dose and dose-rate exposures, validated these changes as markers of radiation dose and exposure (Amundson et al. 2001b; Amundson & Fornace 2001). This resulted in a strong database relating dose to responses that was used to develop molecular biomarkers of radiation exposure and dose (Amundson et al. 2001a; Coleman & Wyrobek 2006).

Such research was validated by measuring changes in gene expression using human radiotherapy populations exposed to known doses of radiation during their medical treatment (Amundson et al. 2004). It was found that gene expression changes as a function of the genes measured, as well as the radiation dose, dose-rate, and time after exposure. This makes it difficult to use the change in a single gene as a marker of radiation exposure. Thus, clusters of genes and selected gene changes and times after exposure were used to develop biomarkers for rapid detection of radiation exposure in large populations. There were, however, suggestions that permanent signatures in the genome and changes induced in the expression profiles could potentially be used as biomarkers (Hande et al. 2003).

A large project was focused on determining the usefulness of gene expression in predicting survival of cells in culture. In this study, 60 cell lines used at the NIH were evaluated for both survival and changes in gene expression. It was determined that different subgroups of genes may provide an indication of the radiation sensitivity of cells and tissues (Amundson 2008).

As this type of research was expanded, it has been funded by other government agencies that were interested in rapid biomarkers of radiation exposure in the event of nuclear war or terrorist attacks. Research from the program continues to be important in the development of molecular biomarkers of radiation exposure and dose, but was not a central focus.

3. Changes in Proteins and Metabolites

Scientists recognized early on that changes in gene expression measured by changes in mRNA levels often may not reflect changes in physiological function. Proteins produced by the RNA are the important variable to determine changes in cell function. Therefore, it was important to measure protein changes induced by ionizing radiation and study the potential use of protein changes as biomarkers of radiation exposure. Some of the early work in this area was the result of basic research on the proteins involved in apoptosis or cell death called clusterin (Leskov et al. 2003). It was determined that changes in clusterin could be used as an indication of past radiation exposure (Klokov et al. 2003). With the development of protein microarrays (Coleman et al. 2003a), it became possible to rapidly measure radiation-induced changes in several proteins. This was an important factor in the development of protein changes as biomarkers of radiation exposure (Marchetti et al. 2006). Currently, many of these changes are detected only after doses of >2.0 Gy. In addition to changes in total protein, it was noted that radiation exposure changed the phosphorylation pattern and degree in human melanoma cells (Warters et al. 2007). Proteomics is another important future area of research.

Finally, it has been suggested that radiation can induce changes in normal metabolism, which results in a change in metabolites in the urine or other body fluids (Tyburski et al. 2008). This has led to the development of the field of metabolomics. Again, because most of these changes are detected after high doses of radiation, this was not a major emphasis in the program.

4. Chromosome Aberrations as Markers of Radiation Exposure

Because the DOE Program focuses on low doses, a primary biomarker of radiation exposure studied by the program was the induction of chromosome aberrations. In the past, chromosome aberrations were evaluated as structural changes in metaphase cells, with a focus on the induction of dicentric chromosomes, where two chromosomes are joined and have two centromeres and a pair of fragments (Bender et al. 1988). This endpoint was relatively easy to score by trained cytogeneticists and provided a very good estimate of radiation dose regardless of whether the exposure was *in vivo* or *in vitro*. This made it possible to construct very good dose-response relationships for a wide variety of different exposure types, tissue types, and species.

However, the program's focus has been on understanding the mechanisms involved in the induction of chromosome aberrations.

The major advance in this area has been the development of a technique called fluorescence in situ hybridization (FISH) that make it possible to paint each chromosome a different color. Without FISH it was not possible to detect complex chromosome rearrangements, and many of the radiation-induced aberrations went undetected. Using such techniques, it became possible to detect complex chromosome aberrations where multiple chromosomes are involved *in vivo* (Rithidech et al. 2007a) and *in vitro* (Cornforth 2001) and to construct dose-response relationships.

Researchers determined that the frequency of complex chromosome aberrations changes as a function of dose following exposure to gamma rays (Loucas & Cornforth 2001). Other exposure variables, such as dose-rate, radiation type, and LET of the radiation exposure, also impacted the frequency and type of complex chromosome aberrations produced. The frequency was higher following exposure to alpha particles than was seen for exposure to gamma rays and was not as dependent on dose-rate (Cornforth et al. 2002a).

However, it was determined that even at low doses and dose-rates, a single track of low-LET radiation can induce complex exchanges with up to four breaks involved. This observation may suggest a different mechanism for the induction of multiple chromosome aberrations in a single cell. Such a mechanism could involve the induction of genomic instability and the loss of genetic control. It may also help explain the presence of rogue cells, cells with multiple aberrations seen in a number of different studies of internally deposited radioactive material. These observations may represent the earliest signs of the induction of genomic instability.

Additional chromosome-painting techniques have been developed that make it possible to paint and produce multi-color banding of the chromosomes so that many local regions in each chromosome can be marked (Brenner 2004). Using these techniques, small losses and insertions and inter- and intra-chromosome exchanges can be detected

(Brenner 2004). Such chromosome-banding techniques make it possible to measure small, subtle chromosome alterations that cannot be detected in individual chromosomes with other techniques.

The interchromosome inversions in Figure 20 show where a piece of the chromosome has been inverted and reinserted into the same chromosome. Figure 20c demonstrates that an individual small piece of a chromosome can be lost or added to an individual chromosome. None of these aberrations would have been detected using either normal staining or chromosome painting.

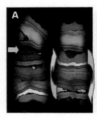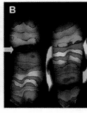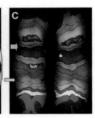

Figure 20. Intrachromosome rearrangements detected by multicolor chromosome banding. A and B are intrachromosal exchanges, C is normal (Brenner 2004).

However, the technique required to prepare the cells, equipment needed, and scoring of the aberrations is expensive and both time- and labor-intensive. This makes the technique useful in mechanistic studies, but very difficult to apply to accidental exposures with multiple exposure types and conditions.

Nevertheless, techniques such as mFISH have been critical in understanding the location of genes, the types of rearrangements induced by radiation, and how the chromosomes interact to produce the observed structural changes. Such research is important and has increased understanding of the microscopic distribution of the chromosomes in the nucleus and how chromosome location influences the frequency and type of chromosome aberrations (Plan et al. 2005). It was demonstrated that homologue chromosomes interact more frequently than others in the production of chromosome exchanges.

5. Micronuclei

The other form of chromosome damage that was investigated in the DOE Program and used as a biomarker is the induction of micronuclei. Micronuclei are detected as small round pieces of chromatin material in the cytoplasm of interphase cells. After cell division, these chromosome fragments are not incorporated into the nucleus and are readily detected in the cytoplasm after the cell divides. The use of Cytochalasin B to prevent cytokinesis results in binucleated cells that make it possible to score for the production of micronuclei only in the cells that have undergone a cell division. Micronuclei are very easy to score and provide a quick measurement of radiation-induced chromosome damage. Extensive studies have been conducted to standardize the assay and make it very reproducible (Fenech et al. 1999). Because of this, recently micronuclei have been applied to many systems as a biomarker of radiation dose and exposure in human populations, as well as in other animals exposed to radiation. Additional research is being conducted to develop methods to automate the scoring of micronuclei, which will make them very useful in biodosimetry.

A wide range of different human populations has been monitored for radiation-induced micronuclei. This includes such groups as hospital workers (Kryscio et al. 2001), nuclear power plant workers (Hadjidekova et al. 2003), uranium miners (Kryscio et al. 2001; Muller et al. 2004), and populations exposed to low dose rates of ionizing radiation (An & Kim 2002). Measuring chromosome aberrations in exposed populations using modern techniques provides a useful indication of radiation dose, even after complex exposures, which include both internally deposited materials and low-dose-rate external exposures (Bauchinger et al. 2001). Many molecular epidemiological studies have been conducted outside the program that help link the frequency of micronuclei to the induction of cancer. In these studies, which are not reviewed here, the risk for the induction of cancer is higher in individuals that have an elevated frequency of micronuclei.

6. Influence of Exposure Variables on Biomarkers

6.2 Dose

As is detailed throughout this book, biological alterations change as a function of dose. In the low-dose region, many cellular changes do not increase as a function of dose but in fact have been demonstrated to decrease, as described earlier as an adaptive response. For acute radiation exposure, other changes such as DNA damage and the formation of γH2AX foci are linear all the way down to the low-dose region. In the high-dose region, biomarkers provide a very useful measure of radiation exposure and dose. The types of biological changes described in this section can provide biomarkers of exposure and dose at all levels of biological organization. Measurement of many biomarkers can be automated and applied to large populations that may have been exposed to accidents, terrorist activities, or nuclear war; such changes represent a useful area of research.

6.2 Dose-Rate

During the development of biomarkers, several studies measured not only the influence of dose on the endpoint, but also how these molecular and cellular endpoints changed as a function of dose-rate. When radiation was delivered at a low dose-rate, there was no increase in the frequency of γH2AX relative to the control level (Ishizaki et al. 2004). This demonstrates the importance of DNA repair of radiation damage as a function of exposure time. Thus, dose-rate is very important in evaluating biomarkers.

DNA repair has been seen for both high- (Asaithamby et al. 2008) and low-LET radiation (Sutherland et al. 2002b; Sutherland et al. 2002a). The dose-rate effect has been demonstrated in most biological systems evaluated over the many years of research on the health effects of radiation (Brooks et al. 2009). There was a decrease in the frequency of micronuclei and an increase in apoptosis in cells exposed to low-dose-rate radiation, suggesting a protection as a function of dose-rate (Boreham et al. 2000). A marked decrease in the frequency of simple exchanges involving two chromosomes was seen as the dose-rate was decreased (Loucas et al. 2004b).

Studies of the cleanup workers at Chernobyl using chromosome aberration suggested that the dose reconstruction and recorded doses on these workers may have overestimated the true radiation doses estimated using chromosome aberrations detected with FISH techniques (Littlefield et al. 1998). An alternative interpretation of the data would suggest that radiation exposure at low dose-rates, which many of these workers received, was less effective in producing biological damage than calculated from the current models used to correct for dose-rate effects. Thus, the measured doses may be accurate, but the effective dose may be reduced. All these cellular and molecular observations support the need for a dose-rate-effectiveness factor (DREF) and the potential for a decrease in risk and effectiveness of low dose-rates relative to high dose-rate exposure (Brooks et al. 2009).

6.3 Radiation Type

Use of chromosome-painting techniques suggests that exposure to high-LET radiation can produce a unique and stable signature of radiation-induced damage (Brenner et al. 2001b). Mitchell (Mitchell et al. 2004a; Mitchell et al. 2004c) suggested that such aberrations do not exist following exposure to low-LET radiation. Others expanded this research to investigate the production of chromosome aberrations in individuals that had body burdens of ^{239}Pu contamination from occupational or accidental exposures (Anderson et al. 2000; Anderson et al. 2003; Okladnikova et al. 2005). This research also suggested that alpha particles produced unique types of complex chromosome aberrations. The frequency of direct interaction and the deposition of energy from the alpha particles emitted by ^{239}Pu in the lymphocyte population are calculated to be limited. Thus, additional research is needed to determine if these aberrations are the result of the direct alpha exposure, bystander effects, or other factors in the environment of these workers.

Extensive research co-funded by NASA and DOE demonstrated that high Z (HZE) particles also produce unique biological changes. One of the most important changes can be measured as different types of chromosome damage. Using FISH techniques, complex chromosome aberrations were observed following HZE exposure. This complex damage suggests a useful biomarker of high-LET exposure (George et al. 2002; Hada et al. 2007; Kurpinski et al. 2009). Another chromosome change identified as being induced by HZE particles was chromatid type of exchanges in G_0 cells. These aberrations are not induced when G_0 cells are exposed to low-LET radiation. The frequency and type of aberrations that could be detected in G_2 cells using the premature chromosome condensation techniques were also found to be unique following exposure to HZE particles (Kawata et al. 2000).

Research identified other changes following high-LET radiation exposure that have different biological properties and may provide a biomarker of exposure to this type of radiation (Brenner et al. 2001b). Because of the dense ionizations produced along the track of a HZE particle, it was of interest to measure the induction of γH2AX foci. Individual cells and tissues were studied, and special imaging techniques (Costes et al. 2006) demonstrated the pathway of the traversal of HZE particles through the cell. The nature and size of the track indicated that a very large amount of energy was deposited in any cell that was traversed by an HZE particle (Leatherbarrow et al. 2006). HZE exposures result in different biological effects, which can also act as biomarkers of this type of radiation exposure (Kurpinski et al. 2009).

6.4 Use of Biomarkers in Contaminated Environments

Biomarkers identified from techniques developed in the program have been applied to environmental and accidental radiation exposures and provide useful information in evaluating radiation hazards. The use of biomarkers to evaluate the radiation dose and damage at Chernobyl is an example of the importance of biomarker research funded by the program. Early studies of the native bank voles in the zones around the accident suggested that 1) the radiation dose and dose-rate were very high (up to 10 mSv/day), 2) there may be extensive genetic damage induced in this population, and 3) the accumulation of damage in the population may have a negative impact in terms of litter size, sex ratio, and other measures of damage on the animals that were living in the highly contaminated areas (Baker et al. 1996b; Baker et al. 1996a). The radiation doses and dose-rates from both the external exposure and the internally deposited radioactive materials such as ^{90}Sr and $^{134, 137}$Cs were accurately measured in these rodent populations and related to the biological damage measured at the cell and molecular levels (Chesser et al. 2000; Chesser et al. 2001). The frequency of micronuclei and other genetic damage in the animals' bone marrow were evaluated and shown to have little increase relative to the levels seen in control populations (Rodgers & Baker 2000).

Because of concerns that the bank voles were radiation resistant, laboratory mice with known differences in radiation sensitivity—BALB, a radiation-sensitive strain, and C57Bl, a radiation-resistant strain—were taken into these highly contaminated zones and maintained in this radioactive environment for an extended period of time. Very limited indication of genetic damage occurred in either strain of mice under these chronic radiation exposure conditions (Rodgers & Baker 2000; Rodgers et al. 2001). The rodent populations at Chernobyl were followed for 3-5 years, and there was no indication of an increase in mutation frequency or other genetic impact on them (Wickliffe et al. 2003b; Wickliffe et al. 2003a). Thus, no measurable genetic impact of the Chernobyl accident occurred in these animal populations, even though they were exposed to very large radiation doses delivered at a low dose-rate. This suggests that dose-rate is very significant in reducing the frequency of radiation-induced damage.

The area around Chernobyl has been made into a wildlife preserve. The ecological diversity, number of species, sex ratio, and other measures of the health of the ecosystem indicated that the ecological damage to this area is very minimal (Baker et al. 1996b; Baker & Chesser 2000).

Although there have been other extensive studies conducted in humans in other contaminated environments, the DOE Low Dose Radiation Research Program was not involved in funding this epidemiological research. Nevertheless, the program provided the data, development, and application of many of the techniques that were useful in this research. The spinoff value of the basic science conducted in the program has had practical applications in the field of biodosimetry and biomarkers that reach far beyond the outlined goals of the program.

7. Major Points: Biomarkers

Many biomarkers are designed to detect radiation exposures in the high-dose range, and although the program did not specifically focus on biomarkers, we nonetheless learned a lot.

- Research from the DOE Program has resulted in advances in the field of radiation biology that were applicable in development of biomarkers of exposure, dose, radiation sensitivity, and disease.
- Chromosome damage provides one of the most reliable estimates of radiation dose. Modern techniques to stain and rapidly measure chromosome aberrations provide a rich source of mechanistic information on the interaction of radiation with cells.
- Molecular techniques have been developed to measure radiation-induced changes in DNA over a wide range of doses. The damage increases linearly with dose and thus provides a good biomarker of radiation exposure.
- The change in gene expression is potentially a great biomarker of radiation exposure since the techniques to measure these changes are very rapid, can be automated, and can be applied to evaluate large populations exposed to ionizing radiation.
- Exposure variables (dose, dose-rate and radiation type) all have a marked influence on the response of biomarkers and must be considered in their application.
- Biomarkers from techniques developed in the program have been applied to environmental and accidental radiation exposures and provided useful information in evaluating radiation hazards.

CHAPTER 7

Mechanisms of Action

1. Need for Mechanistic Studies

When the DOE Low Dose Radiation Research Program issued its first request for proposals in 1999, many scientists suggested that they could not measure biological changes in the dose range of interest, <0.1 Gy (10 rads). Dr. Marvin Frazier, the program's scientific director, replied that in that case, they need not apply. This was good advice, and many new techniques and technologies were developed and applied by program-supported scientists that made it possible to detect previously undetectable radiation-induced biological changes in the low-dose region. These new observations suggested the need for major paradigm shifts in the field of radiation biology.

Before the program, the proposed mechanisms of action for radiation in the low-dose region were easy to understand radiation produced DNA damage, mutations, chromosome aberrations, and cancer. The induction of DNA damage had been observed and measured only in the high-dose region and damage increased linearly with radiation dose. Mutations were observed in many different biological systems ranging from single-cell organisms to plants and mammalian cells in culture.

At high dose rates, mutations also increased linearly with radiation dose. Chromosome aberrations increased in all biological systems exposed to radiation in a well-defined function of dose, which at low doses seemed to be linear. Carefully conducted studies demonstrated that after high-dose, high-dose-rate radiation exposure, cancer frequency increased in experimental animals and humans. As the dose-rate decreased, cancer frequency decreased markedly in animal models. However, in the low-dose region, the epidemiological tools available were not sensitive enough to detect increases in cancer frequency in humans.

The uncertainty in the human data generated was so great that little could be said about the shape of the dose-response curve or if any increase in cancer frequency occurred in the low-dose region. However, together these high-dose observations suggested that radiation caused damage in the genetic material, the amount of genetic damage was linearly related to the radiation dose, and the genetic damage was the primary cause of radiation-induced cancer. Therefore, the radiation dose was directly and linearly related to cancer induction. This was a useful mechanistic approach for high-dose and high-dose-rate radiation exposure.

As described in the previous chapters, technologies and techniques now exist and have been used to evaluate the radiation response in the low-dose region, and scientific observations suggest that radiation risk needs reevaluation. Scientific committees such as the National Academy BEIR VII committee and ICRP acknowledged these phenomena, but suggested it was difficult to determine how the new data could be used to impact the understanding of radiation risk in the low-dose region. When the BEIR VII report "Health Risks from Exposure to Low Levels of Ionizing Radiation" and the ICRP 103 were written in 2006, it was suggested that the observations of genomic instability, adaptive responses, and bystander effects were not sufficiently supported by data on molecular mechanisms to be useful in risk assessment (NRC 2006; Valentin 2007).

The BEIR VII committee suggested "…until the molecular mechanisms responsible for genomic instability and its relationship to carcinogenesis are understood, extrapolation of the limited dose-response data for genomic instability to radiation-induced cancers in the low-dose range <100 mGy is not warranted." For adaptive responses the committee said, "Thus, it is concluded that any useful extrapolations for dose-response relationships in humans cannot be made from the adaptive responses observed in human lymphocytes or other mammalian cellular systems. Therefore, at present, the assumption that any stimulatory effects of low doses of ionizing radiation substantially reduce long-term deleterious radiation effects in humans is unwarranted." Finally, for bystander effects, "Until molecular mechanisms of the bystander effects are elucidated, especially as related to an intact organism, and until reproducible bystander effects are observed for low-LET radiation in the dose range of 1-5 mSv, where an average of one electron track traverses the nucleus, a bystander effect of low-dose, low-LET radiation that might result in a dose-response curving either upward or downward modification should not be assumed."

At the time of the BEIR VII report, the lack of mechanistic data suggested that the new science could not be used in risk estimates. However, since then, the DOE Program has generated additional mechanistic data on each of these responses. The rest of this chapter summarizes the mechanistic data and provides a source of information for future research. To put all these data together using a systems approach will require additional research and funding, but it is essential to be able to use the information produced to date to understand risk, the mechanisms involved, and the shape of the dose-response relationships in the low-dose region.

It is imperative to examine the new radiation biology produced by the program to summarize the data; describe the cell and molecular pathways triggered and altered by low-dose radiation exposure; review the important genes, proteins, chemicals, and pathways involved; and suggest potential approaches to use these mechanistic data to evaluate their impact on the shape of the dose-response relationship in the low-dose region. In light of the new complicated and interactive responses observed following the exposure to low doses of ionizing radiation, it is no longer acceptable to suggest that there is a single mechanism of action.

As mechanistic data continue to be developed, it is essential to use systems biology approaches to develop models that describe important pathways, and provide relationships and weights for each pathway and information on the ultimate shape of the dose-response relationships in the low-dose region. Such evaluations will help to understand and evaluate radiation risk and incorporate these mechanistic data into the calculation of risk factors. This approach will require extensive interactions between research groups, a serious commitment of resources, methods to accumulate and share data, and a commitment to developing and applying the models developed. Early models that pave the way for these efforts have been developed and suggested approaches for systems approaches that can develop more mechanistic models will be discussed in chapter 8.

2. Dose and Energy Deposition

When ionizing radiation deposits its energy in biological systems, molecules are altered. These initial alterations are the basis for all subsequent biological responses. It is critical to be able to relate the proper energy metric to the subsequent biological response as the basis for developing models to describe these responses (Brooks 2005). The deposited energy can be described in several ways. A report recently developed by the ICRU (2011) concludes that the energy metric is very important in conducting and understanding radiation biological experiments. A major point is that the energy distribution and deposition pattern in cells, tissues, and organisms is very non-uniform, especially in the low-dose region. It is important to understand how this non-uniform deposition of energy influences the biological outcome. Microbeam studies have been important in evaluating the influence of dose-distribution on biological effects and suggest that the radiation response target is much larger than that of energy deposition (Braby & Ford 2004).

The primary descriptor for deposition of energy in a tissue has been energy/unit of mass or dose. The development and use of dose as the most important metric in radiation biology is widely accepted. The universal use of dose in radiation biology suggests that the concentration of the interactions and ionizations produced by the radiation plays a central role in the observed biological response. This makes dose-rate important because the concentration at any time also plays a central role in radiation biology in the types of responses and pathways activated. Most of radiation biology, medicine, and risk assessment has been based on these two metrics of radiation exposure.

However, Bond et al. (2005) suggested that if the LNT model is important, the total number of ionizations produced (energy deposited in the system) should be the biological response metric. If this were the case, the radiation-induced biological effects would relate not to dose but to the total energy deposited in the system of interest. Bond et al. (2005) also suggested that total energy deposited in the biological system be used as the basis for setting radiation protection. Using total energy in the system rather than energy concentration (dose) as a metric for describing radiation response would have major impact on radiation biology.

First, it can be demonstrated that for any biological system there is an amount of energy deposited below which it is not possible to detect a biological response. This threshold or energy barrier must be exceeded to detect a significant biological response (Brooks et al. 2000; Brooks et al. 2006a).

Second, if energy is used as the metric for evaluating biological responses, it becomes obvious that very large amounts of radiation (energy) are required to produce a detectable excess cancer above the high background level of spontaneous cancer (Brooks et al. 2007). Evaluation of the atomic bomb data using this concept demonstrated that if this large amount of energy were deposited in a small population it would be lethal. However, because the energy was distributed over a very large population, an increase in responses cannot be detected. This provides a very different view of radiation exposure and risk than when dose is used as the metric for radiation- induced cancer.

The current concept is that every ionization increases the risk for the induction of cancer (NRC 2006). Thus, low doses of radiation, delivered to a large population, increase the calculated risk. Using total energy deposited rather than radiation dose as the metric for radiation exposure and risk provides a very different perception of the risk associated with low doses of radiation. It takes a very large amount of radiation (Energy in Joules) to produce excess cancer. Such information should be communicated to the public, regulators, and those concerned with accidental radiation exposure or terrorist events.

3. Biological Mechanisms of Low-Dose Responses

3.1 Radiation-Induced Changes in DNA

Years of radiation biology research have demonstrated that DNA damage and repair are the earliest and most important changes induced by ionizing radiation, as it plays an important role in the induction of mutations and cancer. The huge database from this research has been used as the mechanistic basis for development of the hit theory, the biophysical models used to predict the induction of cancer, and the LNT (NRC 2006). The research was also used to describe the relationship between radiation dose and cancer development risk. These data remain a critical part of radiation biology.

Because of the extensive database in place at the time the DOE Program began, the program's focus was not on the role of DNA damage on cancer risk, but on the differences in DNA damage induced by radiation and DNA damage from other sources, including the endogenous DNA damage that is present in large amounts. The other focus of the program has been on the role of DNA alterations as a signaling mechanism to alter gene and protein expression, metabolic pathways, and cell function. The information developed has been beneficial in understanding the responses seen in the very-low-dose region where the frequency of mutations is very low and the major impact of the radiation-induced DNA alterations seems to be related to other important biological changes.

3.2 Differences between Radiation-Induced and Spontaneous DNA Damage

Program-funded research has made progress in identifying the differences between radiation-induced damage and that produced by endogenous factors, as reviewed in chapter 4. The mechanisms involved in the production and repair of clustered DNA lesions has been reviewed (Prise et al. 2001) and remains an area of continuing research effort. Early in the program, it was suggested that clustered lesions were unique to radiation and that such damage would be very useful as a biomarker of radiation exposure (Sutherland et al. 2001b). Pinto et al. (2005) suggested that the clustered lesions had unique structure and may be very difficult to repair. These lesions, which have a different distribution in space (Rydberg et al. 1994), were induced more frequently by high-LET radiation than by low-LET X-rays or gamma rays (Paap et al. 2008). Studies on the repair of these lesions using human abasic endonuclease (Ape1) suggested that cellular responses to complex damage may be carried out by multiple processing mechanisms, and that the more complex nature of high-LET-induced damage would have more serious consequences in terms of risk and cellular damage than simple single- or double-strand DNA breakage, which is consistent with other data on high-LET radiation. However, further research determined that there were low levels of clustered lesions induced by endogenous oxidative damage in the normal cells, and that the lesions may not be unique to ionizing radiation (Sutherland et al. 2003c). These observations were extended to human skin and further studied in human skin models (Bennett et al. 2004; Bennett et al. 2005).

Such research suggests that clustered lesions are part of normal human biology and may be produced by endogenous and other environmental insults in addition to ionizing radiation (Bennett et al. 2005). In CHO cells it was demonstrated that increased fgp protein lowered clustered damage and the frequency of Hprt mutations (Paul et al. 2006). The induction and repair of these lesions by both high- and low-LET radiation and their role in radiation risk remains an important area for future research.

3.3 Signaling Molecules

3.3.1 DNA Alterations as Signaling Molecules

Radiation-induced DNA damage is detected by cells and results at the start of a number of cellular events, many of which are involved in the repair of this damage. Double-strand breaks are thought to be the most important primary lesion induced by ionizing radiation (Jeggo 1998) involved in cell killing (Olive 1998). The repair of DNA damage has been reviewed and is available on abcam's website as a flow chart, which was produced in collaboration with James Haber and Farokh Dotiwala of Brandeis University. The genes and proteins involved in the different types of DNA repair are well defined in this flow chart.

3.3.2 The Role of ATM in Signaling

An important breakthrough in the study of DNA damage and repair was the observation that individuals with the disease ataxia telangiectasia have a deficiency in DNA repair. People with this disease have a mutated gene that produces an altered protein called ataxia telangiectasia mutated protein (ATM). The cDNA for ATM was first characterized by (Zhang et al. 1997). This gene has been carefully studied, and the protein has been isolated and characterized (Lavin 1999). The normal protein associated with the mutated gene was able to correct the phenotype of cells that had the mutated ATM gene (Lavin 1999). Lavin et al. (2005) also determined that DNA damage altered ATM signaling in a way that resulted in genomic instability. These observations suggest a link between DNA damage, ATM, genomic instability, and the induction of cancer. Studies of cells, tissues, and organisms with this mutated gene and protein have provided an essential tool for study of radiation-induced DNA damage and repair. This protein, kinase ATM, is the primary transducer of cellular responses initiated by DNA damage caused by ionizing radiation.

Phosphorylation has been shown to be very important in the repair of DNA lesions (Whalen et al. 2008). The phosphate forms of important proteins such as γH2AX are found to localize at the site of DNA DSBs (Rogakou et al. 1998) and have been strongly correlated with and sometimes used as markers of the frequency and repair of DNA breaks (Rothkamm & Lobrich 2003). These lesions have also been suggested as potential markers of cell killing induced by drugs that produce DNA DSBs (Banath & Olive 2003).

It has been shown that ATM, HDM2, p53, and DNA-PK are all involved in H2AX phosphorylation at the site of radiation-induced DNA DSBs (de Toledo et al. 2000; Stiff et al. 2004). It was also demonstrated that phosphorylated p53 directly binds to radiation-induced DNA breaks (Al Rashid et al. 2005) and that p53 is involved in a global regulation of genomic repair genes (Amundson et al. 2002). As a function of radiation type, phosphorylation is very dependent on radiation quality, further emphasizing that the response to radiation is unique as a function of radiation type, and correcting for the increased effectiveness of high-LET radiation by a simple number may depend on the endpoint and the biological processes activated by different radiation types (Whalen et al. 2008).

However, these DNA repair foci also form during many normal cellular processes, so care must be taken to control many biological variables when using foci as measures of the amount of DNA damage induced by ionizing radiation (Tanaka et al. 2007). Studies have been conducted to compare the disappearance of DNA DSBs and the loss of γH2AX foci. The loss or repair rates between these two endpoints do not have a one-to-one correlation. This suggests that they are related but not identical (Kato et al. 2008). Like most biological responses to radiation, there is a difference in individual sensitivity to the induction of γH2AX, showing that genetic background is of prime importance in radiation responses (Hamasaki et al. 2007).

Cells and organisms that are heterozygous for ATM have been shown to be lacking in DNA repair as measured by γH2AX (Kato et al. 2006) and to be sensitive to radiation-induced cell killing (Kuhne et al. 2004), oncogenesis (Smilenov et al. 2001; Hall et al. 2005), and formation of cataracts (Worgul et al. 2005). Studies of the role of ATM on the change in the responses to low doses of radiation from sensitivity to the induction of low-dose radiation resistance suggest that this process is independent of modification of ATM at the ATM ser1981 site (Krueger et al. 2007b).

It has been demonstrated that even a transient inhibition of ATM kinase is sufficient to enhance radiation-induced cell killing (Rainey et al. 2008) and that ATM knockout heterozygous mice display a marked adaptive response for the induction of mutations (Day et al. 2007a). Both observations support the hypothesis that ATM plays an important role in both mutation and in cell sensitivity to ionizing radiation. However, other studies on mice with haploinsufficiency suggested that this condition does not affect the frequency of mutations in solid mouse tissues (Connolly et al. 2006). Haploinsufficiency for ATM and Mrad9 can increase the effects of radiation in the production of cataracts (Kleiman et al. 2007). This condition also changes some of the early events to initiate cell signaling and changes in the function of the cells (Smilenov et al. 2005). The signaling cascade induced by induction and repair of DNA damage has also been shown to influence a number of measurable cellular events such as cell cycle progression, gene activation and expression, changes in ROS status of the cells, senescence, and apoptosis (Smilenov et al. 2005).

3.3.3 *The Interactions between ATM and p53*

Radiation exposure can be linked to ATM activity and p53 is downstream from ATM. Thus, the data have demonstrated that radiation exposure to low doses causes a marked change in gene expression as well as a complex cascade of signaling events. Biological pathways in this cascade have been studied, and many involve the p53 gene (Amundson et al. 2005), the "guardian of the genome," and it is involved many repair processes (Amundson et al. 2002) as well as in the induction of cell death and apoptosis (Slee et al. 2004). The induction of apoptosis is the most conserved function of p53 and seems to be vital for tumor suppression (Slee et al. 2004). However, when this gene is mutated, it can become the "fallen angel" and result in many biological problems and diseases. A well-defined database (http://www-p53.iarc.fr/) has been developed that summarizes the involvement of mutated TP53 genes in the induction of human cancers. The mutational status of the TP53 gene is critically important in influencing the tumor phenotype (Deppert 2007).

The importance of p53 in radiation-induced effects is well established regarding changes in gene expression, radiation damage, control of chromosome instability (Schwartz et al. 2003), and cell survival (Schafer et al. 2002; Amundson et al. 2008). The individual variability in p53 and CDKN1A has also been linked to radiation sensitivity (Alsbeih et al. 2007). The expression of p53 shows that it is a key gene in controlling chromosome integrity (Honma et al. 2000), chromosome instability (Schwartz et al. 2003), and the genomic stability of cells (Perez-Losada et al. 2005). Such studies show that p53 is at the center of a complex web of incoming stress signals and outgoing effector pathways. Understanding these pathways and signals provides links between p53, environmental exposures (Medina et al. 2002), stress (Amundson et al. 2005; Horn & Vousden 2007), genomic instability (Mao et al. 2005), and cancer induction (Yang et al. 2000b; Yang et al. 2000c) as well as many other diseases.

It is important to study the relationships between radiation-induced changes in p53 and changes in other genes and proteins to understand how early genetic events are related to later changes in cells and tissues. This provides insight into how these early changes can be related to animal and human models of cancer. An example of this is found in a study by Williams et al. (Williams et al. 2008b) in which the genotype and radiosensitivity of 39 human tumors were evaluated. It was determined that the normal and mutant TP53 status of the tumors was critical in determining both the radiosensitivity of the cells and histology-dependent variations in radiosenstivity of the cancers. This research suggested that radiosensitivity can predict responses of human tumors to radiotherapy protocols.

Changes in metabolic pathways associated with p53 are important and provide some of the early biological linkages to diseases, as well as play a role in the induction of bystander effects (Prise et al. 2006a; Burdak-Rothkamm et al. 2007; Burdak-Rothkamm et al. 2008). Program research has also identified many chemicals and factors that modify the radiation-induced changes in these pathways.

As of 2007, PubMed has about 3600 citations per year on p53 research (Deppert 2007); therefore, the database cannot be covered here in detail. However, large databases are available that summarize the knowledge about the role of p53 (http://p53.bii.a-star.edu.sg) in radiation response and mechanisms of cancer induction (Lim et al. 2007; Amundson et al. 2008).

Activity of p53 is modulated by MDM2, and the dose of radiation delivered is important in regulating this interaction. At low doses, there is little induction of p53 or MDM2. As dose increases, the level of p53 peaks followed by a down-regulation as MDM2 peaks. This cycle repeats as a function of time after exposure with the number of cycles and the magnitude of the responses depending on the radiation dose. At low doses, there are few cycles. As dose increases, the magnitude of the peaks remains constant but the peak replications increase. Following very high doses, the interaction continues for a prolonged period of time (Tyson 2004). This critical interaction shows a threshold below which no measurable changes occur. Using this approach, it has been possible to show that p53 signaling and the interaction with MDM2 antagonists are modified in some cancer types. This may have implications and applications in potential cancer therapy (Tovar et al. 2006).

3.4 Summary of Biological Response Mechanisms

Once the cascade of responses has been initiated by the radiation-induced DNA damage, many questions remain with major areas of research as follows:

- Many genes are activated by low doses of ionizing radiation. These genes modify many chemical factors and metabolic pathways within the cell that are responsive to low doses of radiation. These pathways have been studied in the program, and it has been determined that they influence the responses in cells both with and without energy deposited in them. These effects are important in the subsequent development of biological changes measured in a wide range of systems, because they may modify response that either decreases or increases radiation-related risk. For many of these pathways, it is not possible to determine which way (beneficial or detrimental) these influences will go. However, the weight of the evidence currently favors the induction of protective effects induced by low doses of ionizing radiation.

- Considerable research has been done on the chemicals and factors involved in transmitting the signals from cells that have energy deposited in them to other cells. This transmission is manifested by the extensive cell-cell, cell-matrix, and cell-tissue interactions. Research on chemicals involved in signaling is one of the program's major areas. It is essential to identify the signals, the targets, and the responses generated.

- The molecular changes result in cellular changes critical to the total response. Such changes depend on interactions between cells and can result in marked changes at the tissue and organ level. New techniques and methods in proteomics and metabolomics are being developed to evaluate all these changes. With these techniques, it has become possible to understand more of the mechanistic basis of the radiation responses to low doses of radiation. Such research will help determine the path forward to consider the mechanisms involved in the radiation-related changes and provide important data for modeling the responses induced by exposure to radiation in the low-dose regions.
- It is important to review the influence of all these mechanisms on inducing genomic instability, bystander effects, and adaptive responsive, as suggested in early observations. Research has demonstrated that these responses are linked by common cellular mechanisms. Because of the large amount of data generated in each area, it is not possible to review it completely. This book focuses on the program-generated publications and research in these areas. It is important that all these factors be considered to construct models that not only describe cellular and molecular events but use systems approaches to identify how such information can be used to evaluate radiation risks.

3.4.1 Transmission of the Signals to Other Cells

Chapter 4 covered the experimental studies that established that bystander effects are a well-defined response to low doses of radiation. The bystander effect is in reality a measure of the communication that occurs within biological systems. Studies have been conducted to determine how radiation-induced changes can be communicated as cell-cell, cell-matrix, and cell-tissue interactions. The bystander effect studies demonstrated that the cells that have energy deposited in them communicated the changes induced by the radiation to the neighboring cells (Shao et al. 2003a; Laiakis & Morgan 2005; Sandfort et al. 2007; Shao et al. 2007).

Such responses as DNA damage and homologous recombinational repair influence the induction of bystander responses (Nagasawa et al. 2005). In cells that are defective in homologous recombination, there is no bystander effect seen for the induction of SCE or chromosomal aberrations (Nagasawa et al. 2008). The induction of cell killing (Schettino et al. 2003; Schettino et al. 2005; Baskar et al. 2007) and apoptosis (Vit

& Rosselli 2003; Vines et al. 2009) in bystander cells has been well documented, as has the induction of micronuclei (Konopacka & Rzeszowska-Wolny 2006). Bystander effects could be produced by either high- or low-LET radiation using microbeam technology (Baskar et al. 2008). Also, Ponnaiya et al. (Ponnaiya et al. 2007) determined that changes in gene expression occur in cells that are not directly "hit" by ionizing radiation. As outlined in chapter 4, cell transformation was one of the first and most important observations made for the induction of a biological effect in cells without direct energy deposition in them (Mitchell et al. 2004c). As research has continued, the term "bystander effects" has been the subject of extensive research to define it in more mechanistic terms (Hei et al. 2008). This communication can alter biological responses to radiation in ways that could be considered both protective (adaptive responses) and harmful (genomic instability) to the body.

After the discovery of bystander effects induced by low doses of alpha particles (Nagasawa & Little 1992) using the induction of SCE as an endpoint, there were many discussions as to how and if these responses were a general biological phenomenon. With the development of the microbeam by DOE Program investigators, it was possible to know which cells had energy deposited in them and to measure the responses in these cells as well as in cells with no energy in them. It is necessary to identify both the "hit" cells and the "non-hit" cells and to determine the pathways involved in the cell-cell communication. Finally, the chemicals and factors released must be identified, and their interaction with the target cells characterized. Experimental techniques must be developed that can determine how the bystander cells are modified and provide measurements of the factors released and the responses induced (Hill et al. 2006; Pyke et al. 2006). These studies demonstrated that bystander responses were very general and have now become well accepted. Research is being conducted to address these scientific questions and determine the mechanisms involved in the cell-cell communication.

It has been known that gap junctions play an important role in cell-cell communication in many cell types and that the protein connexin 43 was an important molecule in this communication. This information was used early in the program to determine that direct cell-cell contact and the presence of gap junctions were critical for the cell-cell communication induced by low doses of high-LET ionizing radiation (Glover et al. 2003). Studies also demonstrated that connexin 43 was sensitive to low doses of ionizing radiation and other environmental stresses. The cell-cell communication induced by low doses of radiation can be modified by blocking connexin 43 using Lindane. Bystander effects were eliminated by this treatment (Azzam et al. 2003a). Such studies demonstrated that gap junctions and direct cell-cell contact is essential in some types of bystander effects.

The molecules released and thought to be important in the induction of cell-cell communication and bystander effects following low-dose radiation exposures have been an important area of research in the program. It was thought that the molecules should be small and be able to travel quickly between cells to account for the rapid responses observed in the cells that did not have energy deposited in them.

An early candidate for the induction of bystander effects was nitric oxide (NO). This molecule has a long enough life >7 minutes (Hakim et al. 1996) to move some distance from the cell exposed to the ionizing radiation (Belyakov et al. 2005) and is involved in oxidative metabolism (Azzam et al. 2002; Ridnour et al. 2005). It has also been linked to transforming growth factor beta (TGFβ1) in the induction of bystander effects in glioma cells (Shao et al. 2008a). This molecule and oxidative metabolism seemed to also provide mechanistic links between gap junctions, bystander effects, and adaptive responses (Azzam et al. 2003b). Ridnour et al. (Ridnour et al. 2005) showed that nitric oxide treatment can induce resistance to oxidative stress induced by hydrogen peroxide, apparently through a glutamate cysteine ligase activity-dependent process. This observation supports the other low-dose responses, where it has also been shown that low doses of radiation activate genes involved in the production of MnSOD (Guo et al. 2003) and glutathione, which are both involved in the protective process against damage from radiation-induced free radicals and oxidative stress (Guo et al. 2003). Many other genes involved in adaptive responses have been identified that are also involved in protective processes (Tomascik-Cheeseman et al. 2004; Chaudhry 2006). It has been suggested that NO radicals choreograph the adaptive response and provide a strong link between the induction of adaptive responses and bystander effects (Matsumoto et al. 2007). Such research provides examples of how bystander effects could result in a protective response, as opposed to the induction of potentially harmful changes such as the micronuclei and other genotoxic responses observed in bystander cells (Azzam et al. 2002).

Shao et al. observed that calcium flux was modulated by low doses of radiation and was an important change induced in bystander cells (2006). This flux was postulated to be critical to the induction of bystander effects in glioma cells and fibroblasts. The molecular pathways activated by the calcium seemed to involve the MAP kinase signaling pathways (Lyng et al. 2006) and were shown to be involved in the induction of bystander responses in cells without energy deposition. Additional studies using glioma cells helped define the role of other signaling factors induced by ionizing radiation (Shao et al. 2008b).

Media transfer experiments showed that materials released from the cells produced paracrine-signaling pathways involved in cell transformation in response to low doses of low-LET radiation (Weber et al. 2005). Many other media transfer experiments demonstrated bystander effects induced in cells exposed to conditioned media (Mothersill et al. 2006; Maguire et al. 2007). These studies were done in JB6 cells that are easily transformed by a number of factors. Other studies showed that autocrine signaling was also important in cell-cell communication and that epidermal growth factor (EGF) was involved in this signaling (Chen et al. 2004b).

Research to identify the signaling pathways suggested that many complex pathways exist and play an important role in cell-cell communication. For example, both protein kinase C (Baskar et al. 2008) and cyclooxykgenase-2 (Zhu et al. 2005) pathways were demonstrated to be important in cell signaling resulting in biological changes in cells that did not have direct energy deposited in them. The magnitude of these indirect changes could be increased following exposure to either alpha particles (Han et al. 2007) or low-LET (Zhou et al. 2008) radiation exposures by simply changing the NaCl concentration in the media. Such studies emphasize the need for control and reproducibility of research in studies of cell-cell communication and the potential for artifacts to be introduced in the data sets.

Additional research using proteomic and other biochemical approaches linked the calcium changes to phosphorylation-dependent calmodulin complex in mammalian cells, both of which are important in cell-cell communication (Jang et al. 2007). It may be that this is an additional pathway associated with bystander responses. As is often the case for these molecular and cellular responses, the same factors do not seem to apply for different types of radiation exposure. Heavy ion exposure, such as that encountered in space, failed to activate the calcium pathways and/or modify early calcium flux (Du et al. 2008).

3.5 Metabolic Pathways

The radiation-induced pathways are interlinked, so it is not possible to discuss them independently. They have been broken down here and discussed by the chemicals and pathways involved. As the signaling pathway continues through many different critical locations, research has demonstrated that alterations from a wide range of factors can modify the signaling and the biological outcome. Elaborate methods have been designed to link all the various interactions with radiation exposure using several different techniques. Since this research was not conducted by the program, these will not be reviewed here.

3.5.1 TGFβ

Early in the program, Barcellos-Hoff and Brooks (2001) recognized that transforming growth factor beta (TGFβ) was important in the expression and modification of direct and bystander-induced radiation-induced damage both *in vitro* and *in vivo*. Important relationships were established between the induction of DNA damage, TGFβ, and different types of cellular damage (Ewan et al. 2002). These relationships were extended from the role of TGFβ at the single-cell level to TGFβ interaction at the tissue level. Barcellos-Hoff (2005a) demonstrated that this compound played an important role in orchestrating the interactions and outcome of radiation with tissues. Further research showed that inhibition of TGFβ blocked ATM activity to genotoxic stress (Kirshner et al. 2006). This again supported a direct link between DNA damage and TGFβ. Other pathways were also demonstrated to play an important role in signaling and controlling DNA repair, including the signaling from epidermal growth factor receptor (EGFR)(Rodemann et al. 2007).

Jobling et al. (2006) established that the ROS status of cells and tissues were directly linked to both radiation exposure and the treatment with TGFβ. Changes in ROS status have also been linked to cell-cell communication and the bystander effect. TGFβ was shown to modify NO activity as a possible mechanism for its role in cell-cell communication (Shao et al. 2008b). NO also played a major role as a signaling molecule during the induction of bystander effects as gleoma cells communicated with fibroblasts (Shao et al. 2008b). Many of these changes in cell signaling and the ROS status of cells have been shown to modify the response to genotoxic agents including radiation.

Other important cellular changes directly related to TGFβ may play a role in cancer development. Andarawewa et al. (2007b) showed that treatment with TGFβ was critical in the transition of epithelial cells to mesenchymal cells. Such a transition could influence the progression of normal epithelial cells to cancer cells. TGFβ was also implicated in modification of the immune response. The regulation of immunological responses during the development of skin cancer suggested an important role of this compound in the carcinogenesis process (Glick et al. 2008).

Ewart-Toland et al. (2004) demonstrated that a gain of function of TGFβ1 polymorphism plays a role in late-stage prostate cancer and may act as a potential biomarker for the progression of this disease. Because of these observations, it was suggested that TGFβ could be used in conjunction with radiotherapy. This hypothesis is being tested (Andarawewa et al. 2007a).

The central role of TGFβ in radiation responses has thus been well established. Additional research in the low-dose region to determine the impact of this compound on cancer risk is essential.

3.5.2 NF-κB

NF-κB (nuclear factor κB an enhancer of activated B cells) is a protein complex that controls the transcription of DNA. It is found in almost all animal cell types and is involved in cellular responses to stimuli such as stress, cytokines, and free radicals produced by ionizing and ultraviolet irradiation. NF-κB is important in maintaining genomic stability and is modified in cells that display genomic instability. Incorrect regulation of NF-κB has been linked to cancer, inflammatory and autoimmune diseases, septic shock, viral infection, and improper immune development.

NF-κB is important in the biological response to ionizing radiation. It is directly linked to the signaling pathway from radiation-induced DNA damage through its interactions with ATM. This interaction has been thought to be important in generating radiation-sensitive cells (Ahmed & Li 2007). The ErbB2 pathway is also involved in radiation-induced activation of NF-κB (Guo et al. 2004). Radiation-induced pathways such as NF-κB that modify radiation response are an important part of the intercellular signaling triggered by low doses of radiation that play a role in split-dose repair.

When radiation exposures are split into multiple fractions with a time period between each of the dose fractions, cell survival increases. This is thought to be made possible by repair of damage produced by the first fraction before it can interact with that produced by later fractions. The interaction between NF-κB and other molecules involved in cell signaling plays an important role in repair of damage between the fractionated doses that is critical to cell survival (Mendonca et al. 2007).

Studies have defined the signaling pathways involved in NF-κB using a variety of different techniques. Low doses of gamma rays activated NF-κB in the bone marrow of mice (Rithidech et al. 2005). Knock-out and transgenic mouse models were used to define the role of NF-κB in cancer induction (Gerondakis et al. 2006). Such studies provide mechanistic pathways to be integrated into the total radiation responses. NF-κB is involved in cross-talk critical in sensitizing cells to cell killing by ionizing radiation (Bubici et al. 2006). It has also been shown that it inhibits mitogen-activated protein kinase signaling in radio-resistant cancer cells, which modifies cell proliferation and increases survival. Such activity also suggests an important role in decreasing radiation-related cell killing (Ahmed et al. 2006). These cytotoxic responses seem to be mediated through interactions with mitochondria. This interaction with the mitochondria and the ROS status of the cell plays an important role in signaling for the induction and promotion of radiation-induced bystander effects (Zhou et al. 2008).

Research suggests that the potential for increasing cell killing through NF-κB is involved in the induction of the adaptive response (Ahmed & Li 2008). The cell killing decreases the frequency of abnormal cells as a function of radiation dose and results in adaptive responses. As these studies have provided a more mechanistic understanding of NF-κB's role in cell killing, it has been suggested that NF-κB could play an important role in cancer therapy (Ahmed & Li 2007).

The role of NF-κB in the induction of genomic instability is also a critical factor and requires additional research. Maintaining the stability of the genome is critical both in cells that are directly hit by radiation and in cells where the signals from hit cells induce changes. Research by Moore et al. (Moore et al. 2005) indicated that in addition to NF-κB, tumor necrosis factor alpha (TNFα) is modified in cells that have energy directly deposited in them from alpha particles but is not altered in bystander cells. It has been suggested that modification of oxidative stress in cell signaling and the importance of the oxidative status of the cells are critical in maintaining and modulating genomic instability induced by TNFα. Modification of these factors has been reported to increase radiation-induced genomic instability (Natarajan et al. 2007).

The concern about atomic war and terrorism has resulted in extensive research on the role of chemicals in modifying radiation responses to protect cells and tissues from ionizing radiation. The DOE Program has not funded extensive research in this area. However, as adaptive response became a recognized result of exposure to low doses of ionizing radiation, it was important to determine the mechanisms involved in this adaptive response and to evaluate some of the chemicals that may protect against radiation.

3.5.3 WR-1065

Research in the program determined that changes in the levels of reactive oxygen and Mn Superoxide dismutase (MnSOD) protected against radiation damage and were involved in adaptive response. It was thus important to understand how other chemicals could modify and protect against radiation-induced damage.

A few projects were funded to evaluate the role of the radioprotection product amifostine (WR-1065), one of the few compounds that had been used in clinics to modify radiation responses. Murley et al. (2002) demonstrated that mouse sarcoma cells could be protected by the enhancement of MnSOD gene expression when exposed to the active metabolite of WR-1065. Oxidative status of the cells and cytoprotection were also modified by both amifostine and TNFα through changes in levels of manganese superoxide dismutase (SOD2) (Murley et al. 2007). This produced radioprotection through changes in the free thiol form of aminfostine (Natarajan et al. 2007), which had been recognized as being important in activating NF-κB gene expression. Such research provides one potential mechanism for the radioprotective action of WR-1065 (Grdina et al. 2002; Kataoka et al. 2002).

The active metabolite of amifostine was recently shown to mitigate radiation-induced genomic instability (Dziegielewski et al. 2008). Because genomic instability has been suggested to play a critical role in the induction of cancer, these studies provide potential pathways for modification of the cancer response and linking these molecular studies to cancer risk.

3.5.4 Clusterin

Clusterins are stress-inducible polypeptides that play an important role in cell survival, proliferation, and apoptosis. Through such actions, they have been implicated in the induction of multiple organ dysfunction following high doses of radiation (Araki et al. 2005). Nuclear clusterin has been synthesized, and its function as an inducer of cell killing defined (Leskov et al. 2003). However, clusterin's activity is modified by its cellular location because clusterin(s) can be located and processed either in the nucleus or can be secreted into the cytoplasm and outside the cell. Clusterin expression has been shown to be induced by low-dose ionizing radiation. It can act to either kill cells or improve the survival, depending on its type and cellular location (Leskov et al. 2001a). It is one of the molecular switches that controls the fate of cells, as it interacts with many factors including insulin-like growth factor 1 receptor/Src/MAPK/ Egr-1 signaling pathway that results in prosurvival (Criswell et al. 2003a; Criswell et al. 2003b). The activity of radiation-induced clusterin can be suppressed by p53 suppressor proteins, which shows that it is linked into the p53 pathway (Criswell et al. 2003a; Criswell et al. 2003b). Nuclear clustetin/X1P8 can bind to Ku70, modify cell proliferation (Yang et al. 2000b; Yang et al. 2000a), and signal the induction of cell death (Yang et al. 2000b; Yang et al. 2000a).

Secretory clusterin protein is implicated in aging, obesity, heart disease, and cancer. Regulatory signaling transduction processes control secretory clusterin expression. Secretory clusterin plays an important role in many of the radiation responses (bystander effects, genomic instability, and adaptive responses) induced by low doses of radiation (Klokov et al. 2004). Radiation exposures and the signal protein, TGF-ß1, increase the production of secretory clusterin. Radiation exposure causes stress activation of signaling pathways that regulate the clusterin cascade pathway and can result in apoptosis and antiproliferation signaling, which can result in killing of prostate cancer cells *in vitro* (Shannan et al. 2007). However, TGF-ß1 can override this suppression and allow massive expression of secretory clusterin. Furthermore, adding secretory clusterin to medium suppresses TGF-ß1-induced cell growth. It has even been suggested that the level and type of clusterin present in a system after exposure to ionizing radiation can be used as a marker of radiation exposure (Klokov et al. 2003).

(Kalka et al. 2000) suggested that secretory clusterin can play an important role in modifying responses to tumor therapy by inducing apoptosis and helping to eliminate skin tumors following phthalocykanine 4 photodynamic therapy. It has been shown to be an excellent candidate for changing radiation resistance in prostate cancer and may be useful in other types of cancer therapy. Research is under way to get clusterin into clinical trials (Shannan et al. 2006). Understanding the mechanisms involved in the action of clusterin in both high and low dose ranges has increased its usefulness in cancer therapy as well as its role in estimating radiation-related cancer risk

3.6 Alteration in metabolic pathways

Data from recent publications indicated that low doses of radiation can alter metabolic pathways (Lall et al. 2014). It was established that doses as low as 0.1 Gy could result in a large increase in glucose consumption while large doses (4.0) Gy caused either little change or a decrease in this responses.

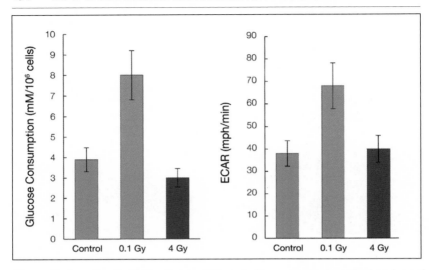

Figure 21. Low-dose Radiation Induces Glycolysis. Demonstrates that after low doses of ionizing radiation there are marked changes in glucose metabolism not seen after high doses. The figure suggests different mechanisms of action in the low dose region which requires additional research.

These low-dose-induced changes were related to increased expression of glucose transporters (ECAR). There is limited evidence that glucose transporters may play an important role in development of cancer (Lall et al. 2014). The potential importance of these changes in radiation-induced cancer still remains to be determined and is an important area of future research. Research to determine the role of radiation variables on these responses is needed. For example, what is the impact of changing dose rate, dose distribution, and radiation type on these responses?

3.7 Changes in Gene Expression

Given that the spectrum of changes induced by low levels of radiation differs from those induced by high radiation doses (Robson et al. 1997; Hande et al. 2003; Tomascik-Cheeseman et al. 2004), the major questions that remain concern the biological impact of these changes. Are the changes induced an indication of an increased cancer risk induced by low doses of radiation, or are they changes that protect the biological system, which would lower the cancer risk in the low-dose region?

The rapid advances in the fields of genomics and proteomics both within and outside the program have provided the tools to understand the biological responses induced by low doses of ionizing radiation and the significance of these changes. Program research determined that many pathways controlling the fate of cells were modified by a range of different chemicals and that low doses of radiation changed the expression of the genes and proteins associated with these chemicals (Azzam et al. 1998; Amundson et al. 2002; Coleman et al. 2003a; Yin et al. 2003; Lyng et al. 2006; Amundson et al. 2008).

As discussed earlier, the observations that radiation can produce DNA damage, mutations, chromosome aberrations, and genomic instability suggest that radiation is harmful at all doses. However, recent research demonstrated that many mechanisms are involved in cancer induction that influence the shape of the dose-response relationships. Extending these recent observations on non-DNA-damage-related changes into the low-dose region and the relationships that exist between changes in gene expression and the initial changes in DNA damage remain critical research areas. The importance of radiation-induced genomic instability and changes in gene expression has not been widely researched. Snyder and Morgan (2004a, b) determined that changes in gene expression profiles change rapidly after radiation exposure, making it hard to relate these early changes to the development of late-occurring diseases. When studies were conducted to determine unique changes in gene expression in an attempt to understand the mechanisms involved in the induction of genomic instability, few clues were found as to the initiation and perpetuation of chromosomal instability (Snyder & Morgan 2004b, a). Expansion of these types of studies may be necessary to link genomic instability to mechanisms of actions for radiation-induced changes in the low-dose region, and to understand the net effects of low doses of radiation on risk.

Research suggests that even though DNA damage is induced linearly as a function of radiation dose, the processing of this damage, the signaling induced by the damage, and the biological consequences of the damage change as a non-linear function of dose. Low doses of radiation were thought to be involved in modification and repair of DNA damage. Early studies with gene expression failed to demonstrate that modified expression of DNA repair genes was altered by low doses of radiation (Tomascik-Cheeseman et al. 2004). It was also determined early in the program that DNA-dependent protein kinases do not play an important role during the induction of adaptive responses (Odegaard et al. 1998). Thus, direct induction and repair of DNA damage may not be as important in the total risk from low doses of ionizing radiation as other cellular processes.

Radiation-induced changes in gene expression suggested that many cellular processes were influenced by exposure to low doses of radiation. Modification of genes involved in control of the cell cycle, changes in oxidative metabolism, and modification of signaling pathways were altered by low doses of radiation. These changes are reflected in adaptive responses produced by both low total doses and low radiation dose-rates. Such changes have been shown to result in protection and sometimes even in the decrease in biological damage below the level observed in control cells (Elmore et al. 2008).

Several genes have been identified that are induced by low doses of ionizing radiation and suggest a potential mechanism involved in the protective effects. Okazaki et al. (2007) showed that the genes involved in p53 and p53-related pathways are modified by low doses of radiation and play an important role in the production and modification of apoptosis. Apoptosis has been linked to radioadaptive responses and to the elimination of transformed cells.

Genes involved in the production of Mn(SOD), known to be involved in radiation protection, are up-regulated by low doses of radiation and play an important role in the induction of the adaptive responses (Guo et al. 2003). This provides another area of research that suggests that low doses of radiation can be protective. Up-regulation of Mn(SOD) has been suggested to reduce the level of damage observed in the systems exposed to low doses of ionizing radiation to levels below that observed in the non-exposed controls.

The level of glutathione can also reduce the frequency of chromosome aberrations, modify cell cycle kinetics, and protect mouse bone marrow cells following ionizing radiation exposure (Ray & Chatterjee 2007). Glutathione is also up-regulated in the low-dose region. Modification of energy metabolism has recently been linked to changes in cellular responses to ionizing radiation. It was demonstrated that treatment of cells with 2-Deoxy-D-glucose, an important marker of modification of energy metabolism, causes cytotoxicity, oxidative stress, and radiosensitization in pancreatic cancer cells (Coleman et al. 2008). PARP-1 and PARP-2 are changed by low doses of ionizing radiation and play a role in radiation-induced resistance and protection (Coleman et al. 2004).

All of these studies need to be integrated in a meaningful way and used in modeling studies as we continue to evaluate the influence of low doses on cancer risk. These types of studies provide early insight into the mechanisms involved in low-dose radiation effects.

Studies with microarray techniques have also provided mechanistic information on how cells communicate with each other following low-dose radiation exposure. The activation of communication pathways seems to be one of the major responses to low doses of ionizing radiation. Such studies show that cells, tissues, and organs communicate and respond to radiation in an integrated fashion rather than as individual cells. Suggestions on current pathways involved in communication as well as insight into future research directions are needed to better understand how such communication will influence radiation responses (Chaudhry 2006).

Studies on radiation-related changes in gene expression have also been applied to understanding the cancer process. Such studies have not been a major focus in the program, so they are not reviewed in detail here. However, several research studies funded by the program have suggested strong correlations between changes in gene expression and cancer induction.

Park et al. (2005) conducted studies determining that the susceptibility of cancer cells to B [beta]-lapachone is enhanced by ionizing radiation. The same research team also noted that up-regulation of NAD(P)H: quinine oxidoreductase by radiation exposure potentiated the effects of bioreductive [beta]-lapachone on cancer cells (Choi et al. 2007). This suggests potential pathways for cancer treatment and control using combined radiation and chemical treatment. It may be possible for some of these combined treatments to be moved into the clinic and provide potential methods to improve cancer therapy.

Gene expression profiles were also used in association with studies that identified chromosome translocations using SAGE that were important in the development of myeloid leukemia (Lee et al. 2006). Such studies are important in providing direct pathways between molecular studies, cellular changes, and the development of disease. With this type of information it will become possible to follow the early changes in gene expression and link them to the cellular outcome and, finally, the late-occurring diseases such as leukemia. Such studies on gene expression suggest that there may be radiation-induced genes that may be exploited in the development of gene therapy (Greco et al. 2005).

Molecular and cellular biology continue to provide powerful tools for understanding the biological responses to radiation, especially in the low-dose region. The potential protective effects of such responses, especially in the low-dose region, have been reviewed and their impact on risk evaluated (Tapio & Jacob 2007). These studies also play an important role in developing methods for radiation protection and were reviewed in a workshop designed to evaluate their use as biomarkers as well as to more fully understand mechanisms involved in radiation protection (Coleman et al. 2003a; Coleman et al. 2003b). Using these techniques to obtain data and relating the results to biological outcomes, especially late-occurring diseases, is an important area for continued research.

3.8 ROS Status of the Cell

Research conducted within the program emphasized the important role of the reactive oxygen species (ROS) of the cells and oxidative stress in the development of radiation-related disease (Azzam et al. 2002; Murley et al. 2002; Azzam et al. 2003a; Gius & Spitz 2006). In the low-dose region, it was determined that protective mechanisms were activated that involved changes in the mitochondria, the ROS status of the cells, and the modification of radioprotective chemicals, including the well-known SH containing radioprotective chemicals. This research provided an excellent link between low doses of radiation and the observed biological responses such as adaptive response, bystander effects, and genomic instability.

Free radicals play an important role in the induction of cancer and other diseases of aging, and radiation produces free radicals and oxidative stress. Research on the impact of radiation-induced free radicals has focused on the alterations in DNA and the induction of chromosome damage and mutations as the result of radiation exposure. A major goal of the program was to determine the differences between biological alterations produced by free radicals during normal oxygen metabolism and those produced from ionizing radiation, as discussed in chapter 3. It was suggested that DNA clustered lesions were unique to ionizing radiation and could act as a marker of radiation-induced damage. Some research suggested that this may not be the case and that factors such as endogenous oxidative damage produces clustered DNA lesions in unirradiated viral DNA and in human cells (Sutherland et al. 2003c; Sutherland et al. 2003b).

Oxidative stress, ATM deficiency, and normal cellular metabolism all seem to be important for the repair of DNA damage (Barzilai et al. 2002). It was determined that repair of DNA damage was defective when the cells were ATM deficient illustrating again that genetic background is one of the most important elements of the responses associated with radiation exposure. The ROS status of the cells was also determined to be critical in the induction of both DNA damage and error-prone DNA repair. This was put forward as a model to link genomic instability and the progression of myeloid leukemia (Rassool et al. 2007). In contrast to these studies, it was also determined that radiation can induce a number of proteins important in DNA DSB repair (Leskov et al. 2001a; Leskov et al. 2001b).

Thus, radiation can both induce DNA damage and help repair it. Such observations support data that demonstrate the fact that DNA damage increases as a linear function of radiation dose but the processing and repair of that damage is non-linear and seems to result in total dose-response relationships for many endpoints that are sub-linear.

3.8.1 ROS and Genomic Instability

There are strong relationships between environmentally induced stress, cellular oxidative stress, chronic inflammation, and the induction of cancer. Diseases such as chronic esophageal acid reflux syndrome result in an inflammatory disease of the esophagus, a major risk factor in the production of esophageal cancer. Thus, the role of ROS in maintaining genomic stability has been the subject of program research.

Large doses of ionizing radiation cause extensive damage at the molecular level, which causes cell killing and then major tissue disruption. A primary outcome of these high doses is alteration of the ROS status of the cells and tissues with marked increases in radiation-induced free radicals. Limoli et al. (1988) determined early in the program that changes in the ROS status of cells in culture were important in the production of apoptosis and reproductive failure of cells. These changes were evident in cells with a compromised genomic integrity (Limoli et al. 1998) and resulted in chromosomally unstable cells (Limoli et al. 2003). The dose-response relationship for such changes has a rather steep slope, and in the high-dose region seems to increase linearly with radiation dose.

Because of the high frequency of radiation-induced genomic instability (Limoli et al. 2003), early studies suggested that individual genes were not one of the major causes. However, Slane et al. (2006) demonstrated that mutations play a role in genomic instability. In cells with mutations in succinate dehydrogenase subunit C, there was a high frequency of genomic instability. In studies of the molecular parameters associated with genomic instability, Pichiorri et al. (2008) determined that genes and cells with fragile sites are important in the induction of genomic instability.

In addition to changes in DNA-associated mutations, other factors play an important role in the induction of genomic instability. Interleukin 8 produced a pro-mitogenic and pro-survival role in radiation-induced, genomically unstable cells (Laiakis et al. 2008). This change in cell kinetics, which resulted in survival of damaged cells, was previously suggested as a mechanism for the selection of radioresistant cell variants seen in genomically unstable cell lines (Limoli et al. 2001a). Together, these results support the role of cell selection in the development of radiation-induced genomic instability.

In addition to selection of cells in genomically unstable lines, cell senescence is an important player. Studies demonstrated that senescence, cell transformation, and genomic instability were all mediated by platelet/megakaryocyte glycoprotein Ib alpha, indicating the importance of this factor in monitoring and maintaining the stability of the genome (Li et al. 2008a; Li et al. 2008b).

Radiation-induced changes in ROS acts as a trigger to initiate cross-talk between the ROS status of the cell and NF-κB. This cross-talk results in a number of molecular changes of biological significance (Bubici et al. 2006). Some of the suggested biological changes produced by the interaction of NF-κB with oxidative stress signaling are the production of genomic instability (Natarajan et al. 2007). It has also been demonstrated that in chromosomally unstable cell lines there is a differential induction and activation of NF-κB complexes (Snyder & Morgan 2005a, b). This again provides additional information regarding the mechanisms involved in radiation-induced genomic instability and, potentially, radiation-induced cancer. As discussed in chapter 4, additional research is needed to link the induction of genomic instability to the formation of cancer.

The relationships between normal metabolism, mitochondrial dysfunction, levels of reactive oxygen species, and radiation-induced genomic instability have been carefully reviewed **(Kim et al. 2006)**. At high doses, these relationships have been well established. It is critical to extend and expand the research on the association of genomic instability and the ROS status of the cells into the low- and very-low-dose regions. Such studies will be crucial in determining if low doses of radiation result in protective mechanisms that stabilize the genome, while high doses result in genomic instability. The important question that remains to be addressed is "do low doses result in responses that increase or decrease the ROS status of the cells?" The answer to this question would provide a potential mechanistic basis for understanding the differences in the biological responses to high and low doses of radiation and an explanation about how radiation can either protect against or enhance the induction of genomic instability.

3.8.2 ROS and Adaptive Responses

The role of cell ROS status on the induction of adaptive responses remains an important area of research because it could have a major impact on the shape of the dose-response relationship and risk in the low-dose region. As research has progressed from single-cell *in vitro* studies to more complex cell and tissue relationships, it has become evident that both tissue architecture and oxidative metabolism are a critical part of the induction of adaptive responses (de Toledo et al. 2006). These protective responses are linked to mitochondrial function.

It has also been demonstrated that mitochondrial DNA repair is important in the induction of cellular resistance to oxidative stress induced by a number of environmental and experimental conditions (Grishko et al. 2005). The role of the mitochondria in the total radiation response also has been found to be very important. Ionizing radiation alters cyclin B1, which is involved in control of cell cycle. This alteration seems to be regulated through NF-κB and the antioxidant enzyme MnSOD, which can modify the oxidative status of the cells and act as a protective mechanism against radiation-induced damage (Ozeki et al. 2004).

Using mouse skin epithelial cells, Fan et al. (Fan et al. 2007) showed that the adaptive response depends on the interaction between NF-κB and MnSOD, producing a decrease in the ROS status of the cells. The relationships between MnSOD, NF-κB, and the adaptive and protective responses have been advanced through research directed toward developing radioprotective compounds. Radioprotective drugs have been used during cancer radiotherapy to protect normal cells against radiation injury. Murley et al. (2004) determined that the production of MnSOD was one of the major pathways altered by the treatment with drugs such as the free thiol form of amifostine. By using experimental protocols that gave repeated administration with this form of amifostine, radioresistance could be maintained, and the level of MnSOD was elevated and seemed to be important in this continued radioresistance (Murley et al. 2007; Murley et al. 2008).

Other research has demonstrated that by altering the ROS status of the cells, it is possible to provide radiation protection against a wide range of different radiation types with a range of different LET such as that which would be encountered during space exploration. In these studies, both antioxidants and Bowman Birk proteins resulted in marked reduction of the free radicals in cells and protected against radiation-induced damage (Kennedy et al. 2006). Such studies suggest that radioresistance observed in MCF-7 breast cancer cells is related to the ROS status of the cells as well as the level of peroxiredoxin II in the cells (Wang et al. 2005). These are experimental variables that may be altered by experimental treatments.

Proteomic and transciptomic analyses have helped determine that mitochondrial dysfunction results in the induction of oxidative stress in cells leading to cell killing through apoptosis (Chin et al. 2008). Additional research demonstrated that apoptosis can be modified by many factors that modify the oxidative stress and ROS levels in the cells and tissues. These studies were conducted in mouse models of Parkinson's disease, but the results may be applicable to other forms of diseases associated with increases in oxidative stress, such as radiation-induced cancer in the high-dose range.

Dong et al. (2007) found that treatment of cells with vitamin E analogues altered oxidative status levels and can induce selective apoptosis in proliferating endothelial cells and stop angiogenesis, which is critical in cancer growth and spread. Such research highlights the role of normal oxidative metabolism and suggests that alterations of this metabolism by any type of stress can be either protective or detrimental in the risk of cancer development. It was also demonstrated that any form of stress may result in stress-induced premature senescence (SIPS) (Suzuki & Boothman 2008), which may play an important role in aging as well as the loss of genomic stability and cancer development. Research in this area could provide mechanistic data to link some of these biological observations and help understand how they may be altered by experimental treatment. Andringa et al. (2006) suggested that altering the metabolism may potentially desensitize cancer cells to radiation and to the toxicity of 2-Deoxy-D-Glucose.

The observations described here demonstrate the critical role of the redox status of the cell in cancer biology. This subject has been reviewed by Gius and Spitz (2006), as has the role of stress and how it alters gene expression, senescence, redox status of the cells, and the risk for cancer. Links between these factors have also been established (Denko & Fornace 2005). All this research makes it clear that the ROS status of the cells is critical during cancer development and that the responses to radiation in altering this status is very dependent on the radiation dose. High doses increase the stress and reactive oxygen levels in the tissues and cells and increase risks, while low doses seem to increase the level of MnSOD, which would protect cells and possibly reduce cancer risks.

3.9 Cellular Changes

3.9.1 Chromosome Aberrations

Radiation responses need to be extended and coordinated across different levels of biological organization from the initial changes in gene and protein expression to changes in the ROS status of the cell and morphological changes in cells. Finally, these changes need to be linked to disease. The first of these cellular changes to be discussed is chromosome aberrations. Measurement methods and the impact of scoring chromosome aberrations were discussed in chapter 3, and it has been well established that scoring chromosome aberrations is very useful in biodosimetry. Studies on chromosome aberrations also supply critical new information on the mechanisms of action of radiation as a function of radiation type, dose, dose-rate, and dose distribution. Most tumors have abnormal chromosomes with either translocations, duplications, losses, or changes in total chromosome number, and studies of these changes have helped evaluate their role in cancer induction. Continued research is needed to understand radiation-induced chromosome aberrations and their role in the development of cancer, particularly leukemia.

3.9.2 Telomeres

Another recent discovery about radiation-induced changes in chromosomes is the role of the telomere in radiation-induced biological damage. New staining techniques made it possible to mark the different telomeres on each chromatid and detect radiation-induced changes in the telomeres. With these techniques, it was possible to demonstrate unique differences that depended on the DNA strand associated with them. Bailey et al. (2001b) performed post-replication processing of the telomere that depended on the DNA strand so that it could be determined which strand each telomere was associated with. Zou et al. (2004) determined that replication of each telomere's DNA was asynchronous and again depended on the strand of the DNA involved. These processes made it possible to identify each telomere and to follow changes induced in the telomeres by experimental procedures, thereby providing a very useful tool for extensive mechanistic research on cellular radiation effects.

Telomeres play a critical role in the proliferative life of cells. As cells and organisms age, the length of the telomere decreases. The protein telomerase is involved in the process of maintaining telomere length, and in early research, it seemed to be essential. Telomere length and maintenance also changes in transformed and cancer cells. Thus, it is important to understand how telomere length is maintained to control the fate of the cells. Nugent et al. (1998) demonstrated that telomere maintenance is also closely linked to end repair of double strand-breaks in DNA. Further work demonstrated that DNA DSB proteins were required to cap the ends of mammalian chromosomes during the formation of telomeres (Bailey et al. 1999). The activity of DNA-PK kinase was found to be essential in protection of the mammalian telomeres (Bailey et al. 2004b; Bailey et al. 2004d).

Even though telomerase is essential in the maintenance of telomere length, Lee et al. (1999) determined that many of the DNA repair enzymes could act through independent DNA repair pathways to maintain telomere length in the absence of telomerase. This research was extended to Saccharomyces cerevisiae where the role of DNA repair on the telomere length was supported (Lee et al. 2002). Bailey et al. (2004b, d) also demonstrated that frequent recombination in telomeric DNA could extend the life and maintain telomere length in cells that were telomerase negative. Such observations provide strong links between DNA repair and the maintenance of the telomere (Bailey et al. 2001b; Bailey & Goodwin 2004). When there was a defect in the telomeres that was present at the same time as DNA DSBs induced by ionizing radiation or other environmental insults, there was interaction between the DNA DSBs and the telomere (Bailey et al. 2004c). This interaction has been reviewed, and it seems to be a very general biological interaction and plays an important role in maintaining the stability of the genome (Bailey & Cornforth 2007).

It has been postulated that radiation-induced genomic instability is important in the generation of cancer. Since genomic instability occurs at a very high frequency, it was critical to look for targets larger than traditional radiation-induced gene mutations for the induction of genomic instability. As research on telomeres has advanced, it has been linked to the induction of genomic instability induced both in cell systems and in animal model systems. The telomere provides a larger target and the higher frequency needed to explain radiation-induced genomic instability. Studies with a mouse model (K-ras p53) on the induction of lung cancer indicated that telomere dysfunction promotes genomic instability as well as increasing the metastatic potential for the cancers (Perera et al. 2008). Reviews of the literature on the inter-relationships between genomic instability and telomere dysfunction suggest that telomere dysfunction is one of the major driving forces in radiation-induced genomic instability (Bailey et al. 2007).

It is important to link radiation-induced changes in telomeres to radiation exposures. Durante et al. (2006) demonstrated that, following radiation exposure with heavy ions that result in very high deposition of energy per unit of distance traveled in the tissue, there are chromosomes that lack telomeres. This would suggest the breakage and loss of genetic material and alteration of the telomeres that could be important in loss of genomic stability.

Zhang et al. (2005) demonstrated that by suppressing DNA-PK activity using RNAi, both telomere dysfunction and mutation frequency were altered. This effect depended on the type of radiation used in the studies, with HZE particles found in space being more effective that gamma rays in producing these effects.

Radiation exposure produces DNA damage, changes in telomere function, and genomic instability. An important review of these effects on telomeres, chromosome instability, and cancer suggested strong causal links between these observed cellular and subcellular changes and the induction of cancer (Bailey & Murnane 2006). Such studies provide a potential mechanism for the induction of cancer through a radiation-induced telomere dysfunction and genomic instability. Because the induction of genomic instability by higher radiation doses increases linearly with radiation dose, such a model supports the LNT hypothesis at these doses.

3.9.3 Cell Cycle
Early in the field of radiation biology, scientists recognized that radiation could cause blockage of the cell cycle at specific stages. It was postulated that the blockage of the cell cycle in the G_2 stage allowed additional time for DNA repair before the cell progressed through mitosis and "set" the damage. This was thought to be a protective mechanism that decreased the damage and risk from radiation. Ku and CHK1-dependent radiation-dependent G_2 blockage were evaluated to define the pathway and mechanism involved in initiating the cell cycle delay (Wang et al. 2002).

Research focused on other radiation-induced biological changes determined that the cell cycle plays an important role in several of the newly observed low dose biological phenomena such as adaptive response and low-dose radiation hypersensitivity. Ahmed and Li (2008) determined that Cyclin D1 was a critical actor involved in altering cell cycle and the induction of the adaptive response. This research contrasted with much earlier research that failed to show that the cell cycle was an important variable in adaptive response. Alterations in the cell cycle were very critical during the induction of low-dose radiation hypersensitivity and the population of G_2 cells was a primary factor that resulted in cell killing in the very-low-dose region of the dose-response relationships.

Research on the role of reactive oxygen on low-dose responses determined that normal regulation of cell cycle progression was altered by flavin-containing oxidases that may be influenced by low doses of radiation (Venkatachalam et al. 2005). Studies using compounds such as isobuosilactone A, which alters ROS and induces apoptosis, also suggest that radiation may alter cell cycle progression. Such compounds modify the kinase 1 signaling pathway that is critical in the induction of human breast cancer (Kuo et al. 2009). The direct association between these observations and the low-dose radiation responses has not been established and is another area of needed research.

New techniques have been developed to follow cell proliferation and clonal expansion using integrated one- and two-photon imaging. Such studies suggest that clonal expansion may be a major factor in determining mutation load in cells and tissues and may be important in the development of radiation-related cancer (Wiktor-Brown et al. 2008).

Wilson et al. (2004) published a useful review on the role of radiation and cell cycle progressions. As more mechanistic information becomes available on the role of radiation in cell cycle progression, clonal expansion and cell cycle changes, the influence of cell proliferation kinetics on cancer risk may be forthcoming.

3.9.4 Hyper-Radiation Sensitivity and Radiation Resistance

Past research at higher radiation doses suggested that cell killing was a simple decreasing function with dose. However, it was determined that cell killing increased rapidly as a function of low-dose exposures (hyper-radiation sensitivity, HRS). As the dose increased the cells became radiation resistant, and the cell-killing slope decreased as dose increased resulting in induced radiation resistance (IRR). Subsequently, numerous studies were conducted to help define the mechanism of action involved in these unique low-dose radiation responses. Early studies focused on the influence of external conditions on the induction of the response, such as oxygen tension (Marples et al. 1994) and dose fractionation (Short et al. 2001). Other studies focused on the potential genes and proteins involved in the responses (Marples et al. 1997; Marples & Joiner 2000; Marples et al. 2002; Chalmers et al. 2004). As is often the case, many mechanistic studies related to HRS and IRR induction were negative: that is, the endpoint studied did not influence the shape of the dose-response relationship in the low-dose region. Further studies determined that ATM Se1981 had no influence on the shape of the cell-killing curve in the low-dose region (Krueger et al. 2007b; Krueger et al. 2007a), and that the recognition of DNA DSBs was not related to the induction of HRS (Wykes et al. 2006).

Krause et al. (2005a,b) determined that low-dose hypersensitivity did not influence the cure of cancers in experimental animals and did not translate into an improvement outcome in ultra-fractionationed radiotherapy *in vivo*. On the positive side, it was suggested that low-dose hypersensitivity is associated with p53-dependent apoptosis. In addition, extensive studies determined that the stage of the cell cycle at the radiation exposure was very important in low-dose hypersensitivity. Cells in the G_2 stage of the cell cycle were most sensitive to low-dose hypersensitivity followed by radiation-induced resistance (Marples et al. 2003; Short et al. 2003; Marples et al. 2004).

The literature on low-dose hypersensitivity and radiation-induced resistance has been carefully reviewed at different times (Joiner et al. 2001; Joiner 2004; Marples & Collis 2008). These phenomena are very important observations relative to the shape of the dose-response relationships in the low-dose region. If low doses of radiation increase cell killing, this treatment could eliminate cells from the population that may be at higher risk for the induction of cell transformation. Conversely, low-dose hypersensitivity and radiation-induced radiation resistance to cell killing could increase cell proliferation in the low-dose region and protect cells that are sensitive to radiation-induced cell transformation in the medium-dose range. It could be postulated that this combination could result in an increase in cancer risk in the low-dose region. Such increases above that predicted by the LNT have not been detected in any animal or human studies and do not seem to be a viable postulate.

3.9.5 Apoptosis

The process of apoptosis, or programmed cell death, has been recognized for a long time and plays a critical role during embryonic development. As cells differentiate and form organs, many of them are programmed to die. For example, in the formation of the hands, the cells between the fingers die on a pre-programmed schedule, allowing the fingers to separate. During the early days of radiation biology, it was not widely recognized that radiation produced apoptosis. Cells were thought to be killed by radiation through either the processes of mitotic death or necrosis. However, it was recognized before the initiation of the program that apoptosis played an important role in cell killing of lymphoblastoid-type cells following exposure to radiation (Schwartz et al. 1995).

This radiation-induced apoptosis had been called interphase death, and lymphocytes seemed to be the major cell type that suffered from this type of death. As the research continued, it became obvious that death of the cells during the G_0 stage was indeed apoptosis and that radiation was a major factor. Studies were conducted to determine the dose-response relationships for the induction of apoptosis. These suggested that for non-genotoxic insults a threshold existed for both radiation-induced cytotoxicity and apoptosis, below which little apoptosis could be observed in human lymphocytes. Schulte-Hermann et al. (Schulte-Hermann et al. 2000) suggested that "Applications of non-genotoxic carcinogens at doses too low to interfere with life-death decisions of cells or for time periods too short to cause irreversible transitions in cell populations may therefore be considered below the biological threshold for a carcinogenic effect."

Because radiation is classified as a genotoxic carcinogen, it was suggested that such responses do not apply to ionizing radiation. Studies using radiation demonstrated that many systems are very sensitive to radiation-induced apoptosis, and no threshold could be detected below which there was no response. It was also determined that, at very low doses, the induction of apoptosis may be protective for the induction of cell transformation and other endpoints of interest for cancer risk (Mendonca et al. 1999). Dose-response studies were conducted in lymphocytes for the induction of apoptosis, and it was determined that the RBE for 280 keV neutrons relative to X-rays was approximately one (Ryan et al. 2006). This suggests that neutrons were similar in effectiveness in the production of damage or the triggering of the signaling pathways associated with apoptosis.

Because the process of apoptosis was well recognized in the embryo, it was a logical extension to evaluate radiation's role in the induction of apoptosis during embryogenesis. Such a role could be important in normal development and suggests a potential role of radiation-induced apoptosis on the induction of birth defects that has been well documented at set stages of development (Hall 2000b). This sensitive stage of embryogenesis is when the limbs and organs are undergoing critical changes during development. Radiation induces apoptosis during embryogenesis, and this could provide a basic mechanism for induction of radiation-related developmental abnormalities (Bladen et al. 2007b).

As is the case for most biological responses, the genetic background of the individuals or cells is critical to determine the magnitude of the radiation response. Early studies of the induction of apoptosis in lymphocytes demonstrated that the p53 status of the cells played a major role (Schafer et al. 2002). These observations were extended to whole animals, and it was determined that the p53 status of mice played a major role in radiation-induced changes in gene expression. These radiation-related changes in gene expression seemed to alter the frequency of apoptosis and could play an important role in repair of DNA damage and removal of cells with damaged DNA (di Masi et al. 2006). Modification of the genetic background of mice to produce a haploinsufficiency for ATM and RAD9, a DNA repair gene, resulted in marked changes in radiation-induced repair of DNA damage. These changes in repair modified other biological processes that were reflected in changes in the frequency of apoptosis and cell transformation (Smilenov et al. 2005).

Additional studies in human thymus and spleen were conducted on radiation-induced transcriptional activity of genes that determined an increased level of expression of radiation-induction of genes involved in the two major apoptotic pathways (Alvarez et al. 2006). This provides links between radiation-related gene induction and the production of apoptosis. Apoptosis plays an important role in cancer risk. It has also been demonstrated that experimental modification of p53 such as with acetlyation at lysine 317 can negatively regulate apoptosis and modify cell survival (Chao et al. 2006). Even though the p53 status of the cells and animals seemed to play a major role in the induction of apoptosis, Takahashi et al. (2005) determined that for high-LET radiation the p53 status did not influence the outcome. Exposure to high-LET radiation may create such a marked biological alteration that the signaling and damage induced act independent from the p53 status. For low-LET radiation exposure there is no question that the p53 status of the cells and organisms is extremely important to trigger the induction of apoptosis following DNA damage. Thus, it is well established that p53 is important in the signaling pathways that are activated by DNA damage and result in apoptosis.

As previously discussed, another important component of the radiation-induced signaling pathway is NF-κB. Radiation-induced NF-κB is directly associated with apoptosis (Meng et al. 2003), and it has been suggested that blocking the NF-κB pathway can alter the frequency of apoptotic cells. Using different types of DNA-damaging agents, it was possible to further link NF-κB to radiation-induced DNA damage and apoptosis. The type of DNA damage influenced the frequency of apoptosis (Strozyk et al. 2006).

A major goal of the program was to determine if a difference exists between DNA damage from endogenous factors and that induced by radiation. Li et al. (2008a) tested the hypothesis that there was a difference between the frequency of apoptosis induced by radiation-related DNA damage and that induced by DNA damage from endogenous sources. They demonstrated that complex DNA damage was more effective in production of apoptosis than simple DNA double-strand breaks. Such observations suggest that radiation-induced complex DNA lesions may play a critical role during radiation-induced apoptosis.

Following radiation-induced DNA damage, many repair genes and processes determine the signaling initiated and fate of the cells with DNA damage. Many of the signaling pathways that result in apoptosis have been identified, characterized, and modified by experimental protocols. It was determined that activation of the TNF-related, apoptosis-inducing ligand (TRAIL) gene was important in the induction of apoptosis (Kagawa et al. 2001). Aravindan et al. (2008) measured the length of time that the signaling processes and molecules involved in apoptosis exist following radiation exposure. The protein-kinase signaling pathway resulting in radiation-induced apoptosis can be modified by both H-ras and Ki-ras. This modification can result in an increase or decrease in the frequency of apoptosis depending on the genes involved and the pathway that is induced (Choi et al. 2004).

It was also possible to determine how the signaling pathways induced by radiation change as a function of time after exposure (Aravindan et al. 2008) as well as a function of the type of radiation exposure (Ryan et al. 2006). Many molecules and DNA repair proteins also play important roles in apoptosis. For example, staurosporine modulates radiation-induced apoptosis, and ceramide is also required for radiation-induced apoptosis (Guo et al. 2006; Deng et al. 2008). It is important to recognize that some of these studies were conducted *in vitro* and others in non-mammalian systems. Additional research is required to determine if such compounds will be important in radiation-induced apoptosis in humans.

To better relate apoptosis to cancer, studies have measured the induction of genomic instability and related it to the induction of apoptosis. Radiation-induced apoptosis profiles in cells that were genomically unstable were different from the induction of apoptosis in normal cells (Nagar et al. 2005). Other studies relating genomic instability to apoptosis determined that both could be induced by exposure to either carbon ions or X-rays (Ryan et al. 2006).

Genomic instability is a step in the carcinogenesis process and may be directly related to the induction of radiation-induced cancer. It has been demonstrated that the frequency of radiation-induced genomic instability increases linearly with high doses. The role of apoptosis in this process is complicated because the removal and loss of damaged cells by apoptosis may reduce cancer risk.

Apoptosis has been demonstrated to be a frequent event following exposure to low doses of ionizing radiation (Bauer 2007a; Portess et al. 2007). Experimental conditions that decreased the frequency of apoptotic cells increased the frequency of APRT mutations in mice repeatedly exposed to ionizing radiation (Liang et al. 2007). Radiation resistance can also be increased by factors that modify cell cycle and reduce apoptosis (Park et al. 2000). Such studies suggest the potential for apoptosis to be protective against late-occurring diseases such as cancer. This observation has been related to the induction of protective or adaptive responses in the low-dose region of the dose-response curve.

One of the most important observations of apoptosis in radiation biology is the suggestion that low doses of radiation can trigger biochemical and signaling pathways in bystander cells that result in selective apoptosis of cells that are transformed and may be in the process of changing from normal to cancerous cells (Bauer 2007a; Portess et al. 2007). If low doses of radiation can selectively cause transformed cells to undergo programmed cell death, then it has been postulated that the cancer risk can be directly reduced (Portess et al. 2007). This would help explain experimental results in the study of cell transformation where low doses of ionizing radiation decrease the frequency of transformed cells below the levels seen in controls (Redpath 2006b, a). Similar results on the induction of mutations could be explained by this apoptosis-related process (Sykes et al. 2006b; Sykes et al. 2006a; Zeng et al. 2006). As will be evaluated in greater detail in chapter 8, such biology could result in non-linear low-dose responses with low doses producing less cancer risk than is present in a non-exposed population (Scott 2007).

Research has demonstrated that many environmental and other factors influence and modify the induction of apoptosis in the low-dose region. Survey studies of gene expression in several cell lines demonstrated that the sensitivity to cell killing could be related to changes in gene expression (Amundson et al. 2008). It has also been shown that ubiquitin can up-regulate apoptosis proteins in cancer cells as a protective mechanism against cancer progression (Zhang et al. 2004). The p53 and p53-related genes are associated with protection from apoptosis during the initiation of adaptive responses (Okazaki et al. 2007). The DNA repair gene gadd45a has sensitized epithelial cancer cells to cell killing, which can change the long-term potential for survival of the patients (Lu et al. 2008). It has even been suggested that changes in the diet can modify apoptosis and result in a "suicide solution for the delay of cancer growth" (Khan et al. 2007). All these studies are important in relating cellular changes to risk. They are critical in using systems biology to relate early cellular changes such as apoptosis and cell killing to radiation risk for the induction of cancer.

As is the case for many biological endpoints, the ROS status of the cell is a critical variable in the induction and prevention of apoptosis. Very high doses of radiation-induced apoptosis can be modulated by treatment with compounds that inhibit energy metabolism (Hunter et al. 2007). This suggests a direct link between energy metabolism, the ROS status of the cells, and the induction of apoptosis in the high-dose region, but provides no information on the responses to low doses. Treating cells with antioxidants decreased the frequency of micronuclei, a potentially detrimental effect of radiation, but it did not affect the induction of apoptosis or the viability of the cells, indicating potentially protective effects (Konopacka & Rzeszowska-Wolny 2006). Research has also demonstrated that treatment of cells with Hsp25 directly inhibited the production of radiation-induced apoptosis by reducing programmed cell-death-mediated ROS production.

Changing the ROS status of the cells can be a protective mechanism. Even treatment with vitamin E is thought to inhibit angiogenesis, an essential part of tumor development, by selective induction of apoptosis in proliferating endothelial cells (Dong et al. 2007). Such studies show the importance of apoptosis during cancer development and the role that ROS status and radiation-induced changes in ROS status may have on cancer risk.

The role of apoptosis in radiation oncology has been reviewed (Meyn et al. 2009), as have important factors that control apoptosis such as NF-κB (Dutta et al. 2006) and other pathways that determine whether a cell lives or is killed by apoptosis (Bartek & Lukas 2006). Such information provides for an understanding of the mechanisms involved in radiation-induced cancer in both the high- and low-dose regions. These reviews and the important scientific publications in this area will be critical as the science of low-dose radiation biology moves forward and attempts are made to use new mechanistic information to evaluate radiation protection and risk.

4. Influence of Genetic Background on Cancer Risk

Cancer has long been known to have a genetic component, as many families are cancer prone. The role of genetic background on radiation-induced cancer was carefully reviewed by NCRP (2010). This report demonstrates that genetic differences in many molecular, cellular, and experimental animal systems support the role of genetic background on biological responses to ionizing radiation (Williams et al. 2008b; Williams et al. 2008a). Genetic background also impacts the induction of genomic instability (Pichiorri et al. 2008).

Mutations in many genes that contribute to cancer result in the production of chromosomal instability in cell lines (Grigorova et al. 2004). Mutations in many of these genes such as BRACA1, BRACA2, CHK2, and BUB1 are thought to play an important role in the induction of breast, ovarian, and other forms of cancer (King et al. 2003). The relationship between spontaneous cancer and mutations in these genes and radiation-related cancer has been a major area of extended research.

The relationship between genetic diseases and radiation-induced cancer is well established for a limited number of diseases such as Nijmegen breakage syndrome and Ataxia telangiectasia (Little et al. 2002a; Little et al. 2002b). To aid in the study of the relationship between genetic background and radiation-induced cancer, mouse models of Nijmegen breakage syndrome have been developed (Williams et al. 2002). Such studies provide a tight link between genetic background and radiation-induced cancer.

Radiation sensitivity has also been established in primary fibroblasts isolated from families that have hereditary retinoblastoms as well as in many apparently normal controls (Chuang et al. 2006). The differences in these changes are greater when the radiation exposure is delivered at a low dose rate (Wilson et al. 2008). The dose-rate-dependent nature of this response has been used to suggest that repair of DNA damage may be one of the major pathways involved in this genetically related, radiation-induced damage.

Genomic instability has been carefully related to a number of DNA-repair-deficient mutants in Chinese hamster ovary (CHO) cells (Somodi et al. 2005). Cells defective in homologous recombination DNA repair have been shown to not be sensitive to the induction of sister chromatid exchanges in bystander cells following exposure to low doses of alpha particles (Nagasawa et al. 2008). However, bystander effects were shown to depend on deficiencies in DNA DSB repair. Zhang et al. (2008) suggesting that DNA repair processes are essential in the initiation of bystander responses in chromosomes. Instability in DNA copy number has also been induced by ionizing radiation (Kimmel et al. 2008).

All these studies suggest that genetic background, genomic stability, and radiation-induced DNA damage are closely related. Because ATM is one of the major signaling proteins that responds to DNA damage, the response of genetic background that influenced this protein was evaluated, and ATM heterozygosity did not influence radiation susceptibility to exposure to ionizing radiation (Mao et al. 2008). This was true both in wild-type background as well as in animals that were heterozygous with respect to their p53 background. These studies become important in establishing the relationships between the effects of genetic background and the role of DNA damage on the activation of communication pathways as an important response to radiation exposure.

This discussion demonstrates that genetic background is critical to radiation response and that there is a range of radiation sensitivity in all cell lines evaluated. Surveys of cell killing and changes in gene expression have been conducted on large populations of different cell types by testing more than 60 cell lines used by the NCI in anti-cancer drug screens (Amundson et al. 2008). Such research suggests that many different genetic factors can influence the induction of cell killing and gene expression, and through these pathways, influence many of the radiation-induced biological responses. The role of radiation on cancer risk is thus very dependent on the genetic background of the individuals and populations being exposed. Genetic background must be carefully considered in any study of the biological effects of ionizing radiation.

Since 2007, the program has funded some studies of the role of epigenetic effects on the transgenerational responses to ionizing radiation. Epigenetic changes do not alter the DNA, but are involved in gene expression, changes in the levels of methylation of DNA, and alterations in protein structure associated with the chromosomes and genes. Such alterations have been shown to alter coat color, metabolism, and cancer risk in mice. Early epigenetic changes in radiation of mice during development are currently being funded. These studies are for the most part in the early stages and represent an important area of current and future research. It has been suggested that epigenetic changes may be related to non-targeted effects of radiation and result in biological changes in cells that do not have energy deposited in them (Kovalchuk & Baulch 2008). If radiation can modify epigenetic changes as well as produce direct changes in DNA, this could have a major impact on cancer risk, especially in the low-dose region. These studies must carefully consider dose-response relationships for the induction of epigenetic changes and the implications of such changes on radiation risk.

5. Major Points: Mechanisms of Action

Extensive research has been conducted on the mechanisms involved in the radiation-induced responses in the low-dose region. As the result of this research, the data are now available to explore the magnitude of the risks from radiation-induced cancer in the low-dose region. From this research, many interacting processes have been identified that are triggered by low doses of radiation. It will require a systems damage approach to integrate this information into a useful framework to be applied to risk. Important observations that help understand the mechanisms of action in the low-dose region are summarized here.

- Biological systems can detect and respond to very low doses of radiation.
- Direct damage to DNA is an important part of the radiation response and increases as a linear function of radiation dose.
- The processing of the damage and the signaling that results from it results in many non-linear processes.
- The signaling pathways induced by DNA damage are important and involve modification of pathways that involve ATM and p53.
- There are multiple genes, chemicals, and metabolic pathways induced by low doses of ionizing radiation that have marked influence on the biological outcome of the exposure.
- Many of these chemicals and metabolic pathways are protective against radiation-induced damage.
- Low doses of radiation modify the ROS free radical status of the cells. Such modifications are suggestive of radiation protective effects

seen in adaptive and protective responses. Higher doses increase the ROS status of the cells to produce responses that are known to damage cells and increase cancer risks.
- In the low-dose region, direct radiation effects and the signaling pathways modify cellular responses, including cell transformation, mutations, chromosome aberrations, telomere function, and cell cycle delay that seem to be protective. High doses change all these same endpoints in a way that would be predicted to harm the organism.
- Radiation can induce hypersensitivity in the low-dose region. As the dose increases there is an induced radiation resistance. Hypersensitivity may be protective by eliminating damaged cells, while induced resistance could be detrimental by protecting damaged cells and allowing them to remain in the population.
- There is evidence that low doses of radiation produce selective apoptosis in cells that are transformed. This provides a major mechanism of action in the low-dose region and may help explain many of the adaptive responses observed. Extensive research on the role of apoptosis in radiation risk demonstrates a potential protective role.
- Research has determined that genetic background plays a critical role in all the responses observed in the low-dose region.
- Research has been initiated on the role of radiation-induced epigenetic changes.

There is a need for a more complete view of the relationships that exist between low-dose radiation exposure and the cancer process. Without a complete systems approach, it will not be possible to apply the current research to radiation protection.

CHAPTER 8

Modeling

Models are essential in the process of transferring basic data to the needs of regulators. These models are developed at many levels of physical and biological organization. It is essential to understand dose and energy deposition to define the x-axis as well as the biological response that defines the y-axis. The most important data that influence regulatory bodies in standard setting are those associated with human studies. These models are evaluated by regulatory bodies and used widely in standard setting. The role of dosimetric, molecular, cellular, and mechanistic data on standard setting still has not been well defined. Models have been developed as part of the DOE Program and provide useful direction and information. Thus, basic mechanistic data can be considered during the standard-setting process. Such data and models also can be important in communication of radiation risks. Complex processes can be expressed as models that are easy to understand and communicate.

1. Traditional Cancer Models

To relate biological data to radiation risk, it is essential to develop models that describe the data. Former radiation risk models have been related to the cancer process (Moolgavkar & Knudson 1981) and have made primary use of the A-bomb data using the dose-response relationships (Hoel 1987a, b). It was recognized early on that cancer is a multistage process (Armitage & Doll 1957) and that to fit the radiation-induced cancer data, a model would require several variables. However, in any model fitting the data, it is common practice to limit the number of variables to as few as possible. Early models limited the variables to two and got an adequate fit to the experimental data (Moolgavkar 1983).

The development of the generalized formulation of dual radiation action played a role in the evolution of this type of thinking (Kellerer & Rossi 1978). In all model development, it was recognized early on that it was essential to relate the variables in the models to real biological changes. It was assumed that these changes were mutations, and extensive research was conducted to determine the mutations essential for cells to transform. It was suggested that it requires at least two mutations to transform a normal cell, and two-step mutation models were developed (Moolgavkar et al. 1990).

Early in the DOE Program, projects were funded to continue modeling using the responses observed in the low-dose regions. Such models were called biologically based models and still are the best approach to using experimental data for model development. The use of the two-mutation model resulted in a good description of the induction of lung cancer in rodents (Leenhouts & Chadwick 1994), and this modeling approach was to describe the radiation-induced lung cancer in rats following radon exposure. These studies suggested that radiation exposure was acting not only on the initiation stage of cancer, where mutations in single cells were thought to be important, but that radiation exposure to the lungs was also essential for the development of cancer, as it acts on the promotion or late stages of cancer development. Such studies suggested that mutations were not the only changes that were crucial in radiation-induced lung cancer. These studies were essential in the development of biologically based models (Luebeck et al. 1999). However, the amount of basic biological data available in the program did not lend itself to modeling efforts, and funding for these efforts stopped for about five years. As additional data was developed, funding was again made available to model responses from the molecular to the cellular levels and to develop methods linking these models to risk estimates (Brooks 2000b, a). Extensive research efforts continue to be conducted in this area.

These early models concentrated on the biological response observed, and they related biological average dose to the individual or the organ where the cancer developed. As described in chapters 4 and 5, extensive research on the variables related to radiation exposure was conducted to determine the proper physical properties that needed to be characterized to relate to the biological process. Another concern was the amount and distribution of the energy in very small targets and how this energy distribution would influence biological responses. Research in this area determined that the cell might be the primary target for biological responses; therefore, it would be important to determine the energy and energy distribution in individual cells. This led to the development of the hit-sized effectiveness functions that detailed the amount of energy deposited in individual cells and the distribution of that energy as a function of radiation type and exposure variables (Sondhaus et al. 1996). Such functions were used to understand the associations between the dose-response relationships and the absorbed dose to individual cells (Bond et al. 1995a).

The extension of the hit-size effectiveness factor and other measures of energy deposition to estimate risk were reviewed as a potential exposure metric for ionizing radiation (Bond et al. 1995a; Bond et al. 1995b;

Brooks 2005). Such research was supported by developing Monte Carlo track structure codes for low-energy protons to understand the energy and the energy distribution deposited in each cell (Uehara et al. 2001). Additional modeling of the interaction cross-sections for intermediate- and low-energy ions determined the distribution of energy in individual cells from radiation exposure (Toburen et al. 2002).

The development of microbeams helped define the track structure of low-energy electrons (Wilson et al. 2004). These studies were especially important as the low-LET microbeams were developed, and descriptions of the energy and microdosimetric distribution derived for 25-KeV beams were essential (Mainardi et al. 2004). These microbeams made it possible to expose single cells to known amounts of total energy. A complete description and knowledge of the distribution of that energy within the cells and in neighboring cells was critical (Wilson et al. 2001). This made it possible to develop a useful database on microdosimetry for low-dose, low-LET exposure to ionizing radiation (Wilson et al. 2000). One approach to developing this information was to use Monte Carlo simulations of single cells exposed to known amounts of energy deposited in individual cells from a 25-KeV microbeam (Miller et al. 2001). The information developed from these types of studies made it possible to introduce microdosimetry into the estimation of risk in the low-dose region (Scott & Schollnberger 2000). This will be discussed later in the section on models for dose-responses in the low-dose region.

2. LNT Models

In the higher dose ranges (>100 mGy), the linear no threshold (LNT) models provided a very good fit to the epidemiology data from the A-bomb (Pierce 2003), which is still the prime source of information used in the development of radiation protection standards. However, as the doses decreased to lower levels, the statistical power of the human studies was not adequate, and it was very difficult to apply the information to human data to estimate risk. Models were needed that were based on biological and biophysical data to predict risk in the low-dose range. In the low-dose region, as the dose continues to decrease, the amount of energy in each of the "hit" cells becomes constant, and only the number of cells "hit" will decrease. This thinking is based on target theory where radiation was treated like a gun that shot out energy at the target and deposited it in individual cells (Lea 1955). This was further developed into the "hit" theory. Both theories have been of great historical importance in the field of radiation biology (Zimmer 1961).

If it is assumed that only the "hit" cells are responsible for cancer, and if DNA and mutations represent the prime target for the production of cancer, the resulting model predicts a linear dose-response relationship in the low-dose region—the LNT hypothesis. The history of the development of the LNT hypothesis has been carefully reviewed (Kathren 1996). These assumptions have formed the basis for the models that the EPA continues to use to estimate risk in the low-dose region (Brenner & Sachs 2006) and to set radiation standards (Puskin 2009). The LNT implies that every ionization has the same potential to cause cancer regardless of the radiation dose. The dose-response relationship in the low-dose region for the primary source of human data, the A-bomb survivors, supports the use of the LNT and also suggests that there is mechanistic information to support this hypotheis (Pierce 2003). The scientific basis of the LNT has been reviewed by Upton (1999) and Chadwick and Leenhouts (2005), as well as review groups (NCRP 2001, NRC 2005). In addition, Brenner et al. (2003) extensively reviewed "What we really know" and supported these assumptions using a mixture of mechanistic information and epidemiological studies.

There is little question that the LNT should be used for setting radiation standards and that it provides a useful tool for regulating and limiting radiation exposure. However, when the LNT is applied in the low-dose region and combined with collective dose measurements, it predicts outcomes that may not be scientifically based (Kocher et al. 2008) and results in the public perception that radiation in the low-dose region represents a large risk. Extensive data has been developed suggesting that for many biological systems and endpoints, there are non-linear dose-response relationships, and there is a large body of scientific data that even suggest that in the low-dose region, radiation can protect against genetic damage, cancer, and other diseases. These data suggest that the risk in the low-dose region may be less than predicted by the LNT.

How should these data be considered in setting radiation protection standards? Perhaps LNT is the best we can do today with the information that we have on relating human cancer to radiation exposure in the low-dose region (Preston 2003). Continued discussion and debate suggest that even if LNT is useful for setting standards, it may not be scientifically accurate for predicting cancer frequency in the low-dose region and needs to be constantly re-evaluated. Without further research and data to support the biological responses in the low dose region, this may not be possible.

3. Non-LNT Models

The Program's funding of low-dose research produced a large amount of data that has been modeled using non-linear models. As stated above, the primary basis for the LNT was that it was fit by target theory. If a single ionization in a critical molecule could result in a DNA damage which then could become an important mutation, each and every ionization would increase the risk in a linear manner. New research has demonstrated that a single ionization activates may processes at the tissue level, many of which may be protective. Thus, hit theory is no longer supported by the data, and the LNT needs to continue to be evaluated as the model used to predict cancer risk. As additional data has become available on radiation-induced changes in the low-dose region, it is obvious that target theory no longer fits the data, especially for non-targeted and adaptive responses. Reviews of these data suggest that target theory modelers can no longer describe the responses in the low-dose range (Schwartz 2004).

Many radiation-induced tumor types in humans, such as leukemia and bone cancer, are best fit to linear-quadratic models, and at low doses many responses seem to have a threshold below which biological changes cannot be detected (Brooks 2006). Some studies on chronic exposure to internally deposited radioactive materials also suggest that lung cancer may have thresholds (Brooks et al. 2009). Threshold models were developed and adequately described the human cancer response to ionizing radiation (Hoel & Li 1998). It has also been demonstrated that the linear-quadratic model is the same as the two-lesion models that were proposed early on and that such models also fit much of the radiation-related tumor data (Armitage & Doll 1957).

There is an abundance of models and other data that suggest non-linear or linear-quadratic responses exist, but because it is difficult to use such models in setting standards, their use has been minimized in considering radiation-induced cancer. The dataset that drives the LNT is related to the observation that when all the solid tumors observed in the A-bomb survivors are grouped, the response is consistent with the LNT hypothesis. However, it has been suggested that there are technical and analytical problems associated with evaluating the A-bomb data that force the function to appear to be linear in the low-dose region (Scott 2005a; Scott 2008). Serious debate continues as to whether these really influence the shape of the dose-response in the low-dose region.

It has also been assumed by the radiation research community that because mutations, chromosome aberrations, and cancer are all stochastic effects and that stochastic effects increase linearly with dose, all these effects will be related on a one-to-one basis and increase linearly as a function of dose. Research has suggested a mechanistic basis for non-linear induction of stochastic effects (Scott 2005b) and that by using Bayesian inference, the risk in the low dose range will be non-linear (Schöllnberger et al. 2001).

Extensive research using cell and molecular systems suggest that low doses of ionizing radiation can decrease the responsiveness of cells to subsequent high doses (the adaptive response discussed previously). It has also been demonstrated that low doses of radiation can decrease the normal background frequency of adverse effects. An essential step in carcinogenesis is the transformation of normal cells to cancer cells. The cell transformation system developed by Dr. Les Redpath is a model system that measures the final steps in this process. The data generated from this system has been modeled, and it suggests that low doses of radiation reduce the cell transformation frequency below that observed in the controls. These data would support the use of hormesis as the model system for estimating risk (Redpath & Elmore 2007).

Modeling these effects suggests that there is a decrease in risk in both cell transformation and cancer (Redpath 2005; Redpath 2007; Redpath & Elmore 2007). Research on chromosome inversions has also been modeled and suggests a decrease in risk and a protective effect against this biological endpoint (Hooker et al. 2004b; Sykes et al. 2006a). Modeling these results also suggested the potential for thresholds for all stochastic effects (Scott 2005b, a). In the low dose region, many factors influence the biological responses and suggest that risk may not be linearly related to radiation dose (Feinendegen et al. 2011).

The extension of these types of results *in vitro* suggest that low doses of radiation may be protective and produce hormesis or beneficial effects. Animal studies have been conducted and were summarized in a review article that demonstrated a change in the latency of several different tumor types produced by low doses of radiation (Mitchel 2006). This review also suggested the potential for low doses to result in a decrease in radiation risk. Other studies suggest a beneficial effect of low-dose-rate radiation exposure. Chen et al. (2007) studied a population that lived in residences made with rebar contaminated with cobalt-60, which made it highly radioactive. These studies suggested protective effects for cancer and birth defects. However, later studies of the same population using different epidemiological techniques did not demonstrate such a protection and suggested excess cancers in the population (Hwang et al. 2006).

Other human studies suggested an adaptive beneficial effect for chronic myeloid leukemia (Radivoyevitch et al. 2002) as well as other protective effects as a function of dose-rate (Leonard 2007b, a). These studies do not represent a complete review of the literature, but provide directions for future research and direction.

A problem associated with protective effects of low doses of radiation has been the lack of understanding of the mechanisms involved. As the research in the low-dose region continues, a mechanistic basis is being developed that helps to explain these observations (Schollnberger et al. 2002). Chapter 5 contains reviews of mechanistic data that impact the responses in the low-dose region.

Application of all these low-dose data to radiation protection standards is problematic and very difficult (Curtis et al. 2004). The impact of the program on standards has been reviewed in detail. To date, the data have not had much impact on the standard-setting process (Morgan 2006). With better mechanistic understanding and the use of systems biology approaches, more progress is expected in the future. This is discussed in more detail in this chapter and chapters 9.

The bottom line in much of this discussion is that many scientists question the LNT, and others call for its rejection as the basis of radiation protection standards; for example, Cohen (Cohen 2008), Jaworowski (2008), and Calabrese (2007). Part of the problem of rejecting the LNT is to determine what models and methods can replace it, especially in setting standards. If thresholds exist, they seem to be different for each endpoint, tissue, organ, and species, making threshold models almost impossible to use.

Hormesis has been suggested as the default dose-response model (Calabrese 2004), and it has also been suggested that there has been a dose-response revolution supporting the rejection of LNT and replacement with hormesis (Calabrese & Baldwin 2003). Such a replacement suggests that there is no risk from low doses and that low doses are beneficial and even essential. Such a change in the understanding of ionizing radiation effects in the low-dose region is hard for many in the radiation research community to accept and has been the subject of debate and publications. Thus, debate continues as to which models should be used to estimate radiation dose in the low-dose region (Tubiana et al. 2008). This discussion was triggered again by the release of two major reports on the health effects of low doses of radiation. One from the U.S. National

Academy of Sciences (NRC 2006) supported the use of the LNT and produced a very impressive document to support this position (NRC 2005). The other report, by the French Academy of Science (Tubiana 2005), reached the opposite conclusion and suggested that the LNT needs to be replaced. Dauer et al. (2010) conducted a literature review and found that much new data has become available since BEIR VII that does not support LNT, and that there remains a need to continue to evaluate the usefulness and accuracy of the LNT hypothesis.

4. Low-Dose Models

Using the LNT as the basis for cancer risk extrapolation has generated widespread concern. Using this approach and model has suggested that the risk for cancer in the future may be increased markedly by the use of medical radiation. Medical radiation exposures continue to increase, and currently it has been calculated that there are 70 million CT scans per year (Mettler 2011). The average dose to the bone marrow from a whole-body CT can results in 10 mSv. By combining these two observations, a collective dose that is larger than the natural background radiation can be estimated (Brenner & Elliston 2004; Brenner & Hall 2007; Mettler 2011). With this approach, 70 million persons receiving CT scans that result in 0.01 Sv will result in 700,000 person×Sv each year. From such a dose a large number of excess cancers can be calculated. However, with recent data on effects in the low dose region demonstrate that it is not possible to predict excess cancer using collective dose and the LNT.

Compare this collective dose to that from the nuclear weapons fallout in the downwinders of southern Utah, where a population of about 25,000 people got an average dose of 0.03 Sv to result in a collective dose of 750 person×Sv . Even though the collective dose is low, Congress passed the Radiation Exposure Compensation Act (RECA). If any person develops a radiation-related cancer like those that increased in the A-bomb population, lived in the Utah counties that had the highest fallout, and was in that county at the time of the fallout, he or she was paid $50,000. To date, the RECA program has paid out more than a billion dollars in claims (Ziemer 2009). If a similar program were in place for those that received medical exposures, it would result in unacceptably high payouts. Similar comparisons can be derived from nuclear workers. All this illustrates over-concern about the fallout and the nuclear workers. Because the dose is small from each CT scan, the risk is very small, and the real and immediate benefit from the scan is much larger than any calculated risk.

Combining the LNT model with extrapolated risk estimates and the large collective doses results in a large number of "calculated" excess cancers (Brenner et al. 2001a; Brenner & Elliston 2004). Similar types of calculations were made for routine mammography screening (Brenner et al. 2002). What these calculations lack are the benefits derived from the exposures associated with the diagnosis of disease and a discussion of the potential protective effects of low doses and dose-rate (Scott 2007). Including such information would help the public understand the tradeoffs involved when they undergo a medical procedure that uses radiation and would help them make informed choices associated with those procedures. Brenner (2009) wrote a review that shows how the extrapolation from low doses to very low doses is done and helps understand and justify the LNT procedure.

5. NASA Models

Research has been jointly funded by NASA and DOE to help understand the radiation risk in space travel. Many of the goals of NASA and DOE are the same: to understand the influence of low-dose and low-dose-rate radiation exposure on risk. The differences are that in space radiation levels are higher, and the type of radiation present in space is different than that in the terrestrial environment. In space there are many different types of high Z energy particles (HZE) such as iron ions that have large mass and very large energies. Responses to these HZE particles may be unique because they deposit such a large amount of energy along the track that they travel in tissues. Durante and Kronenberg (2005) wrote a useful review on HZE particle research and how this information could be used to evaluate space-travel risk. Space also contains very large fluxes of protons that do not exist on earth, and the average radiation dose rate is also much higher than that on earth. Studies of this higher-dose-rate environment will aid in extrapolating the effects of higher dose-rates to the lower ones on earth. In the event of a terrorist attack or nuclear accident, dose-rates may be similar to those in space. This further defines the importance of studies conducted on the space radiation environment.

The radiation environment in space has been carefully defined and is measured on each space mission (NCRP 2010). To construct models appropriate for space flight, NASA has conducted a research program using appropriate types of radiation. The biological data collected from this program and data generated from astronaut evaluations are used to limit the uncertainties associated with the risk derived. The risks set for the astronauts are different than those set for workers on earth.

Early modeling of the risk from space radiation was similar to that done for workers on earth. The initiation-promotion models widely used to model radiation effects were applied to space radiation to estimate the tumor prevalence from high-charge and high-energy particles (Wilson et al. 1995). Other models such as the two-stage clonal expansion model (Curtis et al. 2002) that had wide application in defining cancer risk on earth were also applied to space irradiation.

Application of these models assumes that the initiation, promotion, and clonal expansion mechanisms responsible for cancer on earth is similar to those in space. However, as modeling and biological data have improved, the unique responses seen from space irradiation are being incorporated into the modeling efforts. Cucinotta and Durante (2006) evaluated the cancer risks from the complex space environment including galactic cosmic rays, dose, and dose-rate. Such evaluations are essential to put space risk into the proper framework and minimize the uncertainty associated with these risk estimates.

Studies on the hematopoietic (blood formation) system are important because this is one of the parts of the body most sensitive to ionizing radiation. Studies using space-simulated photons (Gridley et al. 2008) as well as other types of radiation in a simulated solar particle event (acute exposure to photons and protons as well as solar particle event protons) were conducted to evaluate their impact on cell killing, the immune system, and other responses associated with the hematopoietic system (Gridley et al. 2008). Additional research included bystander effects that may be produced by the HZE particles in developing models of cancer risk (Brenner & Elliston 2001). Research and modeling studies continue to be funded by NASA and DOE to link the biological risks from space exposure to that on earth.

The debate over the use and appropriateness of the LNT continues (Tubiana et al. 2008) and has been showcased in several national and international scientific meetings. Many debates featured Dr. David Brenner, who supported the LNT, versus a variety of different scientists who were not in favor of the LNT (Brenner & Raabe 2001; Brenner & Mossman 2005; Averbeck 2009; Brenner 2009). One of the most productive debates was at the 2008 NCRP meeting. The debate featured a representative favoring the LNT and one against it (Averbeck 2009; Brenner 2009). The publications that resulted from this debate provided a useful update on the state of the science in this important area. Both sides suggest that they won the debate, but only time and additional science will tell (Tubiana et al. 2008). The problem with such debates is that the public doesn't know which side is "right," so the public perception associated with the risk from radiation exposure remains confused. As more data become available, the debate may generate more light and less heat.

6. Animal Models

The use of molecular, cellular, and animal models has played an important role in understanding the mechanisms involved in radiation-induced cancer. Extensive research has been done on experimental animals to determine the dose-response relationships that exist between different types of radiation exposure and biological damage. These life-span studies will not be reviewed here. Stannard and Baalman (1998) and Thompson (1989) provide an extensive and useful review on the life-span studies conducted in the dog. These well-conducted studies provide an extensive data resource for comparisons to current research and to aid in extrapolation of radiation risk to humans. Extensive whole-life studies have also been conducted in rodents.[1]

With modern technology and the production of mice with specific genetic backgrounds, the development of knock-in and knock-out mice that contain known genes of interest provides great research potential. These possibilities are not be reviewed here, but a resource for genetically defined animals is available at ORNL and funded by the DOE Program. This ORNL program developed a number of strains of mice through recombination of defined strains which were both radiation sensitive and resistant. These "recombomice" that have been developed with radiation studies in mind are very useful in studies to evaluate the role of genetic background on responses to radiation.[2] It has been suggested that *in vivo* recombination following chronic exposure in these mice is decreased below the spontaneous level as a form of adaptive response (Kovalchuk et al. 2004). Hendricks and Engelward wrote a useful review on these recombomice that defines how they are derived and many of their potential uses in scientific studies (Hendricks & Engelward 2004).

Animal models have been developed that can be used to study specific diseases; for example, a mouse strain is available that can be used to study radiation-induced AML (Darakhshan et al. 2006). Another model is available for study of Nijmegen syndrome (Williams et al. 2002). Many such mouse models provide a genetic background that is either sensitive or resistant to radiation-induced cancer. These models have been exposed to low doses and dose-rates, have demonstrated marked adaptive responses, and suggest that low doses might be protective from radiation-induced cancer (Mitchel 2006).

7. Molecular, Cellular, and Tissue Biology Models

The basic mechanistic data on radiation effects in the low-dose region was reviewed in chapter 6. This information was used to develop models at the molecular, cellular, and tissue levels and to suggest how such data might be applied to cancer risk assessment.

7.1 Modeling DNA Damage

In developing models, it is important to evaluate the changes on the basic level of the DNA and to understand how radiation interacts at this level to produce DNA damage and repair. There was a dose-dependent misrepair of DNA double-strand breaks that was modeled for both high- and low-LET radiation (Rydberg et al. 2005). The type of radiation exposure is important in evaluation of such damage, as was seen in the models of space and microbeam radiation. The role of energetic electrons (100 eV to 100 keV) was studied, and models were developed to predict the type of DNA damage produced (Nikjoo et al. 2002).

An observation made about radiation-induced DNA damage was that multiple damage sites occurred in a small area. Methods and results of research conducted on fragment and multiply damaged sites in the DNA were discussed in chapters 4 and 5 (Sutherland et al. 2001b; Sutherland et al. 2001d; Sutherland et al. 2003a; Sutherland et al. 2003b). The size of the DNA fragments was also important because from the size distribution, it was possible to develop models that could predict the different levels of structure in the DNA. Models of DNA breakage based on random walk, a mathematical formalization of a trajectory that consists of taking successive random steps, were developed to understand how chromatin structure influenced the DNA breakage (Ponomarev et al. 2001a). Other models of DNA breakage following high doses of radiation were used to predict the damage induced by low radiation doses (Ponomarev et al. 2001b). Such models will be useful in making the extrapolations between high and low doses of radiation and can help test the shape of the dose-response curves.

These models were used to study the influence of the LET of radiation to help understand DNA breakage induced by HZE particles in space (Ponomarev et al. 2001a). This again provided preliminary information that can be used in risk estimates following exposure to the space radiation environment. In other research on space radiation, DNA DSBs were produced by nitrogen ions with a wide range of LETs. The model predictions for the induction of clusters of DNA DSBs were compared to the measured frequency and distribution to validate the model (Fakir et al. 2006).

Other studies were conducted using a Monte Carlo algorithm to simulate the spectrum of DNA damage induced by ionizing radiation (Semenenko & Stewart 2004). This research was followed up by studies to evaluate the repair of the clustered DNA damage sites. Repair of these lesions is thought to be limited and may be important during the development of cancer. This study evaluated both base and nucleotide excision repair of these lesions and matched the repair to the model properties (Semenenko & Stewart 2005). In the companion paper (Semenenko et al. 2005), this research was expanded, and the measured DNA repair was compared to that predicted with good results. Such studies are essential to gain confidence that the model will predict the real world.

One of the concerns about this research was the potential to produce DNA breakage during the processing of the samples and the difficulty in determining the radiation-induced DNA breakage versus that produced as an artifact. Methods were developed to solve this problem and reduce breakage during preparation of the samples (Ponomarev et al. 2006). With these new techniques in place, additional studies will be needed to determine the role of high-to-low dose extrapolation, the influence on chromatin structure, and the influence of HZE particles in producing DNA damage.

As discussed in chapter 5, DNA damage acts as a trigger for many changes in the cells, and radiation affects the change in gene expression. Model methods have been derived to detect changes in gene expression in the presence of inter-individual variability (Rocke et al. 2005), and these need to be applied widely in the studies described in chapter 5.

7.2 Modeling Chromosome Aberrations

The next level of cell and molecular organization thought to be important in radiation-induced cancer is chromosome aberrations. Many cancers have well-defined chromosome aberrations, some of which seem to be markers of the radiation-induced disease. Studies and development of models of chromosome aberrations provide a useful foundation for radiation-related cancer risk. With the development of chromosome-painting techniques described in chapter 4, it has been possible to identify every chromosome and determine which ones interact to form radiation-induced breaks. These techniques were combined with DNA-damage-processing pathways, which made it possible to link DNA damage to the production of chromosome aberrations (Levy et al. 2004). Such links between different levels of biological organization form the basis for future modeling using systems biological techniques, which will be described in more detail in chapter 9.

Several models were developed to predict chromosome aberration frequency and interaction distances between chromosomes. One of these was the random breakage and reunion model, which suggested that all interactions between chromosomes were random, and the chance of any chromosome interacting with any other depended only on its size. With such a model it was possible to predict interaction distances based on chromatin geometry (Sachs et al. 2000). The development and further application of chromosome-painting techniques made it possible to stain interphase cells and determine the domain of the chromosomes during interphase. Such information can be linked to the production of aberrations that are expressed when the cells have progressed to the metaphase stage of the cell cycle.

Using chromosome domain during interphase, models were developed to predict radiation-induced chromosome aberrations (Holley et al. 2002). It became possible to generate chromosome aberration spectra and use them to predict aberration frequency and interactions between chromosomes (Levy et al. 2007). Before chromosome painting was available, it was thought that all the chromosomes interacted with each other randomly, and the frequency of this interaction depended only on chromosome size. With additional analysis and modeling it was determined that there were an excess of radiation-induced chromosome aberrations between homologous chromosomes (Plan et al. 2005). This suggested that the location of the chromosome in the interphase nucleus played an important role in the induction of chromosome aberrations.

As additional data was generated using chromosome-painting techniques, it was demonstrated that there are cells that contain very complex chromosome aberrations, with multiple chromosomes involved.(Vazquez et al. 2002). Such aberrations were difficult to explain with older models. New methods of biophysical modeling were put in place, and additional insights were gained on how these complex chromosome aberrations could be formed (Hlatky et al. 2002; Sachs et al. 2002). These models were further developed and produced quantitative analyses of radiation-induced chromosome aberrations that were related to the observed number and types of aberrations (Sachs et al. 2004).

Another measure of chromosome damage is the induction of micronuclei. These represent small pieces of chromosome that are not included in the nucleus after cell division. They are easy to score and can be used to relate physical variables to the induction of chromosome damage. One of the mechanistic studies done with micronuclei was to set up culture dishes in a way that energetic heavy ions would traverse the cells with different portions of the ion track being located in different parts of the culture system. With this system it was postulated that it would be possible to detect the influence of the Bragg peak, where there are more ionizations per unit distance traveled, on the induction of micronuclei. There was a suggestion of an increase in this area but it was not as great as would be predicted based on the number of ionizations deposited in that region of the dish and in those cells in the region (Wu et al. 2006). Such studies support the concept that all the cells in the culture dish are responding to the insult, and the bystander effects may be influencing the total response.

7.3 Modeling Cell Killing

It is well established that radiation kills cells effectively, and it is used in therapy because of this characteristic. Models have been developed to describe radiation-induced cell killing. The induction of DNA damage and the failure of that damage to repair have been thought to be a source of cell killing. Two-lesion kinetic models were developed to determine if rejoining of DNA DSBs was directly linked to cell killing (Stewart 2001). These models predicted an association between these two biological processes but suggested that other mechanisms for cell killing are involved following exposure to ionizing radiation. Microdosimetric models were linked to cell killing through bystander effects. The type of bystander effects studied were those that were transferred through the media and did not depend on cell-cell contact. It was demonstrated that the response could be explained using these microdosimetric models and that this interaction with the cells was responsible for the release of soluble substances into the media (Stewart et al. 2006).

The cell cycle and the length of time that cells spend in each part of the cycle varies depending on the tissue. Most epithelial tissues, which are the source of radiation-induced carcinomas such as in the liver, have most cells in the Go or resting phase of the cell cycle. When an insult kills cells, other cells must divide and replace them. This stimulus for cell proliferation plays an important role in radiation-induced cancer, especially following high radiation doses. In developing models for carcinogenesis, it is important to include consideration of the cell cycle and the movement of cells from one stage of the cycle to the next. A multistage carcinogenensis model was developed that used cell cycle as one of the variables, which is a great step forward in modeling (Hazelton et al. 2006). Movement from the Go to the other stages of the cycle may be one of the triggering events in radiation-induced cancer.

Most studies on cell killing have been conducted in tissue culture systems using *in vitro* single-layer cell cultures. Extensive cell-cell interaction can modify the killing of mammalian cells. Studies using a vertebrate embryo were very important in explaining many of these interactions (Bladen et al. 2007b). Such studies also support the role of bystander cells in protecting and modifying the responses to ionizing radiation.

7.4 Modeling Bystander Effects

The data from bystander effects have been modeled to understand the role of this observation on radiation-induced cancer risk. Because DNA DSBs have been shown to be important in producing chromosome aberrations and have been linked to cancer induction, it is important to determine if DNA DSBs are induced in bystander cells that have no direct energy deposited in them. Using a microbeam, Sedelnikova et al. (2007) determined that DNA DSBs could be produced in cells that do not have energy deposited in them. These studies were carried out in a three-dimensional human tissue model and support the theory that bystander effects can be detrimental to cells and organisms.

An early model of bystander effects combined the damage from the Bystander Effects and Direct effects (BaD) model developed by Brenner and Sachs (2002a). Using this model, it was suggested that bystander effects could potentially dominate radon risk (Brenner & Sachs 2002a, b). Additional data was published that suggested the damage from bystander effects may be significant, but the risk from radon would not be influenced or changed significantly by the by these effects (Brenner & Sachs 2002a, b, 2003). Other models suggest that bystander effects can be either detrimental or protective (Schollnberger et al. 2007). The adaptive response has been demonstrated in bystander cells, and after low doses of radiation many protective mechanisms have been shown to be triggered in bystander cells. There is little doubt that bystander effects exist following radiation, and the effects of this were again demonstrated to potentially be beneficial (**Azzam & Little 2004**). The review written by (**Ballarini et al. 2002**) is useful but requires constant updating as additional data accumulate.

The bystander effects are the result of cell-cell and cell-matrix communication. Modeling this communication both in terms of cancer formation and radiation-induced damage has been very useful. Some models have focused on cell cultures and tried to develop a comprehensive stochastic model of these cultures (**Hanin et al. 2006**). However, it is well established that cells in monolayers in culture do not respond to radiation in the same way that cells grown in three-dimensional cultures do. Modeling engineered cultures of breast cells demonstrated that the architecture, function, and neoplastic transformation is very dependent on the culture conditions and the interaction of the cells in three dimensions (Nelson & Bissell 2005). Special models of intercellular interactions that are essential in the cancer process further support the role of bystanders and communication in both increasing and decreasing the cancer frequency following radiation or other environmental insults (Sachs et al. 2005). Many of these interactions that modify the ultimate outcome of the exposure seem to depend on gap junction communication (Green et al. 2005), and blockers of gap junctions such as Connexin 32 can eliminate the bystander effects in many cell culture systems (Green et al. 2002).

7.5 Modeling Genetic Background

The genetic background of any biological system has a marked influence on the response to radiation. This fact needs to be further evaluated as research on the risk and modeling of this risk proceed. For example, biologically based modeling of chronic myeloid leukemia provides a path forward for modeling many other diseases influenced by genetic background (Radivoyevitch et al. 2001). The influence of genetic background on risk has recently been reviewed (NCRP 2010). This document demonstrates that genetic background is important, but currently the tests for identification of individuals at increased risk are inadequate to be applied to impact radiation protection. Several human genes are known to increase the risk for radiation-induced cancer. For example, mutations in BRAC 1 and 2 increase breast cancer risk. A review of the data on genetic background on risk is provided in the NCRP reference for those interested in following up this subject.

8. Risk Assessment Models

A major goal of the DOE Program was to provide a large, well-documented database on the effects of radiation in the low-dose region. Using this database, reviewed here, will make it possible to develop a scientific basis for risk assessment and to justify and identify the models that are used to predict risk in the low-dose region (Brooks 2000a, 2003). As a systems approach is taken to understanding carcinogenesis, the role of genetic background and cell-cell and cell/matrix communications must be considered, because they play an important role in promoting or inhibiting cancer development.

9. Major Points: Modeling

- Traditional models were used to fit human epidemiology data.
- In the higher-dose ranges, both linear and non-linear models have been used to fit the human data.
- In the low-dose range, the biophysics of energy deposition and "hit" theory have dominated the field of radiation biology and resulted in LNT hypothesis and models.
- LNT models are essential in controlling radiation exposure, but may not accurately reflect radiation risk.
- Extensive data has been generated and modeled that incorporate biological data at all levels of biological organization.
- Models suggest mechanisms of action that can be tested by experimentation.
- Many molecular, subcellular, cellular, and animal models have non-linear dose-response relationships suggesting different mechanisms of action for the production of damage in the low-dose region compared to the high-dose region.
- Models of DNA and chromosome damage have been very helpful in understanding how the damage is formed and repaired.
- Models of data that demonstrated a decrease of background levels of damage by low doses of radiation suggest the potential for adaptive and protective effects in the low-dose region.
- Models of cancer must represent all the mechanisms and molecules involved in radiation induced initiation, promotion, and progression of cancer.
- The continued use of biological-based models has played an important role in the evaluation of the data generated by the program.
- LNT models are adequate in the high-dose region of the dose-response relationships, but do not fit the cell and molecular data in the low-dose region.
- Special models have been developed to evaluate cell and molecular responses.
- The influence of radiation type on models has played an important role in the development of models for the risk associated with space flight. These models must consider exposure to HZE particles (very high-mass and high-energy particles) encountered in space.
- Animal models continue to play a central role in transferring information from basic science to epidemiology.

- Many animal models with defined genetic background can be used in radiation studies that provide increased mechanistic understanding when linked to molecular, cellular, and tissue models.
- Because radiation is a very good cell killer, models of killing have been useful in radiation therapy.

Notes

1. For information or tissues from the life-span studies on rodents, contact Dr. Gayle Woloschak at g-woloshack@northwestern.edu.
2. For information on the mice, contact Dr. Brynn Voy at voybh@ornl.gov.

CHAPTER 9

Taking a Systems Biology Approach to Risk

As described in the previous chapters, extensive datasets have been generated on the effects of low doses of radiation. The new technology and techniques developed or applied by the DOE Low Dose Radiation Research Program have been important in defining the biological responses in the low-dose region. These extensive datasets have resulted in better understanding of the mechanisms of action for radiation as well. They have also enabled some unique models of radiation-induced biological changes.

Interactions are complex between the many different mechanisms active after exposure to low doses of radiation and the modeling of these responses. The data generated in the program were a major factor in defining and recognizing these complex responses to low doses of radiation. The observed multiple responses and complex interactions require a new approach to modeling the responses in the low-dose region. The old approach was to assume that DNA damage was the mechanism of action for radiation damage in the low-dose region and linear models were described damage and risk. Ultimately it will be essential to extrapolate responses across different levels of biological organization (Feinendegen et al. 2007) to determine the shape of the dose-response relationships in the low-dose region. The dose-response relationships can then be applied to estimate the risk associated with low doses of ionizing radiation.

One new approach to understanding the biological responses in the low-dose region has been described as systems biology. Systems biology integrates the responses from the molecular to human population studies into complex models. These models are then used to determine how biological responses influence risk at each level of biological organization and are combined as these unique levels of biological organization are reached. Studies from all levels of biological organization need to be linked together. The ultimate goal is to develop a level of understanding of the mechanisms of action of low doses of radiation that makes it possible to predict responses and to use these responses to move up to the next level of complexity. When it is possible to predict responses, it will also be possible to define the shape of the dose-response relationships and appropriately associate the risks with these low-dose exposures.

One of the driving forces in developing a systems biology approach was recognition that DNA damage is not the only important target in production of cancer (Barcellos-Hoff 2005a). The damage also triggers a set of signaling processes that plays an important role in radiation-induced cancer (Barcellos-Hoff 2005b).

The ability to measure changes in gene expression in large numbers of genes made it possible to generate very large datasets that describe radiation-induced changes in gene expression. These data demonstrate that gene expression changes as a function of many different exposure, physical, and biological parameters. The publications on these large datasets generated on radiation-induced changes in gene expression were discussed in chapters 4 and 5. Such information illustrates that changes in gene expression provide a functional genomics approach and entry into systems-level biological studies (Amundson et al. 2008). Such systems biology genomics research provides one of the basic sets of data needed to determine the responses of biological systems to low doses of radiation. However, with the generation of such massive data sets, new methods are required to analyze these databases. Several of these methods to handle data were reviewed in chapters 7 and 8, such as nearest neighbor analysis, cluster analysis, self-organizing maps, and computational methods to evaluate degree and significance of increase or decrease in expression.

The level of change in gene expression does not link directly to changes in protein expression and function, which makes it essential to evaluate radiation-induced changes in protein levels, modifications, and activity. New technologies have used shared peptides in the quantification of different proteins. This label-free technique uses a combination of liquid chromatography/tandem mass spectrometry (LC-MS/MS) to determine the proteins expressed and the levels of those proteins (Jin et al. 2008). Identifying large numbers of protein changes makes a systems approach to evaluating these data and linking them to changes in gene expression essential.

Other technologies were also developed to identify protein modifications, especially phosphorylation and protein localization in the cells. The protein modification was measured using a linear discriminant analysis to accurately identify the modified phosphopeptides (Du et al. 2008). Studies that determine the impact of protein localization on biological function were also developed (Raman et al. 2007). Both phosphorylation and protein localization alter cell and tissue function and must also be considered in systems biology approaches.

Each of these cell and molecular assays must be linked to a functional measure that is important to radiation-induced cancer. Without such links the molecular data are not useful in evaluating radiation risks.

A study by Miller et al. (2008) relating changes in proteins to genomic instability illustrates how changes at the molecular level can be related to functional changes in the stability of the genome. In this study, changes in mitochondrial proteins were defined in genomically unstable cell lines, and the protein changes were related to the oxidative status of the cells and the maintenance of genomic instability. Such studies provide potential methods to extrapolate from genomic instability to oxidative status of the cells and to relate the ROS status of the cells to cell transformation and ultimately to radiation risk. Many steps and much information are required to make these links, but this illustrates how systems biology can be used in radiation risk estimates.

In chapter 5 the interactions between cell and matrix were described. It was determined that exposure of the mammary gland stroma promotes the formation of tumors in unirradiated epithelial cells (Barcellos-Hoff & Ravani 2000). Cell-cell and cell-matrix interactions and signaling are involved in formation of cancer. The interaction of low doses of radiation with the microenvironment suggests new mechanisms of action that become an important part of understanding risk and systems biology (Tsai et al. 2005). These observations and approaches are useful in describing multicellular interactions and how they can modify outcome and perhaps risk. It has also been suggested that similar approaches may be used to evaluate radiation-induced multi-generational responses (Barcellos-Hoff & Costes 2006).

It is important to be able to describe signaling pathways, understand the biological modifications associated with the signaling, and generate models that adequately describe the pathways. As signaling pathways have been described, it became evident that certain "nodes" of activity exist where single proteins or genes play critical roles in controlling multiple pathways. An illustration of this type of interaction is shown in Figure 22. Using multiple types of genomic approaches provides more accurate data on the mechanism of action.

The communication between all these pathways becomes very complex. Important advances in data analysis involved in signaling pathways and how they can be modeled have been made (Miller & Zheng 2004). Such evaluations represent one of the key elements of systems biology and will be used more widely as additional data are provided on the mechanisms of cell-cell communication and the interactions that control the expression of cancer. Petrini (2007) reviewed the role of cell and matrix communication and showed that the interactions between cells are critical to the cellular function.

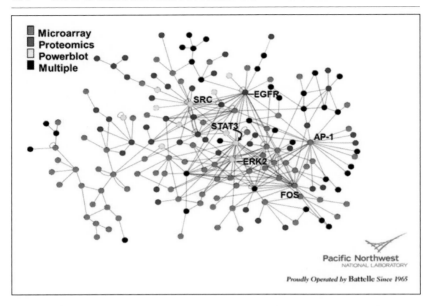

Figure 22. Integrated data provide more comprehensive and accurate network reconstruction (Waters 2012).

The research that generated data demonstrating the phenomena of bystander effects, genomic instability, and adaptive response suggests that whole tissues are responding to radiation exposure. Such tissue and organism responses to radiation imply that radiation-induced cancer is an emergent phenomenon and that many complex processes are involved at all different levels of biological organization (Barcellos-Hoff 2008). These processes seem to change as a function of radiation dose, with high doses initiating a different set of processes than low doses.

It was recently shown that using a systems biology approach, it is possible to make useful risk estimates for astronauts who are exposed to a wide range of different types of radiation (Cucinotta 2008). Such modeling exercises provide guidance that will be useful in determining the slope of the dose-response relationships and the radiation-related risks in the low-dose region. National and international meetings have helped provide future direction for research in the area of systems biology. As additional research is conducted and integrated, understanding of the role of low doses of radiation on risk estimates can be improved, and the uncertainty associated with these risk estimates can be reduced.

1. Major Points: Systems Biology

- Many biological changes and mechanisms associated with risk from ionizing radiation in the low dose region were explored by the program. Systems biology approaches enable all of the new data to be considered and related to risk.
- The program determined that systems biology was a good approach to integrating many of the new findings.
- Simple models of DNA damage, chromosome damage, and mutations do not adequately describe all of the complex biology that is involved in responses to low doses of radiation.
- Models need to be constructed at each level of biological organization. Using a systems approach, models can be linked across levels of biological organization so that mechanistic molecular and cellular models are related to well-defined functional changes.
- It is essential to relate molecular and cellular changes to functional endpoints at the cell and tissue level to be useful in risk estimates as an intermediate step in predicting radiation risks in the low-dose region.
- The very large data bases generated on the biological responses induced by low doses of radiation (changes in gene expression, the epigenome, and proteins) require a systems biology approach to integrate them into risk assessment models.
- Additional research efforts are needed to incorporate systems biology approaches into risk estimates.

CHAPTER 10
Program Communication and Monitoring

During formation of the DOE Low Dose Research Program, the BERAC subcommittee outlined the key question related to communication of the results of the research : "How can the information derived from the low-dose initiative best be communicated to scientists, policy makers, stakeholders and to the public?" Further, they suggested methods and topics to be included in the research plan to accomplish this:

"To communicate with the public about low dose management requires a well developed plan based on strong basic social science research. The goal of the research effort would be to understand the likely public responses to scientific findings from the Low Dose Program's research and responses to the plans for modifying existing standards based upon these scientific findings. The following topics should be included in determining public responses to issues regarding low dose exposures: 1) public perceptions of risk; 2) the perceived importance of the activities and conditions that produce low dose radiation; 3) trust and confidence in risk managers, regulators, and decision makers; 4) the role of the media in characterizing different positions on risk controversies; 5) the role of advocacy groups; 6) the manner by which risk is characterized and assessed; and 7) the procedures by which decisions are made."

With this in mind, several approaches and research projects were funded to carry out this part of the research plan.

1. Advisory Committee

At the start of the program, a science advisory committee was organized as a subcommittee of BERAC by Dr. David Thomassen, who was the Program Coordinator, Office of Biological and Environmental Research, to aid in the communication and direction of the program. The committee was chaired by Dr. Sharon Friedman, Lehigh University. The committee provided valuable guidance during the program's early development and was instrumental in securing funding to help communicate research results and set up mechanisms to keep this communication active.

2. Workshop on Risk Communication

In 2000, a workshop was held at Decision Research Institute in Eugene, Oregon, titled "Workshop on Low Dose Radiation Exposure and Risk Communication." It was organized by Dr. David Thomassen and was under the direction of Drs. James Flynn and Paul Slovic. In this workshop, the communication needs of the program and the needs and problems associated with risk communication in general were carefully reviewed and discussed. The complete results of this workshop can be found in a report to the Department of Energy, Office of Science, Office of Biological and Environmental Research. The topics covered included:

- Underlying Problems of Risk Communication
- The Social-Cultural Context
- Organizations, Institutions, Trust, and Risk
- Guides to Study Social-Cultural Context for Risk Communication
- The Social Amplification Risk Framework
- The Social Geography of Risk Communication
- Value-Based Structured Decision Processes
- Cross-Cutting Research Questions: What should we attempt to learn from our studies?
- Research Tasks: Studies to improve risk communication within its societal context
- Conclusions: The next steps for the risk communication research project

This workshop addressed a number of communication problems that came up repeatedly during the program and provided a backdrop against which DOE could help address these problems. Additional meetings were held with those involved in communication, including the media, who suggested that it would be difficult to get this information to the public because much of it was not considered "newsworthy." Providing the public with information that would change the way that they perceive radiation risk is essential, but such communication may be very difficult because of the media's lack of interest in positive information about the low risk associated with radiation exposure (Flynn & MacGregor 2003).

This has proven to be true as the program has progressed. Only research that tends to raise alarm about radiation is released by the media. Research data suggesting that scientists understand the risks of low doses of radiation, that the risks are well defined, and that radiation risks may be adequately conservative are very difficult to get into the media and placed before the public. This remains a major challenge of the program because the public and regulators have models and hypotheses that they accept and are unwilling to expand their vision beyond (Leonard 2008).

3. Lead Scientist

The first DOE call for proposals in 1998 requested applications for a Lead Scientist to work closely with DOE in facilitating program operations. Among the requirements for the lead scientist: "The Lead Scientist will be funded from the program and will provide scientific leadership to the community of the researchers in the research program. Interested applicants should demonstrate their understanding of the needs for and the uses of the types of scientific information likely to be developed in this research program. They should demonstrate their understanding of previous epidemiologic and experimental studies involving low dose, low dose-rate exposures to radiation or chemicals. Finally, interested applicants should demonstrate their knowledge of research opportunities and capabilities at National Laboratories, universities, and industry in the area of molecular and cellular responses to low dose, low dose-rate exposures." While at Washington State University, I was selected as the lead scientist and served in this position until 2004, when I was succeeded by Dr. Mary Helen Barcellos-Hoff, Professor of Radiation Oncology and Cell Biology at NYU Langone Medical Center, New York.

4. Investigator Workshops

A key venue for communicating research results is the annual Investigator Workshop. All principal investigators funded by the program are required to participate in these workshops by presenting a poster summarizing their scientific progress over the past year. DOE has striven to organize a highly focused symposium on a single theme or issue, in which the current state of the art is reviewed, and potential future directions assessed.

Individual scientists who have made important contributions to the program are asked to make oral presentations that further elaborate on their progress and the general workshop theme. In addition, DOE invited scientists from outside the program who have made important scientific breakthroughs in the field to make presentations. These presentations help keep the program scientists up to date on the research progress in the field. Interactions between the program scientists and these invited experts are vital in developing new scientific direction for the program. Finally, the workshop provides the opportunity for interactions among the program scientists that can lead to collaborations that can become the basis of future research and help guide the program's direction. Workshop participants include BER program staff, program staff at other agencies, BERAC subcommittee members, and scientists from other DOE-funded programs whose research has useful links to the Low Dose Radiation Research Program. In addition, staff from regulatory agencies, such as the EPA and the NRC, actively participated.

Earlier in the program, extensive efforts were made to invite members of the public and political action groups with interest in radiation issues. At these early meetings, members of Indian tribes, the Hanford Advisory Board, the mayor of Rocky Flats, Colorado, political action groups, down-winders, and nuclear workers with interests and concerns about exposures to low doses of radiation attended. These workshops included small discussion groups, with the members of these public groups placed in groups with scientists and regulatory agency representatives to address mutual concerns. These workshops helped develop understanding and trust among the participants. As the invitees got to know the scientists, they learned that the researchers did not have an agenda and that the research produced would have a firm scientific basis. In later years, the workshops became more focused on the scientific issues and future directions, but these groups continued to be invited and to participate actively during the presentations and discussions. However, without direct invitations and help with funding for the meetings there was limited interactions from most of these groups.

5. Presentations

As chief scientist, I gave 76 presentations to the radiation scientific community between 1998 and 2008 to ensure it was aware of the program and its research. I also made presentations to groups outside the radiation community, such as the American Chemical Society, American Pharmacists Association, American Statistical Association, International Consortium for Research on Health Effects of Radiation (ICRHER), DOE national laboratories, the Washington State Department of Ecology, universities, research laboratories, and agencies like RiskRad that fund low-dose research in Europe. I also made presentations to U.S. government agencies, including DOE Washington DE, DOE Richland, EPA, NIH, and NIAID.

The program managers also gave presentations to other agencies using data generated from the program to help these agencies understand the program's importance. They also made presentations to agencies and institutions involved in making recommendations related to radiation standards, such as the National Council on Radiation Protection and Measurements (NCRP), the Electric Power Research Institute (EPRI), and the International Council on Radiation Units (ICRU), as well as to BEIR VII of the National Academy of Sciences.

It was also important to make presentations to the public to facilitate understanding of the magnitude and importance of the program. There were 37 presentations made to different groups like Rotary Clubs, the Nez Perce Tribe, the Navajo Tribe, Downwinders in St. George, Utah, the Hanford Advisory Group, the Boy Scouts of America, and many other public groups. I have PowerPoint representations of these presentations available.

6. Website

The major source of communication of the program's research results to the public and researchers both in and outside the program was its website. The website was originally developed at Oak Ridge National Laboratory under the direction of Dr. John Wassam. Lead scientist Dr. Antone Brooks developed the background information addressing the public's needs and following the scientific progress of the program. The website was transferred to Washington State University Tri-Cities in 2001, under Dr. Brooks' management. When Dr. Brooks retired in 2008, the website was moved to the Pacific Northwest National Laboratory under the direction of Dr. William Morgan. It can now be accessed at http://web.archive.org/web/20150905100153/http://lowdose.energy.gov/, and it contains dose rate charts, research highlights, a database of publications produced by the program, program project descriptions, frequently asked questions, a glossary of radiation-related terms, an inquiry page, and links to other radiation research programs and resources. This site continues to be an important method of communication with the public. With loss of program funding this web site is no longer being updated with recent scientific developments in the low dose and dose rate region.

7. Open Literature Publications on Communications

Program-funded research resulted in many publications in the open literature that followed on previous research on the public's perception of risk from radiation. It is well established that many myths and stories have resulted in a high level of fear of the health effects of radiation exposure among the general public (Slovic 1996). Results of studies of how this fear was generated suggested that the assignment of numbers to risk and the concept that every radiation-induced ionization increases cancer risk have been major contributors (Purchase & Slovic 1999).

In the field of radiation biology the questions asked about low-dose exposure have been "How low is low enough? What dose is acceptable?" These are the same questions asked about any environmental exposure to chemical or physical agents. Early publications from the program focused on the question asked by MacGregor et al. (1999) in their article "How exposed is exposed enough? Lay inferences about chemical exposures." This kind of information is essential as a basis for public discussion and education and to understand how the public views radiation (Flynn & MacGregor 2003; Leonard 2008). Such questions are critical in helping regulators and the public understand the role of low doses of exposure to risk-inducing radiation or any other toxic agent.

A number of open literature articles about the program's progress were written and published in different scientific journals and proceedings (Brooks 2000a, b, c). Also published were papers describing how the results of the program would provide a scientific basis for standards (Brooks 2003; Brooks et al. 2007).

8. Major Points: Communication and Monitoring

- Communication of research results was and remains a priority of the DOE Low Dose Radiation Research Program. This has included communication between researchers, communication of the data to regulatory agencies, and communication of the findings to the public, as well as education.
- An advisory committee was set up to provide input and direction for the program and a lead scientist was funded to provide a vital link between the program and the scientific community.
- The program emphasized open publication of its data to provide public and scientific access to this information.
- The annual contractors' meetings enabled everyone involved in the program to interact and gain valuable information from others.
- The website has become the major repository for information on the health effects of low doses of radiation, a site to store and access data published in the open literature, and an educational resource for the general public on radiation exposure, doses, and risk.

CHAPTER 11

Current and Potential Impact on Standards

1. Standards Setting in the Low-Dose Region

When the DOE Low Dose Program was first being funded in the late 1990s, radiation standards were under much scrutiny. The research associated with the program had substantial input into the basic biology that could influence standards. From the program's start the importance of the data as part of the process in determining the risk from exposure to low doses of ionizing radiation was recognized, and efforts were made at every step to do this (Brooks 2003). From 2005-2008, four reports were published that have directly impacted radiation standards.

In 2006 the National Academy of Sciences published the Biological Effects of Ionizing Radiation (NRC 2006) report. I made a presentation early in the preparation of this report (October 2000) that outlined the research projects being funded by the DOE Program and discussed the potential impact and future data that would be available. The BEIR VII report has a good review of program data generated from 1999 to 2004. As pointed out earlier, the BEIR VII report suggested that while important, the data on bystander effects, adaptive responses, and genomic instability were not developed adequately. A better mechanistic understanding of these processes was required for them to be useful and be included in the standard-setting policy.

BEIR VII continued to use the biophysical LNT model for extrapolation of risk into the low-dose region where it is not possible to gain useful information from human epidemiological studies. With this model, the risk of cancer proceeds in a linear fashion at lower doses without a threshold. Thus, according to BEIR VII, the smallest dose of radiation has the potential to cause a small increase in cancer risk in humans. The DOE Program's results, however, have demonstrated that there are very different biological responses following low doses of radiation than those observed after high radiation doses. Thus, the mechanisms of action in the low-dose region are different than in the high-dose. There is a well-defined transition in responses in the dose range of about 0.1-0.2 Gy where the slope of the line is lower than observed in the higher-dose region. The challenge remains to determine if the slope is zero, greater than zero, or less than zero in this low-dose region.

In 2009 the Electric Power Research Institute funded a review of more than 200 publications published after the BEIR VII report (Dauer et al. 2010). This review demonstrated that additional mechanistic data were available on the responses of biological systems to low doses of radiation that need to be considered in future standard-setting activities. The report supported the need for a low-dose-rate effectiveness factor and reviewed the evidence that the mechanisms of action change as a function of dose and that the slope of the dose-response curve in the low-dose region is lower than that observed in the high-dose region. A summary of this report has been published (Dauer et al. 2010). These data suggest that the current standards are adequately conservative when the LNT is used to predict the risk in this region.

Two parts of a five-part report have been released by The United Nations Scientific Committee on the Effects of Atomic Radiation (UNSCEAR). The report includes five scientific annexes, the last three of which have not yet been released:

1. Epidemiological studies of radiation and cancer
2. Epidemiological evaluation of cardiovascular disease and other diseases following radiation exposure
3. Non-targeted and delayed effects of exposure to ionizing radiation
4. Effects of ionizing radiation on the immune system
5. Sources-to-effects assessment for radon in homes and workplace

They concluded from the report that the risks from cancer and genetic effects previously recommended did not require any change at the time.

Additional reports were issued by the ICRP. Report 99 by Valentin (2006) evaluated the low-dose extrapolation of radiation-related cancer risks. Report 103 updated radiation protection recommendations. This report suggested the continued use of the LNT hypothesis combined with an uncertain dose-dose-rate effectiveness factor (DDREF) to extrapolate risks from high doses of radiation into the low-dose region.

Finally, a report from the French Academy of Sciences raised some serious questions about the validity of using the LNT for evaluating carcinogenic risk in the low-dose region. They suggested, as has been supported by the DOE Program, that the biological mechanisms and responses are different at low doses and high doses. They suggested that the use of the LNT model may lead to an overestimation of risk at low doses.

After organizations such as the NAS, NCRP, and ICRP make recommendations, it is up to the government agencies charged with controlling exposure to determine how to use them. I was a member of the

EPA Radiation Advisory Council (RAC) that was charged with reviewing BEIR VII and recommending how EPA should implement the BEIR VII report in setting radiation risk standards. For the most part, EPA accepted the BEIR VII recommendations with some modifications as recommended by the RAC. A complete review of the cell and molecular data was presented to the RAC and as part of the uncertainty analysis it was included in the recommendations to EPA. The RAC pointed out that one of the largest uncertainties associated with the risk in the low-dose region was the model that was used to extrapolate the risks from high doses to the low-dose region. A summary of the cell and molecular data was prepared for EPA, much of which was generated by the program; it was included in the recommendations as an appendix. Thus, the data was considered, but at the time was not adequate to influence the setting of standards.

2. The Program's Impact on Risk

One of the first reviews of the low-dose data was a book chapter published on data and results from the program that appeared in 2006 in Advances in Medical Physics, edited by A.B. Wolbarst, R.Zamenhof, and William R. Hendee. This chapter, "Biological Effects of Low Doses of Ionizing Radiation," represented a good balance of viewpoints and data from the molecular to epidemiological data (Brooks 2006; Brooks et al. 2006a). Topics included 1) an overview of BEIR VII; 2) the role of cell-cell communication and the bystander effect, which demonstrated that the responding target was much larger than the "hit" cell; 3) a review of the adaptive protective responses and reported molecular, cellular, and experimental animal data supporting it; 4) genomic instability and its role in radiation induced cancer; 5) molecular changes induced by radiation in the high- and low-dose regions with data to support the differences in the biological mechanisms as a function of radiation dose; 6) and an overview of "limits" of detection of biological changes indicating that in the low-dose region there are thresholds or limits below which biological changes cannot be detected. This publication was a good review of major points and datasets generated by program researchers that had an impact on setting radiation standards.

The theme of the 44th Annual Meeting of the National Council on Radiation Protection and Measurements was "Low Dose and Low Dose-Rate Radiation Effects and Models." Several presentations were given on molecular, tissue, and animal responses to low doses and dose-rate radiation, and this meeting resulted in several publications associated with the application of research data to standards. In most cases it is difficult to relate such studies directly to risk, but these studies impact the database for determining the role of radiosensitivity (Kato et al. 2009), dose-rate, dose distribution (Brooks et al. 2009), and molecular factors that modify cancer risk (Kennedy et al. 2006; Barcellos-Hoff & Nguyen 2009; Morgan & Sowa 2009). The current and past epidemiological studies were reviewed, and several studies were presented in which the dose and dose-rate were very low. It was demonstrated that some types of cancer exhibit non-linear dose-response relationships, while when the total solid tumor data is evaluated, the dose-response is linear over the whole range of doses (Gilbert 2009; Shore 2009). The meeting featured a debate and papers from Dr. David Brenner representing the BEIR VII committee and Dr. Dietrich Averbeck discussing the French Academy Report and point of views.

An important session in this meeting was "Low Dose Radiation Effects, Regulatory Policy and Impacts on the Public," in which the problems associated with incorporating mechanistic data from the program into regulatory decision-making were discussed (Locke 2009). The weight-of-evidence approach was recommended, and interactions between scientists working at the molecular, cellular, and epidemiological levels of biological organization were deemed essential for any of the information from the program to impact standards. Each of the government agencies also made presentations about their unique problems and needs for use of mechanistic data in standard setting.

They all agreed that the standards should be based on the best scientific data available, but admitted to a wide range of different problems and reasons why this is difficult to do. Their presentations incorporated other input essential in decision making and demonstrated that science is only one element in standard setting and in many cases does not drive the decisions that must be made to protect the public from the potential effects of radiation (Tenforde & Brooks 2009).

A presentation on how beliefs about radiation influence policy and decision making (Jenkins-Smith et al. 2009) demonstrated that even though most scientists do not accept the LNT as the most scientifically sound method of regulating exposure in the low-dose region, the majority of scientists and the public agree that using it is the prudent policy. The federal programs that reimburse the public for past radiation exposure represent a policy that was instituted through Congressional action. The presentation illustrated that such programs may not be based on the best science. They make many conservative assumptions to determine who should be reimbursed to ensure that those exposed are not neglected. In many cases, there is no attempt to determine a link between the dose, exposure, and disease. For example, the down-winders in southern Utah were exposed to low doses (0.03 Gy) over a protracted period. Using the LNT and making very conservative assumptions of the small population exposed (about 25,000 people), only a very small number (<50) of cancers would be predicted to be induced by this exposure. To be politically correct and to correct the "wrong" of exposure from fallout, it was determined that they would receive compensation if they live in selected areas in Utah and develop a cancer that is related to the types of cancers that were observed to be elevated in the Atomic Bomb survivors (Ziemer 2009).

Because ~40% of the population develops cancer, and many common cancer types were included (bone, renal, leukemia other than chronic lymphocytic, and lung), there will be many of these cancers in any population. The Radiation Exposure Compensation Act (RECA) has approved payment of $50,000 to 11,815 individuals (Ziemer 2009). This illustrates that science does not drive the system, but the need to err on the safe side is very critical in policy making.

Finally, a presentation was made on how to combine science and regulations for decision making following a terrorist incident involving radioactive materials. This presentation pointed out: "It is important that an emergency response is not hampered by overly cautious guidelines or regulations. In a number of exercises the impact of disparate guidelines and training in radiological situations has highlighted the need for clear reasonable limits that maximize the benefit from an emergency response and for any cleanup after the incident" (Poston & Ford 2009). Recommendations must be very clearly defined for the first-responders so that unnecessary anxiety does not impede their ability to quickly respond to the needs created by the disaster. These types of presentations illustrate the importance of a good science background, but also show that other factors are equally important in controlling radiation exposure.

This discussion demonstrates that, to date, the DOE Program has had limited impact on standards setting. However, it also shows that the program has played a critical role in providing data and information on the responses in the low-dose region that, with the development of better methods of using the data, will have an important impact. The program has helped further understanding of the biological responses induced by exposure to low doses of radiation. The low-dose research data has demonstrated that 1) the scientific community understands the biological responses following low-dose radiation exposure, 2) there are no surprises (risks much higher than the current standards) or data that suggest we have underestimated the risk in the low-dose range, and 3) the use of the LNT is useful for controlling radiation exposures, is conservative, and provides an adequate and appropriate safety factor for risks in the low dose region. These low-dose data support the huge database that exists in the high-dose region to control the population risk from radiation-induced damage.

The data produced by the program has resulted in some major paradigm changes in radiation biology and will be very important for future activities associated with understanding and predicting the risks from low levels of radiation exposure.

3. Major Points: Impact on Standards

- A major U.S. report (NRC 2006) on the risk from low doses of radiation acknowledged the research from the program, but did not use it in making risk estimates. The report cited the need for more mechanistic data before the findings from the program can be applied to risk.
- The French Academy of Science reviewed the data on the effects of low doses of radiation including that produced by the program and recommended that the LNT was not valid for estimating risk following low doses of ionizing radiation. They suggested that the use of LNT would overestimate the risk in the low-dose region.
- International organizations, UNSCAR and ICRP, both issued reports that acknowledged the research from the program. Nevertheless, both of these organizations and reports continued to use the LNT to calculate risk from the human epidemiological data. Regulatory agencies (EPA and NRC) with responsibilities for setting radiation standards reviewed the new data from the program, evaluated the BEIR VII and French Academy reports, and accepted the more conservative recommendations for continued use of the LNT.
- The 44th annual meeting of the National Council for Radiation Protection and Measurements was focused on the responses in the low-dose region and resulted in a good compilation of the data generated by the program.
- To date, the basic biology has had little impact on changing standards or regulations used to control radiation exposure.
- The data from the program is widely recognized as important, and continued effort is needed to insure that risk estimates and standards are based on the best scientific data available.

CHAPTER 12
Applying Lessons Learned to Future Direction

1. Introduction/Overview

It may sound cliché, but it is critical we learn from the experiences of the DOE Low Dose Radiation Research Program so that mistakes made in the past will not be repeated in the future. To fail to do so is to ignore overwhelming scientific evidence compiled from humanity's greatest effort to understand low dose radiation and its effects on health. This evaluation of the lessons learned is only my opinion, and it should be read as such. However, as the past chief scientist on the program, I hope my input for future research will be considered to help restore the needed funding, provide a scientific basis for regulatory standards, and yield the needed results for protection of human health. Here's a high-level overview of some of those key lessons, which are discussed in detail later in the chapter:

The scientific advances made by the program put the field of radiation biology in a position to move forward and conduct mechanistic research that can ensure that radiation protection standards are based on the best possible science. To impact standards, and adequately and appropriately protect human health from potential radiation effects, further funding in this area is required.

- Scientific management and funding of the program requires a high-level, oversight advisory committee to keep any scientific program focused in the needed direction to impact radiation standards. Past funding was based on a three-year cycle—a very inefficient use of scientific talent. Funding and management of larger centers could be used and funded for a longer period of time. With this system the bench scientists could invest their time conducting meaningful research and not simply chasing money. This would help provide the research direction needed to support radiation standards and regulate radiation exposure.

- Communication and outreach are essential to any scientific research program if it is to have any impact on regulatory decisions. It is essential to make sure that the scientific information is conveyed to the public, concerned action groups, and regulators. Such communication makes it possible for everyone to have a seat at the table to understand the importance of the science to taxpayers' lives. Effective communication could have decreased disruption, and saved lives and money following both Chernobyl and Fukushima. Effective communication is also needed to help assure adequate funding and to insure that the information generated by the research moves to the level needed to impact standards.

2. Results

2.1 Scientific Advances

2.1.1 Lessons Learned

The program was originally funded based on recent advances in biological techniques such as the sequencing of the genome, development of gene expression arrays, and chromosome painting. These new tools were combined with other new technologies, like the microbeam, to address questions in the low dose region, not possible prior to these innovations. The program used modern techniques to evaluate the influence of radiation dose, dose-rate, dose distribution, and radiation type. Some of these new techniques and tools were just being developed at the start of the program and yielded important contributions to the field of low dose radiation biology.

When the program started, genomics was new. The human genome was declared sequenced in 2003, five years after the program's start. These developments made it possible to easily measure changes in gene expression in thousands of genes in a short period of time and relate these changes to radiation exposure parameters. Such genomics techniques were widely used in the program (Yin et al. 2003, Amundson et al. 2003, Coleman and Wyrobek 2006, Amundson and Fornace 2003). As it progressed, many more omics were developed, and a few of these were applied to the response of low doses of radiation epigenetics (Burnal et al. 2013) and proteomics (Jin et al. 2008, Rithidech et al. 2007b).

A particularly important lesson learned is that when a new observation is made, it is important to have such an observation duplicated by multiple investigators. An example of this is the observation made by Bernal et al. (2013) that low doses of radiation given at the proper time during pregnancy seem to have a protective effect against epigenetic effects, as shown in Figure 23. The figure shows the coat color of the offspring as a function of the radiation dose.

Animals with the yellow coat are overweight, diabetic, and cancer prone. The pseudoagouti animals do not have these traits. Small doses delivered at the proper time resulted in a larger fraction of the offspring having a pseudoagouti coat color and fewer of the mice having the yellow

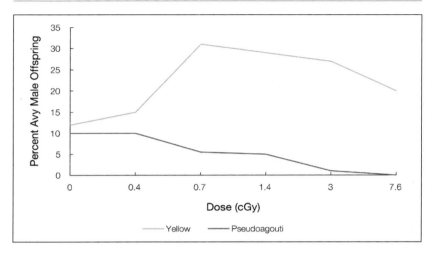

Figure 23. Fetal Radiation Exposure and Coat Color Change in Male Avy Mice (Bernal et al. 2013).

coat. Thus, in this model system, radiation reduced the risk for cancer. The lack of funding for anything that didn't expressly address risk from radiation made it impossible to continue this work in other laboratories to validate this extremely important observation. The lesson learned from the science was that each new research effort opens up new areas of knowledge that can provide information on total risk. It is critical to follow up on new observations *regardless* of the impact on risk. The program's science yielded many potentially vital research directions that were simply never explored and need to be further researched to ensure that the best science is used to address risk.

Like the above example, many other low dose biological responses observed were not known or suspected when the program was initiated. The lesson learned here is that we cannot predict the outcome of any new experiment. Funding is essential to test these new observations and apply them to regulation of radiation exposures. As new technologies become available, they must be applied at the human level to address important past and future problems and questions.

Although the program eventually made many key discoveries using animals in research, it started with studies focused on measuring radiation-induced changes in molecular and cellular systems using cells grown in tissue culture (*in vitro*). With this approach, it was possible to control variables and study one well-defined biological change at a time. As the program progressed, it became obvious that this approach was not viable. Cells in tissue culture did not respond the same as cells in a more realistic environment where there are many controls, feedback mechanisms, and metabolic paths activated and where defense mechanisms are present (Dauer et al. 2010). It is not useful to measure a single change since multiple interdependent changes are induced by radiation exposure. These observations were incorporated into the program, and a change in direction was initiated. Complex cellular systems were studied using skin and lung tissues in culture. The responses in these complex systems did not duplicate cell studies *in vitro*, suggesting that the focus must be further expanded to include experimental animals.

In animals, defense mechanisms like the immune system are in place to limit the progression of mutated cells to the development of cancer. In addition, the genetic background can be modified, resulting in alterations in defined changes in biological repair pathways. This lesson learned is that cell and molecular studies should be conducted in animals. It is essential that such studies be conducted to make it possible to link experimental data from molecular and cellular studies to human epidemiological studies. The lesson learned from the program is that studies *in vitro* provide basic information, but without applying this information to whole animals and humans, there will be little impact on helping protect human health or modify standards. The future direction of the program must consider this lesson learned and link basic molecular and cellular biology to human epidemiological studies. This must be considered for basic cell and molecular data to make any impact regulations that control human exposure.

2.1.2 Future Directions

Much of the data generated by the program did not fit the preconceived ideas in the field of radiation biology; findings were constantly contradicting the textbooks and previous paradigms (Brooks 2005). These changing paradigms influence the basic premise that radiation increases risk as a linear function of dose. Many biological processes are involved in cancer that result in non-linear dose response relationships. Ignoring these processes results in very conservative and costly radiation protection standards. Additional research is needed to help drive future research toward an accurate, mechanistic understanding of radiation's effects. The first paradigm change was driven by the observation of "bystander effects," where cells without energy deposited in them show biological responses to radiation exposure. This observation indicates that the "hit theory," widely used in radiation biology, may not be an accurate description of how radiation induces biological changes. Tissues and organs respond as a whole—a complex network of cells functioning together—rather than as individual cells (Barcellos-Hoff and Brooks 2001, Barcellos-Hoff and Nguyen 2009).

The second paradigm shift is from the mutation theory to genomic instability and tissue interactions theories (Morgan 2002, Limoli et al. 2000b, Barcellos-Hoff 2005b). It has been well documented that radiation is a very good cell killer, and because of this, it is a rather poor mutagen. Transmitted mutations, once thought to be of prime importance in risk assessment, are no longer considered a major risk factor for radiation exposure (NRC 2006). Publications from the program suggest that single mutations may not be the primary cause of radiation-induced cancer, but that the tissue environment and the induction of genomic instability may play an important role. Such data on genomic instability may have a serious impact on risk assessment from low doses of ionizing radiation (Morgan and Sowa 2009).

Finally, low dose, radiation induced, adaptive protection has been demonstrated in many molecular, cellular, tissue, and animal systems, suggesting that the mechanisms of action following low doses of radiation are different from those induced by high doses (Dauer et al. 2010, Feinendegen et al. 2007). Such data require additional study and again suggest that the Linear-No-Threshold model overestimates risk in the low dose region. Because of these data, the concept that each and every ionization (regardless of dose or dose-rate) increases cancer risk must be reconsidered. It has become evident that even if a single ionization produces a single DNA damage and mutation and that this may be linear, there are many non-linear biological processes involved that are responsible for cancer. It is essential to have a more holistic approach to follow the changes needed to take place as a normal cell is transformed and ultimately results in cancer. This approach was first nicely outlined by Hanahan and Wienberg (2000) and updated by Hanahan and Wienberg (2011). This approach was discussed earlier and is shown in Figure 24.

As a normal cell changes its capability to produce cancer, there are also protective mechanisms at the tissue and organ levels that need to be overcome to result in cancer. Many of these changes have not been shown to be induced by exposure to low doses of radiation. Thus, many of the carcinogenic processes are independent of radiation exposures. However, all these changes are postulated as a requirement for the production of cancer. Additional research may be required in many of these areas to really define the role of radiation in the production of cancer.

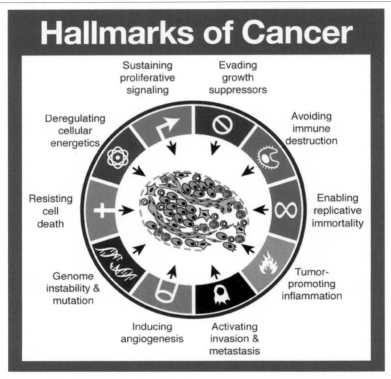

Figure 24. A graphic representation of the "Hallmarks of Cancer" from Hanahan and Wienberg 2011.

Past radiation research has focused on some of these areas (genomic instability and mutation, cell death, signaling, replicative immortality, ROS status, and inflammation), while almost no data are available on the effects of radiation on other key events in the critical pathways (Brooks et al. 2016). Conducting research with broader focuses than the initial DNA and cellular events is essential to define the shape of the dose-response relationship. Early events and late changes as cancers are formed have been studied for years, but the time between the radiation exposure and the induction of the disease—the latent period—has received little attention. Evading growth suppressors, deregulating cellular energetics, inducing angiogenesis, and activating invasion and metastasis may or may not be influenced by radiation exposure. Studies in these areas should be one of the major thrusts for future scientific direction. Additional mechanistic studies using modern tools are needed for each of these areas to provide a predictive basis for setting radiation standards.

Scientific advances have continued at a rapid rate, and as demonstrated above with protective epigenetic changes in mice, many of these were not considered in the past. It is well recognized that it is not necessary to change DNA sequence nor to delete or rearrange DNA in order to change the phenotype or characteristics of offspring. Epigenetic DNA modifications such as acethlylation, methylation, etc. can change gene expression and be reflected in the phenotype in animals in the next generation without changes or loss of DNA. New scientific tools are now available that had not been developed while the program was being funded. Computer power continues to increase rapidly, and the use of the internet and machine learning provides methods for more powerful statistical analysis. Computers also offer more readily available information to communicate science messages to the public and decision makers.

Even materials science has made major advances that will impact research in the low dose region. Bio-nanotechnology is opening ways to follow materials and changes at the single-cell level. Microfluidics has even added information on potential communication between cells. Imaging technology has advanced with new approaches to light microscopy with super-resolution, Raman spectroscopy, and x-ray microscopy. These advances in instrumentation take resolution into the nanometer range. Observing changes at this level in individual cells provides power to future studies on low dose radiation effects. These new technologies make it possible to study radiation effects in single cells in a realistic, complex, well-defined, and controlled biological environment.

Advances in molecular biology are just as rapid as those on the technology side. For example, in the past it was possible to study changes in DNA and gene expression in large numbers of important genes. With scientific advances, it is now possible to measure changes in a large number of different types of RNA, such as micro RNA, long noncoding RNA, and circular RNA. The field of proteomics has made major strides with protein modifications influencing many biological endpoints. The role of ubiquitination, phosphorylation, SUMOylation, and NEDylation in modifying phenotype expression is expanding and provides mechanistic understanding of the responses to both low and high doses of radiation. These can potentially provide means for better understanding the mechanisms involved in low dose radiation responses.

Advances have also been made at the cellular level. In the past it was assumed that cell killing following exposure to low doses of low LET radiation resulted in a plateau with little cell death followed by an exponential decrease. Research in the Low Dose Program demonstrated fine structure in the dose-response relationship for cell killing. In the low dose region, cells were killed much more effectively than predicted previously, resulting in an area of hypersensitivity. As the dose increased, the cells became radiation resistant, and cell killing per unit of dose increased. When the dose increased further, to the area where we were able to measure cell killing in the past, cell killing followed the previously well-defined exponential decrease (Joiner et al 2001). This fine structure in cell killing may have application in standard setting that requires additional research. It was also demonstrated that low doses of radiation-induced protective apoptosis selectively killed transformed cells (Bauer et al. 2007a, 2007b; Portess et al 2007). This response was postulated to differentially eliminate transformed cells and decrease radiation cancer risk. New research in cancer biology has demonstrated that there are many new mechanisms that control cell death, such as autophagy, paraptosis, pyroptosis, and necroptosis. Understanding how low doses of radiation influence these processes, especially in cells injured by radiation, could be very valuable in understanding radiation-induced cancer and may well provide potential direction for protection and treatment.

To have adequate and appropriate standards that minimize fear, optimize financial investments, provide real benchmarks for responses to nuclear accidents and nuclear war, future research must provide a scientific basis for standards. With these questions and needs in mind, important scientific questions can be addresses and resolved with additional research. These objectives and goals represents the charge for future research. With a strong scientific basis, important questions that impact our daily lives and radiation standards can be addressed. One of the most important questions is: What is the shape of the dose-response relationship for cancer induction in the low dose region? At the current time, there are data that support several different dose-response relationships. The energy deposited (physics) increases as a linear function of dose, but the biological response induced by that energy deposition (biology) results in non-linear responses at the cellular and molecular level. In the low dose region, many unique responses are triggered that suggest that these doses result in less cancer than extrapolated from high doses. Some data suggest that low doses of radiation may be protective against a wide range of insults and decrease the cancer frequency below that observed in non-exposed systems (Azzam 1996, Redpath 2007, Sykes et al. 2006). A quick summary of the responses seen after exposure to high and low doses are illustrated in Table 1 (chapter 5, page 95).

As indicated in the table, high doses of radiation increase free radicals, while low doses decrease free radicals by increasing the production of

MnSOD and Gluthione. Apoptosis, programmed cell death, is increased after high doses, while low doses increase selective apoptosis of transformed cells, which eliminates transformed and damaged cells. Mutation frequency and cell transformation frequencies are increased by high doses and are decreased following low doses. High doses suppress immune responses while low doses increase several indicators of immune responses. Better understanding and defining of the mechanisms involved in these important differences represents a critical area for future research.

Using rather high doses it was possible to detect the induction of genomic instability in special cellular systems (Limoli et al. 1999), which brings into question the validity of the single mutation theory of cancer. In fact the role of single gene mutations in cancer risk comes into question. The impact of genomic instability on risk assessment has been reviewed, and additional research is needed in the low dose region for this interesting finding (Morgan and Sowa 2009). The interaction of the microenvironment with radiation and its subsequent carcinogenic response also demonstrate that single DNA damage and mutations may not play the major role in radiation-induced cancer (Barcellow-Hoff 2008). Research that provides scientific data on the role of mutations and microenvironment is needed to determine the shape of the dose-response relationship in the low dose region. Understanding the shape of the dose-response relationship in the low dose region will determine whether low doses of radiation have no detectable effect, a harmful effect, or a beneficial effect. This simple understanding will have an impact on the biggest problem associated with low doses of radiation, the fear of radiation in the low dose region (Walter et al. 2016). The fear of radiation is in part driven by the lack of adequate communication of the science associated with low doses and the over-reaction of both the press and the scientific community to perceived damage.

2.2 Communicating and Outreaching to Inform Policy and Protect the Public

2.2.1 Lessons Learned

The DOE Program was started with adequate funding for several projects focused on outreach and communication of results to the decision makers and the concerned public. Communication specialists were funded from the outset, and, as described in the book, several important attempts were made to build bridges of understanding between the public and scientists. At one of the contractor's meetings, the informed and concerned public and antinuclear advocates were all invited, and funding was provided to cover their costs. At this meeting we invited members of the Hanford Advisory Board, the mayor of Rocky Flats, representatives of the Indian Tribes, just to mention a few of the groups. The people invited almost all came and actively participated. Through these discussions and interactions, connections were established between the public and the scientists conducting the research. This helped increase credibility and trust between the scientists and the concerned public. This meeting was successful in acquainting all parties, and a level of trust was developed. They shared concerns, and discussed research the advocates thought would be helpful for the groups that they represented. From this trust and respect derived by both the scientist and the non-scientists, the data generated by the program would be better accepted and applied to the needs of the concerned groups. However, due to lack of funding, this type of meeting was not duplicated and the potential for further input and interaction was minimized. An important lesson learned is that one-on-one interactions between the scientists doing the research and the concerned citizens is needed to build trust and credibility, both of which are essential for research in this area to impact future regulations. Without this credibility, no scientific discovery can impact public policy and regulatory decisions.

Another important attempt to communicate our scientific information to the public was to schedule a meeting with the news media. Again, by having our best scientists and the news media at the same meeting, we attempted to increase outreach efforts. One of the concerns voiced at by the news media was that research showing radiation is perhaps less hazardous than previously assumed is not front page news. When research results suggest a previously unknown hazard, it is front page news, but data that suggests that radiation is less hazardous than previously thought is hard to get to the public. Lessons learned from this interaction were that scientists have to be proactive even when the data suggest less hazard and make special efforts to make their research results known and understandable to the news media. Scientists must be able to convince the news media that it is good news when they demonstrate that something that generated fear is not as serious a health hazard as previously conceived. It will take a system of rewards to the scientists for outreach efforts and a sustained dedicated, multi-faceted communications effort to respond quickly and accurately to ensure effective publicity. Employing communications professionals to work closely with scientists offers the best way to outreach to the news media, as well as other audiences.

Later in the life of the program, a meeting was organized to provide interactions between the research scientists, bureaucrats, and regulatory agencies. Basic scientists, funding agencies (DOE, EPA, NRC and NIH), regulatory agencies (NIH, NIEHS, OSHA) and scientific advisory groups (NCRP, ICRP) were all involved. It became obvious that much of the science being conducted was not directed to the needs of these people who are responsible for developing regulations. It was also apparent that each of these groups spoke a different language. It was a challenge getting scientists to meet the needs of the regulators because the scientists and regulators weren't communicating effectively. When the scientists made their presentations, the regulatory agencies slept. When the regulatory agencies made their presentations, the scientists slept. And when the bureaucrats made their presentations, everyone slept. Was this communication? Perhaps not. The DOE representatives that sponsored the meeting thought that the meeting was a failure. On the contrary, in my opinion this meeting laid the needed groundwork for the future and increased interactions. These meetings should have been continued. They were not continued, however, because of lack of funding. The lesson learned is that we must all learn to speak the language used in determining regulatory needs. Long term, regular interactions and efforts to understand the needs of the regulators are essential. Ongoing collaboration between scientists and regulators—aided by communications professionals—are necessary to establish effective, ongoing communication to allow regulators to make decisions. Future research efforts in the low dose region should be directed toward projects that can directly impact regulators' ability to set standards and limits.

Unfortunately, all of these important meetings and communications efforts were eliminated due to lack of funding. The program's limited funds were redirected away from communications efforts toward basic scientific projects. The program's researcher meetings, the website, and the chief scientist position were continued, however, and provided the major sources of interaction between the scientists being funded in the program. The outcome of the slashing of communications efforts prevented us from advancing the meaning of the scientific data to the places it needed to go to influence policy. It was not adequate to have the scientists mainly talking with each other. To inspire outreach efforts from scientists, reward and recognition need to be in place to reward scientists for using the data they generate to reach out to a greater audience. Most researchers are simply not professional communicators. Some scientists are better than others, and the skills that these scientists have should be used and rewarded. Without professionals, it's difficult to strategically amplify findings to influence decision makers. Research without communication is dead on arrival. Without communications efforts, no matter how good the science is, it's unlikely to inspire policy.

2.2.2 Future Directions

The scientific data generated must pass rigorous scientific review and be published in the open literature in journals with high scientific impact factors. However, scientific data alone cannot impact standards. The reviewed data must be summarized and published in well-established news outlets and other venues across the rapidly growing media. To be effective in making policy changes, the data and information must be coupled with a communication program that makes the data available and understandable to funding agencies, regulators, the news media, and the concerned public. Any new program should have a strong focus in this area. Funding for experts who understand how to communicate complex information should be required. With adequate funding, it would become possible to take advantage of expertise and modern communication tools. For example, the internet can be used as both a data repository and a hub to informatics. In addition, the internet has become a primary source of information and provides a vital means to get the information out to the concerned public. It is essential that this information has very high credibility and is viewed as a valid source. Technology provides a means for outreach and international collaborations that were not possible in the past. With communications support helping to clarify the "why it matters" for decision makers and the concerned public, it becomes possible to generate a scientific basis for standards that will be accepted. Thus, interaction between scientists and professional communicators is essential to disseminate critical information to protect and benefit public health. This can be done both by providing information directly to the public, but more importantly by providing it to those responsible for determining regulations. With solid science and communications support, future researchers can motivate regulators to make informed, value-based decisions about standards.

2.3 Management, Funding and Scientific Review

2.3.1 Management

2.3.1.1 Lessons Learned

The management of the program was done in a manner similar to that used by the National Institutes of Health. This model for funding and accepting scientific proposal has been used extensively by most agencies that fund basic research. That is, a call for proposals was issued, and acceptable proposals were reviewed by a committee of scientific experts, and the projects were scored on their scientific strengths. These scientific scores were provided to DOE, which reviewed the projects that received fundable scientific scores and evaluated them as to how they addressed programmatic needs. By combining scientific strength and programmatic needs, funding decisions were made, and the research was started.

In the DOE Program funding was awarded on a three-year cycle, and each project was required to submit a new proposal every three years. This approach had advantages and disadvantages. The advantage was that the DOE received a wide range of approaches with scientific credibility that met the call for proposals. This opened the door for some very creative, but not always useful or focused, approaches using new technology to address the problems outlined in the call. The disadvantage was that the direction of the overall program was not adequately focused. Individual scientists sometimes focused on their own research interests and went in different directions. Management then became an exercise in herding cats rather than providing adequate direction up front. With proper focus the total body of research could have more directly impacted radiation standards to provide adequate and appropriate protection of human health. The lesson learned is that to reach the ultimate goal of providing data to influence regulatory decisions, a focused and carefully monitored approach is needed. As was seen in BEIR VII (NRC 2006) and the incorporation of the recommendations into regulations by EPA and NRC, it was not possible to utilize most of the new information in the regulatory process because of lack of mechanistic understanding of the new paradigms.

When the program was first initiated, a science advisory committee was formed. This group with a wide range of scientific backgrounds and expertise provided useful suggestions and early directions for the program. However, because of lack of funding, this committee was disbanded, a serious mistake. Without a doubt, outside advice from other agencies and professional organizations—as well as a well-defined and constituted science advisory group—is essential to keep any program focused and moving in a unified direction. With direction and focus, data can be generated to impact regulations designed to adequately and appropriately control human exposures. As a result of lack of scientific advisory group oversight, the direction for the program was limited to the DOE office and the reviewers selected. The program direction lacked focus. Thus, the data and research needed to impact standards and help regulators make important regulatory decisions was not adequately met, and the program's ultimate impact suffered.

2.3.1.2 Future Directions

Comprehensive scientific guidance and oversight is required for any major research effort. An advisory committee of scientists with a wide range of backgrounds is essential to provide scientific direction and bring new ideas and insights into the field of radiation biology. The committee should have individuals representing a wide range of scientific backgrounds, groups that impact standard setting, other funding agencies, and members of the concerned public. The first charge for such a committee would be to draft a set of recommendations for future directions. It would then be their responsibility to carefully monitor scientific direction and progress towards well-defined goals.

The management of any future research needs to be organized to facilitate interactions between researchers and provide strong oversight to focus the research on the major problem: generating the scientific basis for realistic and appropriate standards to protect the public. One approach to meeting these goals would be to establish centers of excellence at universities or national laboratories for low dose and dose-rate radiation research. Working with the advisory committees suggested above, the centers would have an overall charge to provide sound scientific input and to help maintain focus on providing the interaction and direction needed.

These centers would develop a multidisciplinary scientific team with a wide range of scientific expertise, from physics to molecular biology, with a focus on using the systems biology approach. This approach makes it possible to incorporate data derived at all levels of biological organization (metabolic, molecular, cellular, tissue, experimental animals, and humans) to impact standards. A major focus of the centers would be to link modern technology and biology to human studies to make the information applicable for human risk assessment.

2.3.2 Funding

2.3.2.1 Lessons Learned

During the course of the program, it became obvious that for many of the studies being conducted, a three-year funding cycle was not adequate. This is a universal problem that needs to be addressed. This problem exists for most scientists conducting studies in the life sciences. Scientists found themselves in the middle of important research when it became time to renew their contracts. The scientist, who should have been investing time conducting basic science, was then required to pause the research and invest serious time and energy to meet the next funding cycle. Thus, research efforts were diluted by the requirement to apply for additional funding every three years. This increased stress in maintaining an adequately funded research team, and directly resulted in a marked decrease in scientific productivity during the last year of the contracts. Under these conditions it became almost impossible to derive the data required to address important scientific questions. A longer funding cycle for selected projects would make it possible to more effectively use the funding provided to answer the scientific questions.

Linking molecular and cellular studies to human risk requires long term, expensive, labor-intensive experimental animal and human studies. This needed new direction is not adequately addressed using a three-year funding cycle and cannot support needed scientific goals. Future research programs need built-in flexibility to provide funding for some of the projects over a longer time period to generate the needed data. If some projects that involved animal or human studies, and took longer to collect data, could have been funded on a five-year or longer cycle, it would have made it possible to more efficiently conduct critical studies that were needed to incorporate recent molecular, cellular, and physical techniques available into experimental animal and human studies. Longer funding schedules for selected projects would make better and more reliable data available for strengthening radiation standards.

2.3.3 Future Directions
It is suggested that core grants given to the centers for excellence would be on a five-year cycle that could be renewed if the progress and scientific direction were adequate.

Involvement of scientists outside the center that have tools, research ideas, and technologies that would aid in the overall goal of developing a scientific basis for standards would be essential in providing the scientific breadth. Individual scientists could receive funding to support the research of the centers and could be funded on either a three- or five-year cycle depending on the experimental protocols developed and the time required to meet the goals outlined. It is important to have these outside scientists involved in order to stay abreast of new technologies and techniques, providing a broad scientific base. They would be an integral part of the program and participate in all the communications and directions involved. They would be required to give annual progress reports to the center, and members of the center would be involved in funding decisions as these projects came up for renewal.

For these centers to be successful, it would be important to have a long term and adequate funding commitment. Because of the high level of expense involved in modern biology and human studies, an adequate longer term funding base is needed. Funding would be determined by the funding agencies, political pressures, and the research community. However, it is helpful to look to the past and note that funding at the level of 20 million dollars per year was not adequate to conduct basic science studies in cell culture systems. Much more funding would be required to link modern biology with human epidemiological studies to produce data that addresses the important questions associated with the health effects of low doses and dose-rate radiation exposures.

3. Summary

It is absolutely critical that we continue to pursue a better understanding of biological responses to low doses of radiation, and in the current changing world the need for that information continues to grow. Concerns related to nuclear war, nuclear energy, waste clean-up, terrorist activities, dirty bombs, and the increased use of radiation in medicine and industry make it absolutely necessary to have a scientific basis for the health risk from radiation in the low dose and dose rate region. The DOE Low Dose Radiation Research Program was at one time the world leader in this field and used the best tools available to make a substantial start producing data and developing a mechanistic understanding to determine the biological responses at all levels of biological organization. With the demise of the program, the rest of the world—Europe, China, India, and Korea—are starting up and continuing research programs while the United States is a second-class citizen in conducting this essential research. Areas that remain to be addressed require the linking of cell and molecular data to human health outcomes and the development of science-based radiation protection standards. Many important daily questions and concerns revolve around the lack of knowledge of the risks for cancer and genetic disease associated with exposures to low doses of radiation.

Fear of exposure to low doses of radiation is one of the major problems associated with using nuclear power as a source of affordable energy to help drive our economy. It was fear, not radiation dose, that resulted in thousands of abortions following the Chernobyl accident, and it was fear, not radiation, that resulted in many deaths from moving people too quickly after the tsunami that caused the Fukushima nuclear power plant disaster. This fear must be addressed with the best possible science and communication. Nuclear power is one of the options that must be considered for our future economic well-being and to meet greenhouse gas reduction goals. Understanding the risk associated with the doses associated with normal operation or accidents associated with nuclear power is critical. Can research and communication of results help? This is a question than can only be addressed after adequate efforts have been expended.

What about doses and biological effects associated with nuclear accidents like Chernobyl or Fukushima? Chernobyl was the worst nuclear accident in history. Huge amounts of radioactive material were dispersed into the environment. The program funded several studies to look at the impact of the accident on the animals in the highly contaminated zones. These studies demonstrated that there was little evidence of DNA damage, mutations, litter size, or other measureable effects in the animals in this region. This was an unexpected result based on the high total doses and the level of contamination present and suggested very good repair systems were present to counter the damage done at a low dose-rate. Follow-up studies on the human populations did not demonstrate an increase in leukemia or solid cancers. The only cancers that were elevated were cancers of the thyroid in young people. These people accumulated large concentrations of ^{131}I in the thyroid gland, which resulted in large doses and an increase in cancer frequency.

Data from Chernobyl should have provided a scientific basis to determine the need to evacuate people following the Fukushima nuclear power plant disaster. The tsunami that compromised the Fukushima nuclear reactor resulted in the death of many people and also resulted in the release of radioactive material into the environment. Those people did not die from radiation exposure, however. They died from fear and a lack of knowing that led to a frantic evacuation of large numbers of people. Many of these people were old and had life-threatening diseases. Without proper medical care, housing, and nutrition, it has been calculated that the relocation may have resulted in over 1,600 premature deaths.

Without irrational fear of radiation, additional care and time could have been taken to move these people without the impact on their health. The radiation doses delivered to the public following Fukushima were low and based on our best scientific knowledge were not predicted to cause additional health effects (NCRP 2006, WHO 2012, Brooks et al. 2016). The concerns over radioactive fish, produce, and movement of contamination through the environment have resulted in vast resources being invested with little or no beneficial effect. Additional research and communication efforts on the biological responses and risks are essential to prevent such responses in the future if we have a nuclear war, terrorist attacks, or nuclear accidents. Continued funding of research on the effects of low doses of radiation can help provide this information.

Public fear of low level radiation exposures impacts the quality of our everyday life (Walter et al. 2016). The data derived by continuing research in the low dose region have implications in the use of radiation in medicine. These data from the program help the public address a question faced by over 90 million people a year: Should I have a CT scan or not? What are the risks and benefits from this procedure? Every time we visit a dentist, we get an x-ray. Is this a hazard? When we have a heart problem, radiation is used to help detect and treat the problem. Is this a hazard? Data from the program help people know that in the low dose region, the body responds differently than in the high dose region, and the risk is non-existent. If a medical doctor suggests the use of radiation for either therapy or diagnosis of a disease, it should not be refused because of fear.

The program was initially funded in response to the concerns over the potential health effects and high costs associated with low doses involved during clear up of nuclear waste. Currently billions of dollars are invested each year cleaning up nuclear waste sites. The clean-up guidelines require the doses to be reduced to levels that are below natural background. If the guidelines were based on better scientific data, such as could be derived from continued research in the low dose region, the hazards could be properly controlled, and the investment be more carefully used. With increased understanding, it may be possible to modify the levels required for clean-up and save billions of dollars a year.

Finally, the program uncovered the science needed to start answering these uncomfortable but critical questions needed for protecting public health: what are the effects of a dirty bomb or low doses associated with a nuclear weapon during a nuclear war or single detonation? The impact of a dirty bomb or any type of nuclear terrorism will be as "bad as we perceive it to be." In other words, it is essential that our perception is supported by scientific data. If the event involves radioactive materials, regardless of the levels or risks, the public will be very fearful, and the response may not be justified by the potential health effects. During a nuclear war, the deaths and damage from the blast, shock wave, and fire triggered by the bomb would be extensive, and it would be important to focus on these. However, in addition to the short-term effects, very large populations would be exposed to low doses of radiation. It is essential that the military and the general public have a realistic understanding of the risks associated with potential late effects from these exposures.

Whether in the event of full-scale nuclear war or the detonation of a dirty bomb, the need to disseminate accurate information about the risks associated with low dose radiation would be imperative. For this and the many other compelling reasons articulated in this chapter, it is obvious that a well-organized, funded, and focused research program to address these types of questions is essential.

In the past, one of the major problems of research in the low dose regions was the lack of communication, focus, and direction. Any new program must have focus on the goal of providing a scientific basis for radiation standards. To do this, the goal must be firmly entrenched, orchestrated, and communicated. Without focus and direction it will not be possible to generate adequate mechanistic data to link molecular and cellular data to animal and human data in order to provide a scientific basis for standards. Large data bases from a number of different sources is required to impact standards. The DOE Low Dose Radiation Research Program was important in starting to provide one such data base. Continued research, coupled with a strong, well-focused, and well-communicated program, may make it possible to reach the ultimate goal of having science-based radiation protection standards that are adequate and appropriate. It is important to have the new scientific data base reviewed on a regular basis by expert committees, including those that make recommendations for standard setting such as NCRP, ICRP and National Academy of Science. With these reviews and interactions, the research and database generated may modulate public fear, modify policy, and result in the proper science based radiation standards that we all know are necessary to further the progress and proper use of our valuable resources.

Epilogue

In 2008 it seemed like it was time for me to retire. Thus, I dropped out of science, resigned from WSU, left my position as a member of the Board of Directors at the National Council on Radiation Protection & Measurements, resigned from the EPA science advisory boards, and withdrew from the editorial board of the *International Journal of Radiation Research*. I thought my days as a scientist were over, so I volunteered to serve on a mission for the Church of Jesus Christ of Latter-Day Saints. My wife, mother-in-law, and I were assigned to a girls camp in the Cascade Mountains from 2008 to 2009. At this time, I still had limited funds left on my grant from the DOE Low Dose Program. I talked to Dr. Metting and suggested that I could use these funds to write a book on the history of the program and summarize its scientific accomplishments. Wonderfully for me, she agreed. I was not out of science! Every morning I got up early, put a few hours in on the book and then went about my daily task of running the camp. By the end of our two-year stay as missionaries, I had a draft of the book completed and submitted it to DOE. Dr. Metting and I had several discussions as to what to do with it now that it was written. It was suggested that we publish it as a government report. "That would be a great way to hide it," I thought. I wanted it published as a real book.

I feel blessed to have had a career in science during these special times. During my scientific life, funding was available for those who were able to conduct experiments, publish the data, and make useful scientific presentations. Only those who were good and willing to work hard survived. My job was to do science and follow the progress of science, which is not the case today for many scientists. Today many scientists invest a great deal of their time trying to get money. They write proposals, get rejected, write a paper, get rejected, write again, and do everything they can to chase money. It is such a waste of time and training. You really need to be able to take a beating to be a scientist.

Science provided me with an adequate salary, a chance to get acquainted with scientists from around the world who were important in shaping our radiation protection policy, and an opportunity to serve on a number of interesting committees and to travel to fascinating parts of the world to attend and present my scientific findings at interesting meetings. The major impact of science on my life was the excitement of discovery.

During my sixty-plus years of science I have never encountered a situation where the risk from the internally deposited radioactive material, with non-uniform dose distribution and protracted exposure, presented a greater risk or produced more biological damage than predicted from acute or chronic external exposure. My own experience, the experience of others in the program, and the data itself all suggest that the extreme fear of radiation I had developed back in my years chasing fallout—and that many people have to this day—was unjustified. Radiation is a very useful tool. Proper use and control is possible for nuclear power to provide a major portion of our electrical needs. The detection, handling, and disposal of nuclear waste is based on good science and with proper care can be done without impacting our health or that of our offspring. With care, radiation as we now understand it can serve humankind in many fields. My only major failure has been my lack of ability to communicate that to the public to my satisfaction. My life has been, as Dr. Roger O. McClellan said when he introduced me at the Taylor Lecture, a great ride for me with a job that I love.

APPENDIX A
BERAC REPORT PROGRAM PLAN

Biological Effects of Low Dose and Dose Rate Radiation

Prepared for:
Office of Biological and Environmental Research
By
BERAC Subcommittee

I. Executive Summary

There has always been natural background radiation present our environment. In addition, there is a high frequency of naturally occurring cancers that exist in all human populations. The radiation background and the large spontaneous incidence of cancer make it impossible to determine if low levels of radiation are capable of causing cancer.

Extensive research on the health effects of radiation using standard epidemiological and toxicological approaches has been used for decades to characterize the response of populations and individuals to high radiation doses, and to set exposure limits to protect both the public and the workforce. These standards were set by extrapolation of effects from high-dose studies using modeling approaches because of the inability of science to detect cancer following low doses of radiation. Thus, the historic approach has been the Linear-no-Threshold model that requires that each unit of radiation, no matter how small, will cause cancer. This model predicts effects from radiation in low-dose regions where it is not possible to demonstrate effects. Excess cancer at low doses are thus based on calculations and not on observations.

Most of the projected radiation exposures over the next 100 years will be to low doses and low dose-rates from waste clean-up and environmental isolation of materials associated with nuclear weapons and nuclear power production. The major type of radiation exposures will be low Linear Energy Transfer (LET) ionizing radiation from fission products. The DOE Program will thus concentrate on studies of low-LET exposures delivered at low total doses and dose-rates. The program will use advances in modern molecular biological and instrumentation to address the effects of very low levels of exposure to ionizing radiation. It will concentrate on understanding the relationships that exist between normal endogenous processes that deal with oxidative stress and the processes that are responsible for detection and repair of low levels of radiation induced damage. There is a single major question associated with the radiobiology of low dose exposures, "Are there adverse health effects induced by low dose and dose-rate exposure to ionizing radiation as predicted by the Linear-No-Threshold hypothesis?" To address this major question it is also important to answer:

1. Is the damage induced by ionizing radiation and the repair of that damage different from the endogenous oxidative damage and repair present during normal life processes? High levels of oxidative damage are produced and repaired daily in our bodies. Does this repair extend to oxidative damage from ionizing radiation?

2. Can endogenous repair capability prevent cancer induction following low levels of radiation exposure? Such repair could then result in a threshold of exposure below which there is no increased cancer risk.

3. Can molecular and tissue responses to radiation-induced damage prevent or reduce development of cancer? Such responses could modify the processing of damage and/or determine whether or not damaged cells are eliminated from tissue.

4. Do genetic differences exist that result in the inability of some individuals to repair radiation-induced damage? Such genetic differences could result in sensitive individuals or sub-populations that are at increased risk for radiation-induced cancer.

The major goal of this program is to ensure that human health is being adequately and appropriately protected. It is currently costing billions of dollars to protect workers and the public from man-made radiation exposure that is lower than the natural background levels of radiation. If it is determined that there is no risk associated with these exposures, these resources could be more effectively directed toward more critical health related issues.

The funding for research in radiation biology has decreased because the Linear-no-Threshold models were conservatively protective and the scientific tools and methods available in epidemiology and toxicology were not adequate to address questions associated with cancer risk following low doses of radiation. Research to define the genome, to understand structure-function relationships and to apply molecular biology to medical problems has resulted in the development of a range of new scientific instrumentation. These instruments and methods can be modified and applied to address basic radiobiological problems. In association with the development of instrumentation, there has been an explosion of knowledge in the fields of molecular and cellular biology. For example, it is now possible to identify the genetic basis of many diseases, to clone and amplify individual genes, to grow a wide range of critical cell types associated with cancer, and to develop trangenic animal models. All these techniques help understand and modify the expression and action of many genes. With new molecular techniques and proper application of instrumentation, it will now be possible to increase our understanding of normal processes that repair oxidative and radiation-induced damage at the molecular, cellular and tissue levels, and to determine the role of low levels of radiation in changing these endogenous processes. This research program will take advantage of the modern methods and technologies to address these important national and international issues and follow the leads that are emerging from modern biology.

The overall theme of the new research program will be to understand the endogenous processes that are responsible for maintenance and repair of radiation-induced damage. If the damage and repair produced by normal oxidative endogenous processes is the same as that produced by radiation, it is possible that there are thresholds of damage that the body can handle. If the damage from ionizing radiation is different from normal oxidative damage, then its repair, and the hazard associated with it, may be unique. To understand the relationship between normal oxidative damage and radiation-induced damage, studies will be conducted at very low doses and dose-rates and the perturbation of the normal physiological processes characterized at all levels of biological organization. Under this major theme, there will be three major research goals.

- To determine if there are dose or energy thresholds of exposure below which there is no significant biological change or below which the damage can be effectively dealt with by normal physiological processes. If such levels exist, there should be no regulatory concern for exposures below these thresholds since there will be no increase in risk.
- To determine how unique genetic background may alter individual sensitivity for the induction of cancer from radiation exposures and how genetic make-up influences individual and population risks.
- To communicate the research results to policy makers, standard setters and the public so that current thinking will reflect sound science.

Research conducted under this program will help determine potential mechanisms of cancer induction by low levels of radiation and if human health is being adequately and appropriately protected from these low levels of radiation exposures.

II. Introduction

Estimates of cancer risks following exposure to ionizing radiation are based on epidemiological studies of exposed human populations, principally the Japanese A-bomb survivors. While analyses of these populations provide relatively reliable estimates of risks for high dose and high dose rate exposures it is the effects of low doses (and low dose rates) that presents the greatest health concerns for radiation workers and the general population. The risks of cancer and mutations produced by very low doses remain a critical unresolved issue, because they cannot be directly measured in exposed populations. Conceptually, we are forced to estimate risks for low-doses and for doses received as chronic protracted exposures or low dose fractionated exposures by applying various dose response models. Currently, overall estimates of low dose risks are based on empirical linear fits of the human data that are then adjusted for low-dose and dose- rate exposures. While this approach has generally been adopted by those charged with assessing radiation risks, others have argued that it is inappropriate. Specifically, this approach may greatly overestimate the cancer risks. Among those who believe current protection standards overestimate risks, many argue that a threshold for radiation-induced cancer exists. This is a critical issue because of the potential societal and economic impact of decisions upon which these estimates of risk are based. Epidemiological data by themselves are not capable of resolving the critical questions at hand; moreover, conventional experimental approaches have gone as far as they can toward addressing low dose issues.

Through recent advances in cell and molecular biology and concomitant advances in chemical and biological technology, scientists have now created an extraordinary opportunity to definitively resolve this critical low dose issue. Specific opportunities are now at hand in four interrelated areas which are key to resolving this issue: 1) characterization of radiation-induced damage to cells and tissues and its relation to endogenous damage; 2) characterization of the repair and processing of radiation-induced damage; 3) determination of the molecular and tissue responses to radiation damage and the consequences of these responses; and 4) defining the impact of susceptible subpopulations on low dose risks.

Over the last several years it has become clear that oxidative free radicals produced by normal cellular metabolism are involved in the production of endogenous DNA damage. The types of damages produced by these free radicals overlap with the majority of molecular damage produced by ionizing radiation. Cellular DNA repair mechanisms, that are highly conserved across species, evolved to remove these endogenous oxidative DNA damages and thus preserve genomic integrity. It is precisely because free radical-induced DNA damages are efficiently repaired that cells have low rates of spontaneous mutagenesis. The question then arises as to whether low levels of ionizing radiation can be efficiently repaired by the same or similar repair systems as endogenous damage resulting in a threshold in the dose response curve. There is ample evidence that DNA repair competence can influence radiation effects, including radiation-induced cancer. There is also accumulating evidence that even low doses of radiation can elicit numerous molecular responses that have the potential to influence consequences. The above considerations support the view that a threshold at low doses of radiation may exist. With the development of sophisticated molecular biological approaches, together with new and evolving chemical and biophysical techniques, it is now be possible to readdress the low-dose issue, including the likelihood of a threshold during the next decade. Coupled with the biological program, new technologies will have to be advanced, including new approaches to measure DNA damage in the very low dose range and to determine molecular responses to such damage at the level of single genes or small changes in gene expression. Much of this technology will be facilitated by interactions with other ongoing programs such as the genome and structural biology programs.

Recent epidemiological and genetic studies suggest there may be a large number of genetic polymorphisms with the potential of conferring an increased risk for cancer as a result of interactions with environmental factors, including radiation. If the frequencies of polymorphisms that impact susceptibility to radiation-induced cancer are relatively high, it could significantly impact risk estimates at low doses for the population in general. It is now possible to identify, map, and clone the genes involved in radiation damage response functions, define the polymorphic frequencies of these genes in the population and determine their importance for susceptibility. This will provide the opportunity to directly determine their impact on cancer risk estimates after exposure to radiation.

III. Program Outline

A. Thresholds

Key Question: Is there a threshold for low LET radiation-induced cancer?

The linear-no-threshold model states that cancer risk increases as a linear function of dose. From such a model it follows that even the smallest dose of radiation is theoretically capable of producing at least some cancers. It therefore becomes important to establish the validity of this model at very low doses. At issue is whether there are thresholds below which no excess cancer or genetic damage is induced. This is a difficult issue to approach experimentally because of the inability to actually measure cancers produced by very low doses. There are several types of thresholds that have been suggested. There are statistical or practical dose thresholds below which no increase in cancer can be detected because of the severe statistical limitations imposed by the high background rate of cancer and the low frequency of radiation induced cancer. There are potential energy thresholds related to track length and structure, especially for low LET radiation, where the amount of energy deposited in a biological system is not adequate to cause produce biological damage. Finally, biological thresholds have been postulated to exist that are dependent on biological repair processes acting on radiation induced damage. The prime goals are to determine whether or not biological or energy thresholds exist following very low doses of ionizing radiation.

The existence of thresholds depends on a number of factors. It must first be determined whether the spectrum of DNA damage produced by ionizing radiation at low doses is qualitatively or quantitatively different from those produced from endogenous sources. The majority of damage produced by low LET ionizing radiation is due to the radiolysis of water in the vicinity of the DNA molecule, leading to free radical-induced DNA damages, which is similar to that produced by endogenous free radicals. However, unlike endogenous damage, production of multiple radicals close to the DNA molecule by ionizing radiation can result in highly localized clusters of damage on the DNA molecule that may be difficult to repair. For example, ionizing radiation is more efficient at producing potentially lethal DSBs. The numbers of qualitatively different lesions compared to the overall spectrum of endogenous damage must be assessed with accuracy at low doses. In addition to determining similarities and differences between the lesions initially produced by ionizing radiation and those resulting from endogenous damage, it is important to ascertain if they are inefficiently or efficiently repaired and whether the repair and processing of the radiation-induced damage results in faithful restoration of genomic integrity. Since ionizing radiation induces an stress response, in many ways similar to many endogenous oxidative processes (e.g., inflammation), it is also important to know what genes are induced in response to low dose radiation exposures and how might the induced genes influence the outcome of radiation damage to the cell.

B. Nature of Radiation Induced Damage

Key Question: Is the DNA damage produced by ionizing radiation at low doses qualitatively and/or quantitatively different from endogenous damage?

Description: The majority of radiation-induced DNA damage results from free radical attack on the DNA sugars and bases, producing single strand breaks, sites of base loss (alkali-labile lesions) and a large number of modified DNA bases. A much smaller number of DSBs produced by direct ionization of DNA or possibly by the processing of multiple single lesions produced in close proximity. Protein-DNA cross-links are also formed, but in very low amounts. In spite of the fact that the frequency of DSBs is much lower than that of other types of damage, in mammalian cells, the DSB is considered to be the primary lesion involved in cellular lethality. This is because DSBs are more difficult to repair with fidelity. Clustered DNA damage that, at least at high doses, appears to be unique to ionizing radiation is particularly difficult to repair (Ward 1994). Free radical-induced lesions present on a single strand of DNA have not generally been implicated in cell death because they are readily repaired by the base excision repair system and because a correct copy of the information is present on the complementary strand. Although the impact of unrepaired DNA damage to vital genes cannot be ignored, it is likely that subsequent processing leading to misrepaired DNA damage is largely responsible for chromosomal aberrations, genomic instability and ultimately carcinogenesis.

Decision Making Value: The problem facing scientists and policy makers is that all the information for radiation-induced DNA damage is at high doses where cells are traversed by multiple ionization tracks. There are no data at the low doses normally considered relevant to public health issues where a cell may only be traversed by a single electron track. It is not difficult to imagine that the spectrum of damage at such low doses may be substantially different from that observed at high doses. Because the background of spontaneous damage produced by free radicals derived from oxidative metabolism appears to be fairly high (Wallace 1997), the question arises as to whether low levels of ionizing radiation significantly add to the background level of damage. Thus it is fundamental to the entire low dose issue to determine whether the amount and kinds of DNA damage produced at low doses are different from those produced endogenously. If the DNA damage produced by low doses of ionizing radiation, is qualitatively similar to the damage produced by normal physiological processes then the amount of damage from the radiation is so small relative to the normal damage that it cannot have an impact on cancer risk. This would support a threshold for radiation-induced cancer. On the other hand, if ionizing radiation produces unique types of DNA and cytogenetic damage that are not produced by normal endogenous processes, the linear-no-threshold model may be supported.

Recommendations and Costs: Characterize and quantify the spectrum of radiation-induced damage at low doses and its relation to endogenous damage.

Characterizing and quantifying damage after very low radiation doses and placing it in context of endogenous damage is critical to this program and will require a major effort. For this effort to be successful, a significant investment in technology development will be required to expand capabilities for identifying and quantifying such damage beyond those currently available. Methodologies having high sensitivity as well as high signal-to-noise ratio will be critical in this effort. Coupling laboratories involved in characterization and quantification with groups with expertise in technology development will facilitate progress in both areas simultaneously. Once these new methods are in common use, the ten-year goal for determining the relationships between endogenous and radiation-induced damage and its repair should be realized. The initial investment in technology development will have to be in the order of $ (DOE staff), while the research programs to answer the critical questions should cost $ (DOE staff).

C. DNA Repair and Processing

Key Question: Does efficient repair and processing or radiation-induced damage at low doses create a threshold for radiation-induced cancer.

Description: In mammalian cells, the principal DNA repair pathways that are involved in the repair of damage to DNA resulting from ionizing radiation are base excision repair and non-homologous end-rejoining. Base excision repair, which evolved to protect cells against endogenous genotoxic damage, removes all the radiation-induced single DNA lesions, base damages, single strand breaks and sites of base loss which together account for about 70% of the radiation-induced DNA damage (for a recent review see Wallace, 1997). This is a simple DNA repair pathway that is well understood and is highly homologous between bacteria and humans with many of the proteins exhibiting up to 40% identity. This pathway is relatively error free in most instances. Interestingly, a confounder specific to ionizing radiation is that multiple single lesions formed in close proximity to one another are recognized by the enzymes of the base excision repair pathway and their processing results in a DSB.

Double strand breaks in mammalian cells are generally repaired by non-homologous end-rejoining. This type of repair does not require homology between the two recombining molecules and is distinct from homologous recombination. Although less well characterized than excision repair, this pathway is extremely important with respect to radiation effects. This is because radiation-induced

DSBs, while lower in frequency than most other types of radiation-induced damage, are the major threat to genomic integrity because of the problems associated with their repair. Mammalian cells and mice defective in components of this pathway are hypersensitive to the cytotoxic effects of ionizing radiation. Recent studies of cancer prone human populations have served to underscore the potential importance of this pathway. Cells deficient in ATM (the recently cloned gene associated with the disease Ataxia Telangiectasia) are defective in damage checkpoint controls, are sensitive to ionizing radiation, and have increased levels of spontaneous and radiation-induced chromosome aberrations. More recently, it has been demonstrated that another protein complex associated with non-homologous end-rejoining is defective in patients with Nijmegen breakage syndrome (Carney et al., 1998, Varon et al., 1998). Like patients with Ataxia Telangiectasia, individuals with Nijmegen breakage syndrome are cancer prone, radiation sensitive and demonstrate increased levels of chromosomal instability. Interestingly, the BRCA1 and 2 genes, found to be defective in many patients predisposed to breast and ovarian cancer, also appear to be involved in DSB repair pathways.

Because of the nature of the damage the non-homologous end-rejoining pathway is more error prone. Subsequent processing lead to mutagenesis, chromosomal aberrations and perhaps genomic instability. These consequences can also reveal important information relevant to the low dose question. For example, newer chromosome painting techniques have revealed that an unexpectedly large proportion of radiation-induced chromosome aberrations is due to exchanges requiring multiple breaks and involving multiple chromosomes (Savage and Simpson 1994). By earlier techniques such rearrangements appeared to be simple exchange events between chromosomes. These newer results present a clear challenge to current theories including key aspects that underpin the linear no threshold dose response.

A further challenge to current paradigms comes from recent observations on radiation-induced genomic instability. It has now been clearly demonstrated that radiation can induce changes in cells that result in an increase in mutations, chromatid type aberrations, chromosome translocations, and a decrease in cloning efficiency in the progeny of irradiated cells many population doublings after irradiation. The induction of genomic instability is postulated to be the underlying event that leads to the cascade of genetic changes that results in the genetic diversity observed in most solid cancers. What may appear to be unique about radiation-induced instability is its high frequency and makes a strong argument that it is not produced as the result of a change in a single gene or even a group of genes. Since the target for induction of genomic instability is located in the cell nucleus (Kaplan and Morgan 1998) the high frequency suggests the target size is likely to encompass a large fraction of the genome. Genomic instability has been demonstrated in both *in vitro* systems (Kadhim et al. 1992) and *in vivo* using mice (Ponnaiya et al. 1997).

Decision Making Value: The repair of radiation-induced DNA damage is of fundamental importance to all aspects of a cell and/or an organism's responses to radiation exposure. The fidelity of the repair and damage processing systems will significantly affect the dose response curve for cancer induction, particularly at low doses. Ineffective repair or misrepair of radiation damage and subsequent processing of this unrepaired or misrepaired damage can significantly impact genomic integrity resulting in radiation-induced mutagenesis, chromosomal aberrations, chromosomal stability, and cancer. Quite simply, if radiation-induced damage is faithfully repair and processed, a threshold is expected. On the other hand, if repair and subsequent processing can lead to errors at low doses but not at high doses, an expectation of a threshold is not warranted. Additional understanding of the molecular mechanisms involved and in the closely linked damage signaling pathways will provide information relevant to the faithful repair of specific lesions, the molecular responses of cells to specific lesions and the consequences of cellular processing of radiation-induced damage compared to that of endogenous damage. Many of these consequences can be assessed using rapidly developing molecular cytogenetic technology such as combinatorial FISH. Because cytogenetic effects represent the synthesis of damage induction, repair and processing, these new technologies provide the opportunity to directly test certain key predictions of models of radiation effects at low doses.

Observations over the past few years, demonstrating the delayed radiation-induced genomic instability, are not readily understood using current radiobiological principles and paradigms. These observations have obvious important implications in understanding radiation effects in general and mutagenesis and carcinogenesis in particular. Developing a mechanistic linkage between cellular responses to low doses of ionizing radiation, genomic instability, and cancer risk or susceptibility is an important part of this program. The study of radiation-induced genomic instability provides the opportunity to: 1) identify cellular target(s); 2) clarify the role of DNA, cellular and tissue repair and the role of cell killing, proliferation and apoptosis on the induction of instability and the development of mutator phenotype; and 3) provide a framework for understanding risks following exposure to very low doses and dose-rates.

The mechanistic understanding derived relative to the repair processes at the tissue, cellular and molecular level can potentially impact current radiation paradigms and policy. The effective removal of damaged cells from a population through repair can result in biological and energy thresholds which need to be defined as a function of dose. The existence of such thresholds could modify clean-up goals and help address the question of "how clean is clean enough?" Such thresholds may also impact setting radiation exposure levels for appropriate health protection. If there are damaged cells that escape this repair process, even after very low doses of radiation, the linear-no-threshold hypothesis and current radiation paradigms would be supported.

Recommendations and Costs: Determine the biological significance of simple base damage compared to DNA lesions of higher complexity.

Considering the numbers involved, base damage might be considered the most important pathway in repairing DNA damage. However, virtually all of this damage is repaired efficiently and that the vast majority of such damage is similar to damage produce by reactive oxygen species generated through normal cellular processes. The increase in DNA damage from reactive oxygen species (ROS) produced by low doses of radiation is insignificant compared to endogenous damage produced by ROS associated with normal cellular function. Therefore, research relevant to low dose effects should concentrate on damage that is unique to radiation. If specific lesions can be identified that are of particular significance for subsequent biological effects, by knowing their dependence on energy deposition and patterns of deposition, it should be possible to predict the likely form for the dose response for their production with considerable reliability.

Identify the pathways involved in damage signaling and processing of damage at low doses of radiation and the biological consequences.

The processing of initial damage to DNA often leads to misrepair products that are complex in nature, involving more than simple end-rejoining reactions. Examples include chromosomal exchanges that involve several chromosomes as part of the same event. At the molecular level there is evidence that otherwise simple exchanges involve co-deletions where large fragments of DNA are lost. Substantially more information is need on 1) the underlying repair processes; 2) sequence context and chromatin structure that may conceivably affect radiation response and target size for biological endpoints relevant to cancer; 3) how such processing leads to mutagenesis, chromosomal aberrations, and genomic instability.

Determine the mechanisms and significance of radiation-induced genomic instability for cancer risk.

Current evidence suggests that DNA repair and processing of radiation damage can lead to instability in the progeny of irradiated cells and that susceptibility to instability is under genetic control. However, there is virtually no information on the underlying mechanisms and how the processing of damage leads to instability in the progeny of irradiated cells several generations later. Further, while there has been considerable speculation about the role of such instability in radiation-induced cancer, its role in this process remains to be determined.

The technical ability to measure specific lesions and to create such lesions in genetic material, thereby facilitating studies of their significance has been a limiting factor. However many of these problems appear to be solvable and are likely to be overcome in the next few years with appropriate incentive from a program such as this for development. Progress in understanding the mechanisms of DNA repair, the interactions between repair complexes, and the structure of repair enzymes is progressing at an amazing rate. Understanding radiation-critical target interactions and subsequent DNA repair after low dose, low-dose rate radiation exposure will be facilitated by close interactions with the genome project and structural biology programs. As a result, it is highly likely that the questions outlined above can be addressed within a 10-year time frame at a cost of (DOE staff).

D. Biological Responses and Cancer

Key Question: Do the molecular responses induced by low doses of radiation protect cells against radiation damage or radiation-induced cancer?

Description: Damage signaling and response pathways are key elements in damage repair and processing, cell-cell interactions and cellular microenvironment. While there has been a significant amount of research defining radiation-induced genes and radiation-induced stress responses in mammalian cells, the relative contribution of a particular inductive response to the cellular consequences (survival, apoptosis, transformation) has been examined in detail for only a few genes (such as p53 or PKC). At low doses no relationship between radiation-induced responses and other oxidative stresses have been defined. Most radiation-induced gene changes reported to date are transient events, occurring at a specific time following exposure and then decreasing some time thereafter. The kinetics of these responses appear to vary with radiation dose, radiation quality, and cell type but systematic studies on specific radiation-induced responses have not been carried out. It must be determined which proteins are specifically induced in response to low doses of ionizing radiation, how these relate to other oxidative stresses, and importantly, how the induced proteins affect endpoints relevant to radiation-induced cancer. There is already some evidence that molecular, cell and tissue responses can influence radiation effects. This evidence has served to challenge current radiobiological theory that underpins the linear no threshold model. Over the last decade, a number of studies have demonstrated an apparent adaptive response in cells irradiated with small doses of ionizing radiation which manifests itself as an increased resistance to the induction of radiation effects of subsequent higher doses of ionizing radiation (Wolff 1998). Although the initial endpoint was chromosome aberrations, adaptive responses to mutation, cytotoxicity, and neoplastic transformation have been observed *in vitro* and in mice, induction of resistance to life shortening and the induction of thymic lymphoma have been found. It is likely that radiation-induced adaptation involves DNA repair, signal transduction and/or cell cycle kinetics. Most evidence indicate the adaptive response is related to oxidative stress and is associated with excision repair, although restriction enzymes that produce DSBs have also been shown to induce the adaptive response to ionizing radiation. More recently, several laboratories have demonstrated changes in gene expression (Le et al. 1998), increases in sister chromatid exchanges (Nagasawa and Little 1992) and induction of cytogenetic instability (Kadhim et al. 1992) in cells not directly irradiated but rather in proximity to irradiated cells. Biological changes in cells not traversed by radiation have been called "bystander" effects. The mechanisms involved to induce bystander effects are under investigation and will help understand the mode of action of radiation. To date, bystander effects are limited to high LET radiation. It is important for this program to determine if these effects can be induced by exposure to low LET radiation delivered at low total doses or dose-rates.

Decision Making Value: An essential component of the Low Dose Program is the determination of the functional significance of gene/protein inductive processes at low doses, and the impact on the damage response and processing pathways. Such responses could influence not only cellular responses to radiation damage but also cell-cell interactions and the interaction of cells with their microenvironment (Bissell 1998) by modifying radiation-induced damage and impacting the target size for radiation-induced cancer. Because they represent a clear challenge to radiobiological theory upon which the linear no threshold model is based, understanding the underlying mechanisms of the adaptive response and bystander effects is likely to provide important insights into critical pathways which directly impact cancer risks. All of these outcomes have important implications for effects at low doses and would argue against a linear extrapolation from high to low doses. Rather, a significant role for inductive processes would provide a further basis for the consideration of a threshold. To properly evaluate the potential impact of such responses for cancer risks, it is essential to focus on cells and tissues that are targets tissues for radiation carcinogenesis and on low dose and low-dose rate effects.

Recommendations and Costs: We recommend that research be focused on end-points that are important in cancer formation. A major thrust of this research will thus be directed toward defining such endpoints. This will require that all questions be addressed both in appropriate cell and tissue cultures and in carefully selected experimental animal systems. This *in vitro/in vivo* approach will help insure that radiation induced changes in isolated cell systems are relevant to carcinogenesis.

These questions require a multidisciplinary approach to characterize responses at low doses and to link these responses to cancer risk. This will require better knowledge of key endogenous processes, the influence of low radiation doses on these processes, and the impact of the interaction between these processes in endpoints of direct significance for cancer risk. Measurement and characterization of these responses will require the development and application of new instrumentation and analytic technology. The overall cost of this program will be (DOE staff).

Endogenous Factors

Studies that help define the normal endogenous processes and how low doses of radiation modify these processes are recommended. This will be done by using state of the art instrumentation combined with modern cellular and molecular biological tools to link cellular and molecular changes with radiation-induced cancer.

Dose-Response Relationships

Exposure- and dose-response studies should be conducted to determine if the basic mechanisms of radiation action following exposure to low-LET ionizing radiation change as a function of total radiation dose and dose rate. High doses of ionizing radiation induce matrix and tissue disorganization, cell killing, changes in cell proliferation kinetics, induction of a multitude of genes and growth factors, and extensive chromosome and genetic damage. It is important to determine the dose of low-LET radiation that can induce these biological changes. It will also be important to determine if cancer can be induced by doses that are too low to produce such changes.

Link Low Dose Biological Response Endpoint with Cancer

Recent research has detected a number of unique cellular and molecular changes following exposure to radiation levels where biological changes had not been detected in the past. Changes such as genetic instability, adaptive responses, induction of radiation repair genes, changes in expression of stress related genes, and bystander effects all are detected following low doses of ionizing radiation. Studies to understand the mechanisms involved in these biological changes and their role in cancer induction are required. Such studies may impact many current paradigms associated with the carcinogenic mechanisms of radiation action.

Methods and Instrumentation Development

Application and development of improved instrumentation as well as advances in cellular and molecular techniques are needed to detection unique radiation induced biological changes. Research in this area is recommended. After detection of biological changes induced by low doses of radiation it again becomes essential to conduct studies to link the cellular and molecular changes observed to the induction of cancer *in vivo*.

E. Genetic Sensitivity and Risk

Key Question: Is there a distribution of sensitivities of the human population to the carcinogenic effects of radiation? How does this distribution influence risk estimates?

Description: During the last decade there has been a progressive increase in understanding of the genetic contribution to complex diseases including cancer. Molecular studies examining the genetic component of cancer susceptibility have lead to the identification of a number of genes conferring susceptibility and the number of such genes is continuing to increase. It seems likely that there may also be individual differences in susceptibility to radiation-induced cancer, and recent developments have suggested a mechanistic linkage between cellular responses to ionizing radiation, and cancer susceptibility. Though phenomenologically based, dose response kinetics for the induction of certain types of cytogenetic damage have been shown to correlate with cancer susceptibility. There is clear evidence in mice and humans for genetic control of susceptibility to radiation-induced genomic instability that may extend to cancer susceptibility as well. Further, physical associations between gene products involved in the response and repair of DNA damage as heteropolymer complexes and their apparent disruption in heritable diseases associated with instability and cancer have been recently described (Carney et al., 1998; Patel et al., 1998). Functional associations linking cell cycle, apoptosis and DSB repair have also begun to be defined in detail which offer further possible pathways for cancer susceptibility (Woo et al., 1998). However, except for rare genetic conditions affecting single genes of high penetrance, there is insufficient information to identify such potential susceptibility genes, estimate the frequency of polymorphisms in these genes in the population, and assess the risks they impose. Molecular technologies provide powerful new ways to analyze the mammalian genome and address these issues. As this area of research matures, more complex issues of genetic interactions, including gene modifiers and gene-gene interactions will be able to be addressed.

Decision Making Value: Studies focusing on genetic susceptibility to radiation-induced cancer will improve the understanding of low dose risks, but will also create opportunities for new basic knowledge of potential wide-ranging importance. Such insights will provide a better understanding not only of the basic process of cancer development but also a clearer appreciation of the interactions of both endogenous and exogenous risk factors. This is an area for which little information is available. The extent to which these studies are likely to impact current understanding of low dose risks depends on the frequency of susceptibility genes in the general population, and their ability to significantly influence low dose risks. If there are enough people who are unable to properly respond to and process radiation damage, then any model of radiation risk to the general population suggesting a threshold would appear to be untenable. It is also reasonable to assume that the distribution of such sensitive sub-populations could have a major effect on the response function of any low dose response model. The largest impact would result from changes in the slope of the initial response function. Such information will also create opportunities to specifically identify susceptible individuals as well as provide insight into approaches to modify such susceptibility.

Recommendations and Costs: Determine the frequencies of polymorphisms in genes involved in repair and processing of radiation damage.

The efforts to identify polymorphisms in genes involved in radiation susceptibility and determine their frequencies in the population would strongly complement new initiatives at NIH and NIEHS focusing on genetic susceptibility. The interests of the low dose program effort in DOE will be unique in that the focus is on radiation damage response pathways which means that genes in the less well-characterized end-rejoining pathways will be a major interest. Because many such genes are only now being characterized and many of the known genes are quite large, these pathways have not been given high priority by other programs interested in genetic susceptibility. This aspect of the program will be closely tied and rely heavily on ongoing results and technical developments in the genome project. Like the genome and structural biology efforts this low dose effort should be coordinated with activities in the other agencies to prevent duplicative effort and facilitate rapid progress. Frequencies of identified polymorphisms will require genetic epidemiological analyses based on principle derived from population genetics. The total cost of this part of the effort is approximately (DOE staff)

Determine the biological significance of polymorphisms with respect to cancer risk.

Identification of polymorphisms is only the first (and perhaps the easiest) step in the program to examine genetic susceptibility. The determination of biological significance is the ultimate goal and the more difficult task. The genome project and structural biology program each play an important role in providing guidance on which polymorphisms are most likely to influence gene function. Population genetics and computation biology approaches that are also integral to the genome project will be required to estimate the potential impact on population risk estimates. Genetic epidemiology approaches also would be of value to relate specific polymorphisms and combinations of polymorphisms with cancer risk. Another potential resource for identifying significant polymorphisms is the inbred mouse strains which have well-characterized broad differences in susceptibility to radiation-induced cancer. Direct assessment of the biological significance of candidate "susceptibility genes" can be undertaken using animal models such as knockout and knockin mice. Other animal models which emerge as part of the genome project (e.g., drosophila models) should also be considered since they may allow experiments to be performed more rapidly than with murine systems. The total cost of this aspect of the low dose program is estimated to be (DOE staff)

F. Risk Communication

Key Question: How can the information derived from the low-dose initiative be best communicated to scientists, policy makers, stakeholders and to the public?

Description: The low-dose research program is expected to produce important new scientific data that may modify existing paradigms associated with radiation induced health risk. Since a new risk paradigm has the potential to impact existing standards and methods used in management of low-dose radiation exposures, communication between the scientific community, policy makers and the public about the potential risk associated with radiation induced disease is vital to the outcome of the low-dose program.

Communicating the results of this research program will be a difficult challenge, since simply presenting scientific findings will not automatically impact risk policy or increase public understanding and acceptance. Influencing policy decisions will require a major change in philosophy by stakeholders and policy makers. For this shift to occur, they must have a good understanding not only of the underlying science and its implications, but also confidence that the public will accept any changes. It is well established that the public is extremely sensitive and adverse to the issue of radiation exposure (Slovic 1996). A high percentage of the public believes any exposure to radiation is likely to lead to cancer. The linear-no-threshold hypothesis supports this public conception and fosters the view that no expense is too great to reduce the risks of radiation exposure or environmental contamination. Therefore, it is not surprising to find that radiation controls tend to be associated with extremely high costs per year of life saved (Slovic 1987; Tengs, et al. 1993).

Decision Making Value: The information derived from the DOE Low Dose program must provide input for decision making but also for public acceptance of risk policy. For the decision making process, it is essential that there is adequate communication between the scientists involved in generation of the primary data and between scientists and those involved in risk policy and risk communication. Through this program the policy makers should have timely understandable scientific information which enables them to make good decisions and communicate these decisions to the public. This communication must not be one way, opportunities for public input to the decision making process is essential.

Effective communication of the results from this program should also foster better public understanding of low dose radiation risk. Communication between the scientific community, the policy makers, and the public about the potential risks associated with radiation-induced disease is vital to the outcome of the DOE low dose program. Good communication will solve problems regarding low dose radiation, facilitate the best policy choices, and develop public understanding and support.

Recommendations and Costs: Develop a public communication program based on principles developed through risk communication research.

To communicate with the public about low dose management, requires a well-developed plan based on strong basic social science research. The goal of research effort would be to understand the likely public responses to scientific findings from the Low Dose Program's research and responses to the plans for modifying existing standards based upon these scientific findings. The following topics should be included in determining public responses to issues regarding low dose exposures: 1) public perceptions of risk; 2) the perceived importance of the activities and conditions that produce low dose radiation; 3) trust and confidence in risk managers, regulators, and decision makers; 4) the role of the media in characterizing different positions on risk controversies; 5) the role of advocacy groups; 6) the manner by which risk is characterized and assessed; and 7) the procedures by which decisions are made. The cost for this research effort is (DOE staff).

Develop a public education program based on principles derived from risk communication science: To present the developments from this program in a form that is useful and easily understood by the public, the education program would develop web pages, written resources for public schools, and coordinate multimedia coverage of research results and public meetings. The public meetings would provide opportunities for the public to meet with scientists and regulators involved in policy making, facilitating public input into the decision making process. The cost of this public education program will be (DOE staff).

Develop a communication network between scientists, policy makers, and DOE administrators:

The low dose program is highly dependent on effective interactions and collaborations among scientists with varied scientific and technical expertise. For this to be successful, a communication network must be developed which will ensure adequate communication. This network would encompass not only the scientists directly involved in the conduct of studies as a part of this program but also those involved in the genome and structural biology programs. An expanded network including scientists, policy makers from a variety of agencies, and DOE administrators is required to keep the program focused on critical issues and facilitate the understanding and translation of result into public policy. The costs for development and maintenance of this active network will be (DOE staff).

IV. Programmatic Structure, Monitoring Progress, Direction and Focus

The research themes identified in this program are to address the questions that provide information that meet the DOE needs for improved understanding of radiation risks at low doses. The questions that can impact current radiation paradigms are as follows: 1) Is radiation induced DNA damage different than endogenous oxidative damage? 2) Can normal physiological processes repair the DNA damage and prevent cancer following low levels of radiation exposure? 3) Can organ, tissue and molecular responses to low doses of radiation prevent or reduce cancer risk? 4) How do genetic differences impact radiation induced cancer risk?

To achieve these goals, the proposed research program should, of necessity, be more focused than the NIH R01 model. For example, a systematic analysis of a single link in the chain of arguments that leads to the assumption of dose linearity or a threshold would be a highly appropriate task within the current context. This is not to say that scientific merit should not be reviewed as rigorously. Clearly, rigorous peer review is essential. However, research proposals for this program should explicitly reflect the framework set out here.

A critical component of this research program will be its ability to continue addressing both the original and changing goals over time. As with any basic research program, especially one that is focused on a specific challenge, program needs will change as results are accumulated from this and other research programs. In addition, as interactions between scientists in this program and at regulatory agencies develop and mature (see next section), program goals will be further clarified and new goals will be identified.

Scientific progress, at the individual project level, will be monitored and evaluated through the use of ad hoc peer review panels and occasional ad hoc mail reviews, under the guidance of BER program managers. The results of these peer reviews will be evaluated and used by BER management to make decisions on the funding of individual projects across the program. BER program managers will also evaluate progress among groups of related projects and across the entire program.

A standing BERAC low dose effects subcommittee through interactions with BER program managers will evaluate overall program progress, direction, and focus. This subcommittee should be comprised of scientists with expertise representing the entire range of program goals. In addition, the subcommittee should include individuals with expertise in or responsibility for developing human exposure regulatory policy. This committee should meet with BER program managers to assess the portfolio of grants within this program, and to recommend changes in emphasis and balance. In addition, the committee would also be charged with defining programmatic areas that require increased / decreased emphasis based results of this program and advances in other fields of relevance to this program (i.e., scientific issues), and new issues related to risk management (regulatory issues). Such recommendations may be reflected in the issue of RFP's for this program if warranted and sufficient research funds are available. The subcommittee will also participate in the low dose effects program contractor workshops (see next section) to be held approximately every 18 months. A major review of the program should be scheduled at the end of five years.

In this subcommittee, findings will be reported, in writing, to BERAC for further discussion, comment, and approval. Final reports will be distributed to scientists in the low dose effects program, BER management, the Director of the Office of Energy Research, program staff at other agencies, and interested congressional staff. The reports will be publicly available in hard copy or on the BERAC web site at http://www.er.doe.gov/production/ober/herac.html. The reports will also serve as the basis for future program solicitations and for the development of special research workshops or symposia to help clarify or debate specific program topics or to inform scientists and the public on program progress and future directions.

V. Program Contractor Workshops – Involving Customers and Stakeholders

The ultimate success of this program will depend on the quality of the science produced and the usefulness of that science to the people and organizations charged with using research results to develop public health protection policy. To facilitate the kinds of interactions that will improve both the science and, hopefully, the usefulness of the results for developing public health protection policy, program contractor workshops will be held approximately every 18 months.

All principal investigators funded in the low dose effects research program will be expected to participate in these workshops. BER program staff, program staff at other agencies, BERAC low dose effects subcommittee members, and scientists from other DOE-funded programs whose research has useful links to the low dose program will also be invited to participate. Finally, staff from regulatory agencies, e.g., the environmental Protection Agency, the Nuclear Regulatory Commission, etc., will be invited to actively participate in these workshops. As well as this oversight role, it is recommended that the subcommittee act in conjunction with BER program managers to act as the Scientific Program Committee for this annual meeting. The Committee's principle charge here would be to organize a highly focused symposium on a single theme or issue, in which the current state of the art is reviewed, and potential future directions assessed.

VI. Future Directions and Research

The goal of these workshops will be several fold. They will serve as forums for exchanging research results, for communicating and discussion ongoing or changing program directions, and as opportunities to evaluate the overall balance of the low dose research portfolio. They will serve as opportunities for scientists in the program to broaden their scientific perspectives and their understanding of how their research project fits into and contributes to the low dose effects program. Finally, and perhaps most importantly, it will provide opportunities for people involved in developing public health protection policy to discuss, with research scientists, the types of new or clarifying information that they need or can use from research.

These workshops will change the way that research scientists think about and conduct their research. They will open new lines of communication among program scientists and between those scientists and the users of the research results being developed in the program. Research results will still be published in peer-reviewed scientific journals; however, the dialogues, the exchanges of information, and the new understandings of the relationship between basic research the development of health protection policy that occur at these program workshop may be among the most significant outcomes of this research program.

References

Bissell, M.J., 1998, Glandular structure and gene expression. Lessons from the mammary gland. *Annuals of the New York Academy of Science*, 842, 1-6.

Carney, J.P., Maser, R.S., Olivares, H., Davis, E.M., Le Beau, M., Yates, J.R., III, Hays, L., Morgan, W.F., Petrini, J.H.J. (1998) *Cell* 93, 447-486.

Kadhim, M.A., D.A. Macdonald, D.T. Goodhead, S.A. Lorimore, S.J. Marsden, E.G. Wright, (1992), Transmission of chromosomal instability after plutonium alpha-particle irradiation, *Nature* 355, 738-740.

Kaplan, M.I., Morgan, W.F. (1998) The nucleus is the target for radiation-induced chromosomal instability. *Radiation Research* 150, 382-390.

Le, X.C., Xing, J.Z., Lee, J., Leadon, S.A., and Weinfeld, M., 1998, Inducible repair of thymine glycol detected by an ultrasensitive assay for DNA damage. *Science* 280, 1066-1069.

Nagasawa, H, J.B. Little (1992) Induction of sister chromatid exchanges by extremely low doses of alpha particles. *Cancer Res.* 52 6394-6396.

Patel, K., Yu, V.P.C.C., Lee, H., Corcoran, A., Thistlethwaite, F.C., Evans, M.J., Colledge, W.H., Friedman, L.S., Ponder, B.A.J., Venkitaraman, A.R., (1998) xxx *Molecular Cell* 1, 347-357.

Ponnaiya, B., M.N. Cornforth, R.L. Ullrich (1997) Radiation-induced chromosomal instability inBALB/c and C57 BL/6 mice; The difference is as clear as black and white. *Radiat. Res.* 147, 121-125.

Savage, J.R.K., and Simpson, P.J.,1994, FISH "painting" patterns resulting from complex exchanges, *Mutation Research*, 312, 51-60.

Slovic, P. (1987) Perception of Risk, *Science* 236, 280-295.

Slovic, P. (1996) Perception of risk from radiation, *Radiation Protection Dosimetry* 68, 165-180.

Tengs, T. Adams, M. Pliskin, J., Safran, D. Siegel, J. Weinsteir, M. Grahm, J. (1993) Five-hundred life-saving interventions and their cost effectiveness. Center for Risk Analysis, Harvard School of Public Health.

Varon, R., Vissinga, C., Platzer, M., Cerosalitti, K.M., Chrzanowska, K.H., Saar, K., Beckmann, G., Seemanova, E., Cooper, P.R., Nowak, N.J., Stumm, M., Weemaes, C.M.R., Gatti, R.A., Wilson, R.K., Digwee, M., Rosenthal, A., Sperling, K., Concannon, P., Reis, A. (1998) xxx *Cell* 93, 467-476.

Wallace, S.S. (1997) Oxidative Damage to DNA and its Repair. In: Oxidative Stress and the Molecular biology of Antioxidant Defenses (Sachdalios, J., ed.) pp. 49-90. Cold Spring Harbor Laboratory Press, Cold Spring Harbor, NY

Wolff, S., 1998, The Adaptive Response in Radiobiology: Evolving Insights and Implications. *Environmental Health Perspective*, 106, 277-283.

Woo, R.A., McLure, K.G., Lees-Miller, S.P., Rancourt, D.E., Lee, P.W.K. (1998) xxx *Nature* 394, 700-704.

APPENDIX B

First Call for Proposals

The calls for proposals follow the development of the program and the direction that was provided by DOE to maintain the program's focus. Each call resulted in a large number of applications. After review for relevance to the program, review groups ranked each proposal according to its scientific strengths. After this ranking, DOE applied programmatic needs, and the proposals were ranked and funded. Shown here is the text of first call, which was made in 1998.

I. Summary

The Office of Biological and Environmental Research (BER) of the Office of Energy Research (ER),[1] U.S. Department of Energy (DOE), hereby announces its interest in receiving applications for research for support of the Cellular Biology Research Program. This Program is a coordinated multidisciplinary research effort to develop creative, innovative approaches that will provide a better scientific basis for understanding exposures and risks to humans associated with low level exposures to radiation and chemicals. Using modern molecular tools, this research will provide information that will be used to decrease the uncertainty of risk at low levels, help determine the shape of the dose-response relationships after low level exposure, and achieve acceptable levels of human health protection at the lowest possible cost.

II. Supplementary Information

Current standards for occupational and residential exposures to radiation and chemicals are based on linear, no-threshold models of risk that drive regulatory decisions and estimations of cancer risk. Linear, no-threshold models assume that risk is always proportional to dose, that there is no risk only when there is no dose, and that even a single molecule or radiation induced ionization can cause cancer or disease. However, the scientific basis for these assumptions is limited and uncertain at very low doses and dose rates.

Much scientific evidence suggests that the risks from exposure to low doses or low dose-rates of radiation and chemicals may be better described by a non-linear, dose-response relationship. This evidence includes long term human and animal studies and research at the cellular and molecular level on the DNA repair capabilities of cells and tissues, bystander' effects associated with low dose exposures, the effects of exposure-induced gene expression, the effects of a cell's micro environment on its response to low dose exposures, and studies of the multi-step nature of cancer development. A more definitive understanding of the biological responses induced by low dose, low dose-rate exposures is needed to clarify the role played by these and other cell responses and capabilities in determining risk.

This research program will focus on understanding the mechanisms of molecular and cellular responses to low dose, low dose-rate exposures to radiation and chemicals to improve the scientific underpinning for estimating risks from these exposures. The program will include research to identify and characterize: (1) the genes and gene products that determine and affect these cellular responses induced at low dose and dose-rates; (2) the role played by these genes and gene products in determining individual differences in susceptibility to low dose, low dose-rate exposures; and (3) methods to synthesize or model molecular level information on genes and gene products into overall health risk. The program will also communicate research results to regulators and legislators. The goal of this research program is the development of scientifically defensible tools and approaches for determining risk that are widely used, accepted, and understood.

Research is encouraged in a number of areas including, but not limited to:
- The effects of and reactions to reactive oxygen species at low doses and/or dose rates.
- The role of gene induction, DNA repair, apoptosis, and the immune system in mediating responses to low dose and/or low dose-rate exposures.
- The nature and significance of bystander' effects in determining cell and tissue responses to low dose and/or low dose-rate exposures.
- The role of cell and tissue microenvironments in determining cell and tissue responses to low dose and/or low dose-rate exposures.

Development of computational techniques, e.g., algorithms and advanced mathematical approaches, for use in determining risk, that model new information from cellular and molecular studies together with available data from epidemiologic and animal studies.

A Lead Scientist will be selected from among all investigators who are successful in receiving research funds in this program. This research program will be directed by a program manager from BER, who will be responsible for providing support and overall direction, including determining the relevance of the goals and objectives of the program. The Lead Scientist will provide scientific leadership to the community of the researchers in the research program. Applicants interested in being considered as a Lead Scientist for the low dose research program should indicate their interest in their research application. In addition to the information requested in the Application Guide, applicants should supplement their applications by describing their qualifications to serve as a Lead Scientist for this program. The supplemental information should be provided as a separate appendix not attached to the main application. Interested applicants should demonstrate their understanding of the needs for and the uses of the types of scientific information likely to be developed in this research program. They should demonstrate their understanding of previous epidemiologic and experimental studies involving low dose, low dose-rate exposures to radiation or chemicals. Finally, interested applicants should demonstrate their knowledgeability of research opportunities and capabilities at National Laboratories, universities, and industry in the area of molecular and cellular responses to low dose, low dose-rate exposures.

Notes

1. Name was changed to the Office of Science in 1998.

APPENDIX C

Dose Range Charts

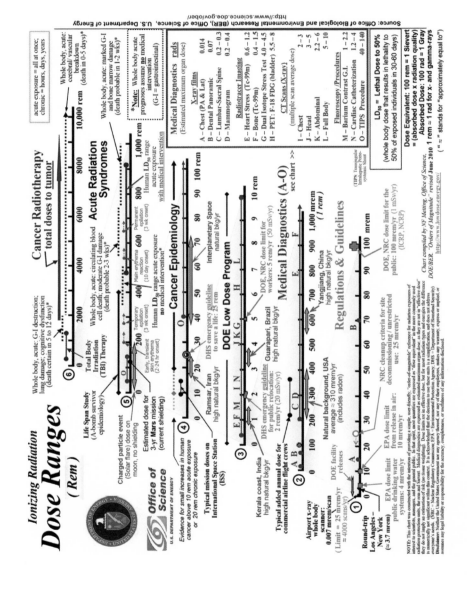

APPENDIX C

Dose Range Charts

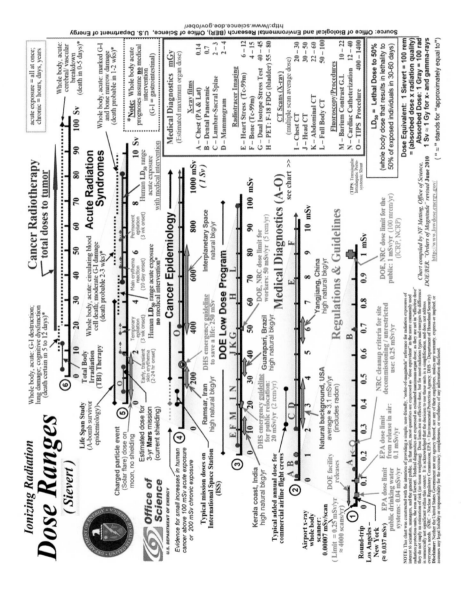

REFERENCES

Ahmed KM, Li JJ 2007. ATM-NF-kappa B connection as a target for tumor radiosensitization. Current Cancer Drug Targets 7(4): 335-42.

Ahmed KM, Li JJ 2008. NF-kappa B-mediated adaptive resistance to ionizing radiation. Free Radical Biology and Medicine 44(1): 1-13.

Ahmed KM, Dong S, Fan M, Li JJ 2006. Nuclear factor-kappa B p65 inhibits mitogen-activated protein kinase signaling pathway in radioresistant breast cancer cells. Molecular Cancer Research 4(12): 945-55.

Ahram M, Adkins JN, Auberry DL, Wunschel DS, Springer DL 2005a. A proteomic approach to characterize protein shedding. Proteomics 5(1): 123-31.

Ahram M, Strittmatter EF, Monroe ME, Adkins JN, Hunter JC, Miller JH, Springer DL 2005b. Identification of shed proteins from Chinese hamster ovary cells: Application of statistical confidence using human and mouse protein databases. Proteomics 5(7): 1815-26.

Akerman GS, Rosenzweig BA, Domon OE, Tsai CA, Bishop ME, McGarrity LJ, MacGregor JT, Sistare FD, Chen JJ, Morris SM 2005. Alterations in gene expression profiles and the DNA-damage response in ionizing radiation-exposed TK6 cells. Environmental and Molecular Mutagenesis 45(2-3): 188-205.

Al Rashid ST, Dellaire G, Cuddihy A, Jalali F, Vaid M, Coackley C, Folkard M, Xu Y, Chen BPC, Chen DJ and others 2005. Evidence for the direct binding of phosphorylated p53 to sites of DNA breaks *in vivo*. Cancer Research 65(23): 10810-21.

Albertini RJ 1999. Biomarker responses in human populations: towards a worldwide map. Mutation Research-Fundamental and Molecular Mechanisms of Mutagenesis 428(1-2): 217-26.

Alsbeih G, Torres M, Al-Harbi N, Al-Buhairi M 2007. Evidence that individual variations in TP53 and CDKN1A protein responsiveness are related to inherent radiation sensitivity. Radiation Research 167(1): 58-65.

Alvarez S, Drane P, Meiller A, Bras M, Deguin-Chambon V, Bouvard V, May E 2006. A comprehensive study of p53 transcriptional activity in thymus and spleen of gamma irradiated mouse: High sensitivity of genes involved in the two main apoptotic pathways. International Journal of Radiation Biology 82(11): 761-70.

Amundson SA 2008. Functional genomics in radiation biology: a gateway to cellular systems-level studies. Radiation and Environmental Biophysics 47(1): 25-31.

Amundson SA, Fornace AJ 2001. Gene expression profiles for monitoring radiation exposure. Radiation Protection Dosimetry 97(1): 11-16.

Amundson SA, Fornace AJ 2003. Monitoring human radiation exposure by gene expression profiling: Possibilities and pitfalls. Health Physics 85(1): 36-42.

Amundson SA, Do KT, Fornace AJ 1999. Induction of stress genes by low doses of gamma rays. Radiation Research 152(3): 225-31.

Amundson SA, Patterson A, Do KT, Fornace AJ 2002. A nucleotide excision repair master-switch: p53 regulated coordinate induction of global genomic repair genes. Cancer Biology & Therapy 1(2): 145-49.

Amundson SA, Bittner M, Meltzer P, Trent J, Fornace AJ, Jr. 2001a. Biological indicators for the identification of ionizing radiation exposure in humans. Expert Review of Molecular Diagnostics 1(2): 211-9.

Amundson SA, Bittner M, Meltzer P, Trent J, Fornace AJ 2001b. Physiological function as regulation of large transcriptional programs: the cellular response to genotoxic stress. Comparative Biochemistry and Physiology B-Biochemistry & Molecular Biology 129(4): 703-10.

Amundson SA, Bittner M, Meltzer P, Trent J, Fornace AJ 2001c. Induction of gene expression as a monitor of exposure to ionizing radiation. Radiation Research 156(5): 657-61.

Amundson SA, Do KT, Shahab S, Bittner M, Meltzer P, Trent J, Fornace AJ 2000. Identification of potential mRNA biomarkers in peripheral blood lymphocytes for human exposure to ionizing radiation. Radiation Research 154(3): 342-46.

Amundson SA, Lee RA, Koch-Paiz CA, Bittner ML, Meltzer P, Trent JM, Fornace AJ 2003. Differential responses of stress genes to low dose-rate gamma irradiation. Molecular Cancer Research 1(6): 445-52.

Amundson SA, Grace MB, McLeland CB, Epperly MW, Yeager A, Zhan QM, Greenberger JS, Fornace AJ 2004. Human *in vivo* radiation-induced biomarkers: Gene expression changes in radiotherapy patients. Cancer Research 64(18): 6368-71.

Amundson SA, Do KT, Vinikoor L, Koch-Paiz CA, Bittner ML, Trent JM, Meltzer P, Fornace AJ 2005. Stress-specific signatures: expression profiling of p53 wild-type and -null human cells. Oncogene 24(28): 4572-79.

Amundson SA, Do KT, Vinikoor LC, Lee RA, Koch-Paiz CA, Ahn J, Reimers M, Chen Y, Scudiero DA, Weinstein JN and others 2008. Integrating global gene expression and radiation survival parameters across the 60 cell lines of the National Cancer Institute Anticancer Drug Screen. Cancer Research 68(2): 415-24.

An M-Y, Kim T-H 2002. Frequencies of micronuclei in peripheral lymphocytes in Korean populations after chronic low-dose radiation exposure. Journal of Veterinary Science 3(3): 213-8.

Andarawewa KL, Paupert J, Pal A, Barcellos-Hoff MH 2007a. New rationales for using TGF beta inhibitors in radiotherapy. International Journal of Radiation Biology 83(11-12): 803-11.

Andarawewa KL, Erickson AC, Chou WS, Costes SV, Gascard P, Mott JD, Bissell MJ, Barcellos-Hoff MH 2007b. Ionizing radiation predisposes nonmalignant human mammary epithelial cells to undergo transforming growth factor beta-induced epithelial to mesenchymal transition. Cancer Research 67(18): 8662-70.

Anderson RM, Marsden SJ, Wright EG, Kadhim MA, Goodhead DT, Griffin CS 2000. Complex chromosome aberrations in peripheral blood lymphocytes as a potential biomarker of exposure to high-LEt alpha-particles. International Journal of Radiation Biology 76(1): 31-42.

Anderson RM, Marsden SJ, Paice SJ, Bristow AE, Kadhim MA, Griffin CS, Goodhead DT 2003. Transmissible and nontransmissible complex chromosome aberrations characterized by three-color and mFISH define a biomarker of exposure to high-LEt alpha particles (vol 159, pg 40, 2003). Radiation Research 159(3): 437-437.

Andringa KK, Coleman NC, Aykin-Burns N, Hitchler NJ, Walsh SA, Domann FE, Spitz DR 2006. Inhibition of glutamate cysteine ligase activity sensitizes human breast cancer cells to the toxicity of 2-deoxy-D-glucose. Cancer Research 66(3): 1605-10.

Araki S, Israel S, Leskov KS, Criswell TL, Beman M, Klokov DY, Sampalth L, Reinicke KE, Cataldo E, Mayo LD and others 2005. Clusterin proteins: stress-inducible polypeptides with proposed functions in multiple organ dysfunction. BJR supplement / BIR 27: 106-13.

Aravindan N, Madhusoodhanan R, Natarajan M, Herman TS 2008. Alteration of apoptotic signaling molecules as a function of time after radiation in human neuroblastoma cells. Molecular and Cellular Biochemistry 310(1-2): 167-79.

Armitage P, Doll R 1957. A 2-Stage Theory of carcinogenesis in relation to the age distribution of human cancer. British Journal of Cancer 11(2): 161-69.

Asaithamby A, Uematsu N, Chatterjee A, Story MD, Burma S, Chen DJ 2008. Repair of HZE-Particle-induced DNA double-strand breaks in normal human fibroblasts. Radiation Research 169(4): 437-46.

Asch BB, Barcellos-Hoff MH 2001. Epigenetics and breast cancer. Journal of Mammary Gland Biology and Neoplasia 6(2): 151-52.

Averbeck D 2009. Does scientific evidence support a change from the LET model for low-dose radiation risk extrapolation? Health Physics 97(5): 493-504.

Azzam EI, Little JB 2004. The radiation-induced bystander effect: evidence and significance. Human & Experimental Toxicology 23(2): 61-65.

Azzam EI, Raaphorst GP, Mitchel REJ 1994. Radiation-induced adaptive response for protection against micronucleus formation and neoplastic transformation in C3H 10T1/2 mouse embryo cells. Radiation Research 138(1): S28-S31.

Azzam EI, de Toledo SM, Little JB 2001. Direct evidence for the participation of gap junction-mediated intercellular communication in the transmission of damage signals from alpha-particle irradiated to nonirradiated cells. Proceedings of the National Academy of Sciences of the United States of America 98(2): 473-78.

Azzam EI, de Toledo SM, Little JB 2003a. Expression of CONNEXIN43 is highly sensitive to ionizing radiation and other environmental stresses. Cancer Research 63(21): 7128-35.

Azzam EI, de Toledo SM, Little JB 2003b. Oxidative metabolism, gap junctions and the ionizing radiation-induced bystander effect. Oncogene 22(45): 7050-57.

Azzam EI, de Toledo SM, Raaphorst GP, Mitchel REJ 1996. Low-dose ionizing radiation decreases the frequency of neoplastic transformation to a level below the spontaneous rate in C3H 10T1/2 cells. Radiation Research 146(4): 369-73.

Azzam EI, de Toledo SM, Gooding T, Little JB 1998. Intercellular communication is involved in the bystander regulation of gene expression in human cells exposed to very low fluences of alpha particles. Radiation Research 150(5): 497-504.

Azzam EI, de Toledo SM, Spitz DR, Little JB 2002. Oxidative metabolism modulates signal transduction and micronucleus formation in bystander cells from alpha-particle-irradiated normal human fibroblast cultures. Cancer Research 62(19): 5436-42.

Bailey SM, Goodwin EH 2004. DNA and telomeres: beginnings and endings. Cytogenetic and Genome Research 104(1-4): 109-15.

Bailey SM, Murnane JP 2006. Telomeres, chromosome instability and cancer. Nucleic Acids Research 34(8): 2408-17.

Bailey SM, Cornforth MN 2007. Telomeres and DNA double-strand breaks: ever the twain shall meet? Cellular and Molecular Life Sciences 64(22): 2956-64.

Bailey SM, Meyner J, Goodwin EH 2001a. Telomeres, DNA Repair Proteins and Making Ends Meet. In: Nickoff JA, Hoekstra MF ed. DNA Damage and Repair. Totowa, Humana Press Inc. Pp. 359-75.

Bailey SM, Goodwin EH, Cornforth MN 2004a. Strand-specific fluorescence in situ hybridization: the CO-FISH family. Cytogenetic and Genome Research 107(1-2): 14-17.

Bailey SM, Brenneman MA, Goodwin EH 2004b. Frequent recombination in telomeric DNA may extend the proliferative life of telomerase-negative cells. Nucleic Acids Research 32(12): 3743-51.

Bailey SM, Williams ES, Ullrich RL 2007. The role of telomere dysfunction in driving genomic instability. International Congress Series 1299: 146-49.

Bailey SM, Cornforth MN, Ullrich RL, Goodwin EH 2004c. Dysfunctional mammalian telomeres join with DNA double-strand breaks. DNA Repair 3(4): 349-57.

Bailey SM, Cornforth MN, Kurimasa A, Chen DJ, Goodwin EH 2001b. Strand-specific postreplicative processing of mammalian telomeres. Science 293(5539): 2462-65.

Bailey SM, Brenneman MA, Halbrook J, Nickoloff JA, Ullrich RL, Goodwin EH 2004d. The kinase activity of DNA-PK is required to protect mammalian telomeres. DNA Repair 3(3): 225-33.

Bailey SM, Meyne J, Chen DJ, Kurimasa A, Li GC, Lehnert BE, Goodwin EH 1999. DNA double-strand break repair proteins are required to cap the ends of mammalian chromosomes. Proceedings of the National Academy of Sciences of the United States of America 96(26): 14899-904.

Baker RJ, Chesser RK 2000. The Chernobyl nuclear disaster and subsequent creation of a wildlife preserve. Environmental Toxicology and Chemistry 19(5): 1231-32.

Baker RJ, VanDenBussche RA, Wright AJ, Wiggins LE, Hamilton MJ, Reat EP, Smith MH, Lomakin MD, Chesser RK 1996a. High levels of genetic change in rodents of Chernobyl (Retracted article. See vol 390, pg 100, 1997). Nature 380(6576): 707-8.

Baker RJ, Hamilton MJ, VandenBussche RA, Wiggins LE, Sugg DW, Smith MH, Lomakin MD, Gaschak SP, Bundova EG, Rudenskaya GA and others 1996b. Small mammals from the most radioactive sites near the Chornobyl Nuclear Power Plant. Journal of Mammalogy 77(1): 155-70.

Balajee AS, Ponnaiya B, Baskar R, Geard CR 2004. Induction of replication protein a in bystander cells. Radiation Research 162(6): 677-86.

Ballarini F, Biaggi M, Ottolenghi A, Sapora O 2002. Cellular communication and bystander effects: a critical review for modelling low-dose radiation action. Mutation Research-Fundamental and Molecular Mechanisms of Mutagenesis 501(1-2): 1-12.

Banath JP, Olive PL 2003. Expression of phosphorylated histone H2AX as a surrogate of cell killing by drugs that create DNA double-strand breaks. Cancer Research 63(15): 4347-50.

Banaz-Yasar F, Lennartz K, Greif KD, Giesen U, Iliakis G, Winterhager E, Gellhaus A 2006. Evaluation of the role of gap junctional communication in mediating radiation-induced bystander effects in Jeg3 trophoblast cells. Radiation Research 166(4): 681-82.

Bao SP, Harwood PW, Wood BH, Chrisler WB, Groch KM, Brooks AL 1997. Comparative clastogenic sensitivity of respiratory tract cells to gamma rays. Radiation Research 148(1): 90-97.

Barcellos-Hoff MH 2005a. How tissues respond to damage at the cellular level: orchestration by transforming growth factor-{beta} (TGF-{beta}). BJR supplement / BIR 27: 123-27.

Barcellos-Hoff MH 2005b. Integrative radiation carcinogenesis: interactions between cell and tissue responses to DNA damage. Seminars in Cancer Biology 15(2): 138-48.

Barcellos-Hoff MH 2008. Cancer as an emergent phenomenon in systems radiation biology. Radiation and Environmental Biophysics 47(1): 33-38.

Barcellos-Hoff MH, Ravani SA 2000. Irradiated mammary gland stroma promotes the expression of tumorigenic potential by unirradiated epithelial cells. Cancer Research 60(5): 1254-60.

Barcellos-Hoff MH, Brooks AL 2001. Extracellular signaling through the microenvironment: A hypothesis relating carcinogenesis, bystander effects, and genomic instability. Radiation Research 156(5): 618-27.

Barcellos-Hoff MH, Costes SV 2006. A systems biology approach to multicellular and multi-generational radiation responses. Mutation Research-Fundamental and Molecular Mechanisms of Mutagenesis 597(1-2): 32-38.

Barcellos-Hoff MH, Nguyen DH 2009. Radiation carcinogenesis in context: How do irradiated tissues become tumors? Health Phys. Health Physics 97(5): 446-57.

Bartek J, Lukas J 2006. Balancing life-or-death decisions. Science 314(5797): 261-62.

Barzilai A, Rotman G, Shiloh Y 2002. ATM deficiency and oxidative stress: a new dimension of defective response to DNA damage. DNA Repair 1(1): 3-25.

Baskar R, Balajee AS, Geard CR 2007. Effects of low and high LET radiations on bystander human lung fibroblast cell survival. International Journal of Radiation Biology 83(8): 551-59.

Baskar R, Balajee AS, Geard CR, Hande MP 2008. Isoform-specific activation of protein kinase c in irradiated human fibroblasts and their bystander cells. International Journal of Biochemistry & Cell Biology 40(1): 125-34.

Bassi L, Carloni M, Meschini R, Fonti E, Palitti F 2003. X-irradiated human lymphocytes with unstable aberrations and their preferential elimination by p53/survivin-dependent apoptosis. International Journal of Radiation Biology 79(12): 943-54.

Bauchinger M, Braselmann H, Savage JRK, Natarajan AT, Terzoudi GI, Pantelias GE, Darroudi F, Figgitt M, Griffin CS, Knehr S and others 2001. Collaborative exercise on the use of FISH chromosome painting for retrospective biodosimetry of Mayak nuclear-industrial personnel. International Journal of Radiation Biology 77(3): 259-67.

Bauer G 2007a. Low Dose Radiation and Intercellular Induction of Apoptosis: Potential Implications for the Control of Oncogenesis. International Journal of Radiation Biology 83(11-12): 873-88.

Bauer G 2007b. Low Dose Radiation and Intercellular Induction of Apoptosis: Potential Implications for the Control of Oncogenesis. International Journal of Radiation Biology. Pp. 873-88.

Baulch JE, Raabe OG, Wiley LM, Overstreet JW 2002. Germline drift in chimeric male mice possessing an F-2 component with a paternal F-0 radiation history. Mutagenesis 17(1): 9-13.

Bedford JS, Liber HL 2003. Applications of RNA interference for studies in DNA damage processing, genome stability, mutagenesis, and cancer. Seminars in Cancer Biology 13(4): 301-8.

Belyakov OV, Malcolmson AM, Folkard M, Prise KM, Michael BD 2001. Direct evidence for a bystander effect of ionizing radiation in primary human fibroblasts. British Journal of Cancer 84(5): 674-79.

Belyakov OV, Folkard M, Mothersill C, Prise KM, Michael BD 2002. Bystander-induced apoptosis and premature differentiation in primary urothelial explants after charged particle microbeam irradiation. Radiation Protection Dosimetry 99(1-4): 249-51.

Belyakov OV, Folkard M, Mothersill C, Prise KM, Michael BD 2003. A proliferation-dependent bystander effect in primary porcine and human urothelial explants in response to targeted irradiation. British Journal of Cancer 88(5): 767-74.

Belyakov OV, Folkard M, Mothersill C, Prise KM, Michael BD 2006. Bystander-induced differentiation: A major response to targeted irradiation of a urothelial explant model. Mutation Research-Fundamental and Molecular Mechanisms of Mutagenesis 597(1-2): 43-49.

Belyakov OV, Mitchell SA, Parikh D, Randers-Pehrson G, Marino SA, Amundson SA, Geard CR, Brenner DJ 2005. Biological effects in unirradiated human tissue induced by radiation damage up to 1 mm away. Proceedings of the National Academy of Sciences of the United States of America 102(40): 14203-8.

Bender MA, Awa AA, Brooks AL, Evans HJ, Groer PG, Littlefield LG, Pereira C, Preston RJ, Wachholz BW 1988. Current Status of Cyogenetic Procedures to Detect and Quantify Previous Exposures to Radiation Mutation Research 196(2): 103-59.

Benjamin SA, Brooks AL 1977. Spontaneous Lesions in Chinese Hamsters. Vet. Pathology. 14: 449-62.

Bennett PV, Cutter NC, Sutherland BM 2007. Split-dose exposures versus dual ion exposure in human cell neoplastic transformation. Radiation and Environmental Biophysics 46(2): 119-23.

Bennett PV, Cintron NS, Gros L, Laval J, Sutherland BM 2004. Are endogenous clustered DNA damages induced in human cells? Free Radical Biology and Medicine 37(4): 488-499.

Bennett PV, Cuomo NL, Paul S, Tafrov ST, Sutherland BM 2005. Endogenous DNA damage clusters in human skin, 3-D model, and cultured skin cells. Free Radical Biology and Medicine 39(6): 832-39.

Berglund SR, Rocke DM, Dai J, Schwietert CW, Santana A, Stern RL, Lehmann J, Siantar CLH, Goldberg Z 2008. Transient genome-wide transcriptional response to low-dose ionizing radiation *in vivo* in humans. International Journal of Radiation Oncology Biology Physics 70(1): 229-34.

Bernal AJ, Dolinoy DC, Haung D, Skaar DA, Weinhouse C, Jirtle RL, 2013. Adaptive radiation-induced epigenetic alterations mitigated by antioxidants, FASEB 67: 665-71.

Bissell MJ, Barcelloshoff MH 1987. The Influence of Extracellular-Matrix on Gene-Expression - is Structure the Message. Journal of Cell Science: 327-43.

Bissell MJ, Aggeler J 1987. Dynamic reciprocity: how do extracellular matrix and hormones direct gene expression? Progress in Clinical and Biological Research 249: 251-62.

Bittner M, Chen YD, Amundson SA, Khan J, Fornace AJ, Dougherty ER, Meltzer PS, Trent JM 2000. Obtaining and evaluating gene expression profiles with cDNA microarrays. Suhai S ed.: 5-25.

Bladen CL, Navarre S, Dynan WS, Kozlowski DJ 2007a. Expression of the Ku70 subunit (XRCC6) and protection from low dose ionizing radiation during zebrafish embryogenesis. Neuroscience Letters 422(2): 97-102.

Bladen CL, Flowers MA, Miyake K, Podolsky RH, Barrett JT, Kozlowski DJ, Dynan WS 2007b. Quantification of ionizing radiation-induced cell death in situ in a vertebrate embryo. Radiation Research 168(2): 149-57.

Blaisdell JO, Wallace SS 2001. Abortive base-excision repair of radiation-induced clustered DNA lesions in Escherichia coli. Proceedings of the National Academy of Sciences of the United States of America 98(13): 7426-30.

Blakely EA, Thompson AC, Schwarz RI, Chang P, Bjornstad K, Rosen C, Wisnewski C, Mocherla D, Parvin B 2006. Radiation-induced bystander effects with a 12.5 keV X-ray microbeam. Radiation Research 166(4): 684-84.

Blakely WF, Brooks AL, Lofts RS, van der Schans GP, Voisin P 2002. Overview of low-level radiation exposure assessment: biodosimetry. Military Medicine 167(2): 20-4.

Bogen KT, Enns L, Hall LC, Keating GA, Weinfeld M, Murphy G, Wu RW, Panteleakos FN 2001. Gel microdrop flow cytometry assay for low-dose studies of chemical and radiation cytotoxicity. Toxicology 160(1-3): 5-10.

Bonassi S, Lando C, Ceppi M, Landi S, Rossi AM, Barale R 2004. No association between increased levels of high-frequency sister chromatid exchange cells (HFCs) and the risk of cancer in healthy individuals. Environmental and Molecular Mutagenesis 43(2): 134-36.

Bond V, Sondhaus C, Couch L, Brooks AL 2005. The Requirement for Energy Imparted in Radiation Protection Practice. International Journal of Low Radiation. Pp. 452-62.

Bond VP, Benary V, Sondhaus CA, Feinendegen LE 1995a. The meaning of linear dose-response relations, made evident by use of absorbed dose to the cell. Health Physics 68(6): 786-92.

Bond VP, Varma M, Feinendegen LE, Wuu CS, Zaider M 1995b. Application of the HSEF to Assessing Radiation Rises in the Practice of Radiation Protection. Health Physics 68(5): 627-31.

Boothman DA, Meyers M, Odegaard E, Wang MZ 1996. Altered G(1) checkpoint control determines adaptive survival responses to ionizing radiation. Mutation Research-Fundamental and Molecular Mechanisms of Mutagenesis 358(2): 143-53.

Boothman DA, Odegaard E, Yang CR, Hosley K, Mendonca MS 1998. Molecular analyses of adaptive survival responses (ASRs): role of ASRs in radiotherapy. Human & Experimental Toxicology 17(8): 448-53.

Boreham DR, Dolling JA, Maves SR, Siwarungsun N, Mitchel REJ 2000. Dose-rate effects for apoptosis and micronucleus formation in gamma-irradiated human lymphocytes. Radiation Research 153(5): 579-86.

Bouffler SD, Bridges BA, Cooper DN, Dubrova Y, McMillan TJ, Thacker J, Wright EG, Waters R 2006. Assessing radiation-associated mutational risk to the germline: Repetitive DNA sequences as mutational targets and biomarkers. Radiation Research 165(3): 249-68.

Bowler DA, Moore SR, Macdonald DA, Smyth SH, Clapham P, Kadhim MA 2006. Bystander-mediated genomic instability after high LET radiation in murine primary haemopoietic stem cells. Mutation Research-Fundamental and Molecular Mechanisms of Mutagenesis 597(1-2): 50-61.

Braby L, Brenner D, Cherubini R, Cremer T, Durante M, Geard C, Kobayashi K, Miller J, Prise K, Taucher-Scholz G 2006. Session 1. In: Extended Abstracts: Proceedings of the 7th International Workshop: Microbeam Probes of Cellular Radiation Response. Radiation Research. Pp. 652-53.

Braby, LA. 2000. Targeting and spatial aspects of microbeams. Radiation Research, ed. Moriarty M, Mothersill C, Seymore C, Edington M, Ward JF, Fry RJM. Pp. 178–81.

Braby LA, Ford JR 2000. Microbeam irradiation patterns to simulate dose. Radiation Research 153(2): 225-25.

Braby LA, Ford JR 2004. Energy deposition patterns and the bystander effect. Radiation Research 161(1): 113-15.

Brenner DJ 2004. Comments on "Chromosome intrachanges and interchanges detected by multicolor banding in lymphocytes: Searching for clastogen signatures in the human genome" by Johannes et al. (Radiat. Res. 161, 540-548, 2004). Radiation Research 162(5): 600-600.

Brenner DJ 2009. Extrapolating Radiation-Induced Cancer Risks from Low Doses to Very Low Doses. Health Physics 97(5): 505-9.

Brenner DJ, Elliston CD 2001. The potential impact of bystander effects on radiation risks in a Mars mission. Radiation Research 156(5): 612-17.

Brenner DJ, Raabe OG 2001. Is the linear-no-threshold hypothesis appropriate for use in radiation protection? Radiation Protection Dosimetry 97(3): 279-85.

Brenner DJ, Sachs RK 2002a. Do low dose-rate bystander effects influence domestic radon risks? International Journal of Radiation Biology 78(7): 593-604.

Brenner DJ, Sachs RK 2002b. Bystander effects may dominate domestic radon risks - But current risk estimates are probably okay. Radiation Research 158(6): 790-91.

Brenner DJ, Sachs RK 2003. Domestic radon risks may be dominated by bystander effects - But the risks are unlikely to be greater than we thought. Health Physics 85(1): 103-8.

Brenner DJ, Elliston CD 2004. Estimated radiation risks potentially associated with full-body CT screening. Radiology 232(3): 735-38.

Brenner DJ, Mossman KL 2005. Do radiation doses below 1 cGy increase cancer risks? Radiation Research 163(6): 692-93.

Brenner DJ, Sachs RK 2006. Estimating radiation-induced cancer risks at very low doses: rationale for using a linear no-threshold approach. Radiation and Environmental Biophysics 44(4): 253-56.

Brenner DJ, Hall EJ 2007. Computed tomography--an increasing source of radiation exposure. The New England Journal of Medicine 357(22): 2277-84.

Brenner DJ, Elliston CD, Hall EJ, Berdon WE 2001a. Estimated risks of radiation-induced fatal cancer from pediatric CT. American Journal of Roentgenology 176(2): 289-96.

Brenner DJ, Okladnikova N, Hande P, Burak L, Geard CR, Azizova T 2001b. Biomarkers specific to densely-ionising (high let) radiations. Radiation Protection Dosimetry 97(1): 69-73.

Brenner DJ, Sawant SG, Hande MP, Miller RC, Elliston CD, Fu Z, Randers-Pehrson G, Marino SA 2002. Routine screening mammography: how important is the radiation-risk side of the benefit-risk equation? International Journal of Radiation Biology 78(12): 1065-67.

Brenner DJ, Doll R, Goodhead DT, Hall EJ, Land CE, Little JB, Lubin JH, Preston DL, Preston RJ, Puskin JS and others 2003. Cancer risks attributable to low doses of ionizing radiation: Assessing what we really know. Proceedings of the National Academy of Sciences of the United States of America 100(24): 13761-66.

Brooks AL 1975. Chromosome damage in liver cells from low dose rate alpha, beta and gamma irradiation: derivation of RBE. Science 190; 1090-92.

Brooks AL 1999. Biomarkers of exposure, sensitivity and disease. International Journal of Radiation Biology 75(12): 1481-503.

Brooks AL 2000a. The Potential Impact of Cellular and Molecular Biology on Radiation Risk Assessment. Technology. Pp. 251-64.

Brooks AL 2000b. Basic biological science in support of radiation risk policy. 112-18.

Brooks AL 2000c. Science in Support of Radiation Risk Policy. Risk Excellence Notes. Pp. 8.

Brooks AL 2001. Biomarkers of exposure and dose: State of the art. Radiation Protection Dosimetry 97(1): 39-46.

Brooks AL 2003. Developing a scientific basis for radiation risk estimates: Goal of the DOE Low Dose Research Program. Health Physics 85(1): 85-93.

Brooks AL 2004. Evidence for 'bystander effects' *in vivo*. Human & Experimental Toxicology 23(2): 67-70.

Brooks AL 2005. From Cell to Organism: The Need for Multiparametric Assessment of Exposure and Biological Effects. The British Journal of Radiology. Pp. 139-45.

Brooks AL 2005. Paradigm Shifts in Radiation Biology: Their Impact on Intervention for Radiation Induced Disease. Radiat. Res. 164, 454-61.

Brooks AL 2006. Where Have all the Thresholds Gone? Wolbarst A, Zamenhof R, Hendee W ed. Journal of Applied Clincal Medical Physics.

Brooks AL 2013. Commentary: What is the health risk of 740 BqL^{-1} of tritium? A perspective. Health Physic 104 (1): 108-14.

Brooks AL, Hui EE, Bond V 2000. Energy Barriers for Radiation-induced Cellular Effects. Yamada T, Mothersill C, Michael B, Potten B ed. Amsterdam, Elsevier Science.

Brooks AL, Lengermann FW 1967. Comparison of radiation-induced chromatid aberrations in the testes and bone marrow of the Chinese hamster. Radiation Research 32: 587-95.

Brooks AL, Lengermann FW 1968. Determining the repair rate for radiation-induced chromatid deletions during the first 24 hours after irradiation. Radiation Research 38: 181-92.

Brooks AL, McClellan RO 1968. Cytogenic effects of strontium-90 on the bone marrow of the Chinese hamster. Nature 219: 761-63.

Brooks AL, Hui TE, Couch L 2007. Very Large Amounts of Radiation are Required to Produce Cancer. Dose-Response. 263-74.

Brooks AL, Carsten AL, Mead DK, Retherford JC 1976. The effects of continuous intake of tritiated water (HTO) on the liver chromosomes of mice. Radiation. Research 68: 480–89.

Brooks AL, Eberlein PE, Couch LA, Boecker BB 2009. The Role of Dose-Rate on Risk From Internally-Deposited Radionuclides and the Potential Need to Seperate Dose-Rate Effectiveness Factor (DREF) from the Dose and Dose-Rate Effectiveness Factor (DDREF). Health Physics 97(5): 458-69.

Brooks AL, Khan MA, Duncan A, Buschbom RL, Jostes RF, Cross FT 1994. Effectiveness of Radon Relative to Acute CO-60 Gamma-Rays for Induction of Micronuclei *In-Vitro* and *In-Vivo*. International Journal of Radiation Biology 66(6): 801-8.

Brooks AL, Bao S, Rithidech K, Chrisler WB, Couch LA, Braby LA 2001. Induction and repair of HZE induced cytogenetic damage. Physica Medica 17: 183-84.

Brooks AL, Coleman MA, Douple EB, Hall EJ, Mitchel REJ, Ullrich R, Wyrobek AJ 2006a. Biological Effects of Low Doses of Ionizing Radiation Advances in Medical Physics. Pp. 255-86.

Brooks AL, Coleman MA, Douple EB, Hall EJ, Mitchel REJ, Ullrich R, Wyrobek AJ 2006b. Bilogical Effects of Low Doses of Ionizing Radiation In: Wolbarst A, Zamenhof R, Hendee W ed. Advances in Medical Physics. Madison, Medical Physics Publishing.

Brooks AL, Bao S, Harwood PW, Wood BH, Chrisler WB, Khan MA, Gies RA, Cross FT 1997. Induction of micronuclei in respiratory tract following radon inhalation. International Journal of Radiation Biology 72(5): 485-95.

Bubici C, Papa S, Dean K, Franzoso G 2006. Mutual cross-talk between reactive oxygen species and nuclear factor-kappa B: molecular basis and biological significance. Oncogene 25(51): 6731-48.

Burdak-Rothkamm S, Rothkamm K, Prise KM 2008. ATM acts downstream of ATR in the DNA damage response signaling of bystander cells. Cancer Research 68(17): 7059-65.

Burdak-Rothkamm S, Short SC, Folkard M, Rothkamm K, Prise KM 2007. ATR-dependent radiation-induced gamma H2AX foci in bystander primary human astrocytes and glioma cells. Oncogene 26(7): 993-1002.

Burma S, Chen DJ 2004. Role of DNA-PK in the cellular response to DNA double-strand breaks. DNA Repair 3(8-9): 909-18.

Burma S, Chen BP, Murphy M, Kurimasa A, Chen DJ 2001. ATM phosphorylates histone H2AX in response to DNA double-strand breaks. Journal of Biological Chemistry 276(45): 42462-67.

Calabrese EJ 2004. Hormesis: from marginalization to mainstream - A case for hormesis as the default dose-response model in risk assessment. Toxicology and Applied Pharmacology 197(2): 125-36.

Calabrese EJ 2007. Threshold dose-response model-RIP: 1911 to 2006. Bioessays 29(7): 686-88.

Calabrese EJ, Baldwin LA 2003. Hormesis: The dose-response revolution. Annual Review of Pharmacology and Toxicology 43: 175-97.

Calabrese EJ, Bachmann KA, Bailer AJ, Bolger PM, Borak J, Cai L, Cedergreen N, Cherian MG, Chiueh CC, Clarkson TW and others 2007. Biological stress response terminology: Integrating the concepts of adaptive response and preconditioning stress within a hormetic dose-response framework. Toxicology and Applied Pharmacology 222(1): 122-28.

Cappelli E, Degan P, Thompson LH, Frosina G 2000. Efficient repair of 8-oxo-7,8-dihydrodeoxyguanosine in human and hamster xeroderma pigmentosum D cells. Biochemistry 39(34): 10408-12.

Chadwick KH, Leenhouts HP 2005. Radiation risk is linear with dose at low doses. British Journal of Radiology 78(925): 8-10.

Chalmers A, Johnston P, Woodcock M, Joiner M, Marples B 2004. PARP-1, PARP-2, and the cellular response to low doses of ionizing radiation. International Journal of Radiation Oncology Biology Physics 58(2): 410-19.

Chandna S, Dwarakanath BS, Khaitan D, Mathew TL, Jain V 2002. Low-dose radiation hypersensitivity in human tumor cell lines: Effects of cell-cell contact and nutritional deprivation. Radiation Research 157(5): 516-25.

Chao C, Wu Z, Mazur SJ, Borges H, Rossi M, Lin T, Wang JYJ, Anderson CW, Appella E, Xu Y 2006. Acetylation of mouse p53 at lysine 317 negatively regulates p53 apoptotic activities after DNA damage. Molecular and Cellular Biology 26(18): 6859-69.

Chaudhry MA 2006. Bystander effect: Biological endpoints and microarray analysis. Mutation Research-Fundamental and Molecular Mechanisms of Mutagenesis 597(1-2): 98-112.

Chen S-T, Pan T-L, Juan H-F, Chen T-Y, Lin Y-S, Huang C-M 2008. Breast tumor microenvironment: Proteomics highlights the treatments targeting secretome. Journal of Proteome Research 7(4): 1379-87.

Chen W, Luan Y, MC S, ST C, HT H, Soong K, Yeh Y, Chou S, Mong J, Wu C and others 2004a. Is Chronic Radiation an Effective Prophylaxis Against Cancer?, Journal of American Physicians and Surgeons. Pp. 6-10.

Chen WL, Luan YC, Shieh MC, Chen ST, Kung HT, Soong KL, Yeh YC, Chou TS, Mong SH, Wu JT and others 2007. Effects of cobalt-60 exposure on health of Taiwan residents suggest new approach needed in radiation protection. Dose-response : a publication of International Hormesis Society 5(1): 63-75.

Chen WNU, Woodbury RL, Kathmann LE, Opresko LK, Zangar RC, Wiley HS, Thrall BD 2004b. Induced autocrine signaling through the epidermal growth factor receptor contributes to the response of mammary epithelial cells to tumor necrosis factor alpha. Journal of Biological Chemistry 279(18): 18488-96.

Chesser RK, Rodgers BE, Wickliffe JK, Gaschak S, Chizhevsky I, Phillips CJ, Baker RJ 2001. Accumulation of (137)cesium and (90)strontium from abiotic and biotic sources in rodents at Chornobyl, Ukraine. Environmental Toxicology and Chemistry 20(9): 1927-35.

Chesser RK, Sugg DW, Lomakin MD, Van den Bussche RA, DeWoody JA, Jagoe CH, Dallas CE, Whicker FW, Smith MH, Gaschak SP and others 2000. Concentrations and dose rate estimates of (134,137)cesium and (90)strontium in small mammals at Chornobyl, Ukraine. Environmental Toxicology and Chemistry 19(2): 305-12.

Chin MH, Qian W-J, Wang H, Petyuk VA, Bloom JS, Sforza DM, Lacan G, Liu D, Khan AH, Cantor RM and others 2008. Mitochondrial dysfunction, oxidative stress, and apoptosis revealed by proteomic and transcriptomic analyses of the striata in two mouse models of Parkinson's disease. Journal of Proteome Research 7(2): 666-77.

Choi EK, Terai K, Ji I-M, Kook YH, Park KH, Oh ET, Griffin RJ, Lim BU, Kim J-S, Lee DS and others 2007. Upregulation of NAD(P)H : quinone oxidoreductase by radiation potentiates the effect of bioreductive beta-lapachone on cancer cells. Neoplasia 9(8): 634-42.

Choi JA, Park MT, Kang CM, Um HD, Bae S, Lee KH, Kim TH, Kim JH, Cho CK, Lee YS and others 2004. Opposite effects of Ha-Ras and Ki-Ras on radiation-induced apoptosis via differential activation of PI3K/Akt and Rac/p38 mitogen-activated protein kinase signaling pathways. Oncogene 23(1): 9-20.

Chu K, Teele N, Dewey MW, Albright N, Dewey WC 2004. Computerized video time lapse study of cell cycle delay and arrest, mitotic catastrophe, apoptosis and clonogenic survival in irradiated 14-3-3 sigma and CDKN1A (p21) knockout cell lines. Radiation Research 162(3): 270-86.

Chu K, Leonhardt EA, Trinh M, Prieur-Carrillo G, Lindqvist J, Albright N, Ling CC, Dewey WC 2002. Computerized video time-lapse (CVTL) analysis of cell death kinetics in human bladder carcinoma cells (EJ30) X-irradiated in different phases of the cell cycle. Radiation Research 158(6): 667-77.

Chuang EY, Chen X, Tsai MH, Yan HL, Li CY, Mitchell JB, Nagasawa H, Wilson PF, Peng YL, Fitzek MM and others 2006. Abnormal gene expression profiles in unaffected parents of patients with hereditary-type retinoblastoma. Cancer Research 66(7): 3428-33.

Cleaver JE, Thompson LH, Richardson AS, States JC 1999a. A summary of mutations in the UV-sensitive disorders: Xeroderma pigmentosum, Cockayne syndrome, and trichothiodystrophy. Human Mutation 14(1): 9-22.

Cleaver JE, Afzal V, Feeney L, McDowell M, Sadinski W, Volpe JPG, Busch DB, Coleman DM, Ziffer DW, Yu YJ and others 1999b. Increased ultraviolet sensitivity and chromosomal instability related to p53 function in the xeroderma pigmentosum variant. Cancer Research 59(5): 1102-8.

Cohen 2008. The Linear No-Threshould Theory of Radiation Carcinogenesis Should be Rejected. American Journal of Physicians and Surgeons. Pp. 70-76.

Coleman CN, Blakely WF, Fike JR, MacVittie TJ, Metting NF, Mitchell JB, Moulder JE, Preston RJ, Seed TM, Stone HB and others 2003a. Molecular and cellular biology of moderate-dose (1-10 Gy) radiation and potential mechanisms of radiation protection: report of a workshop at Bethesda, Maryland, December 17-18, 2001. Radiation Research 159(6): 812-34.

Coleman MA, Wyrobek AJ 2006. Differential Transcript Modulation of Gene after Low vs. High Doses of Ionizing Radiation. Advances in Medical Physics. Pp. 273-76.

Coleman MA, Eisen JA, Mohrenweiser HW 2000. Cloning and characterization of HARP/SMARCAL1: A prokaryotic HepA-related SNF2 helicase protein from human and mouse. Genomics 65(3): 274-82.

Coleman MA, Miller KA, Beernink PT, Yoshikawa DM, Albala JS 2003b. Identification of chromatin-related protein interactions using protein microarrays. Proteomics 3(11): 2101-7.

Coleman MA, Yin E, Peterson LE, Nelson D, Sorensen K, Tucker JD, Wyrobek AJ 2005. Low-dose irradiation alters the transcript profiles of human lymphoblastoid cells including genes associated with cytogenetic radioadaptive response. Radiation Research 164(4): 369-82.

Coleman MC, Asbury CR, Daniels D, Du J, Aykin-Bums N, Smith BJ, Li L, Spitz DR, Cullen JJ 2008. 2-Deoxy-D-glucose causes cytotoxicity, oxidative stress, and radiosensitization in pancreatic cancer. Free Radical Biology and Medicine 44(3): 322-31.

Connolly L, Lasarev M, Jordan R, Schwartz JL, Turker MS 2006. Atm haploinsufficiency does not affect ionizing radiation mutagenesis in solid mouse tissues. Radiation Research 166(1): 39-46.

Cornforth MN 2001. Analyzing radiation-induced complex chromosome rearrangements by combinatorial painting. Radiation Research 155(5): 643-59.

Cornforth MN 2006. Perspectives on the formation of radiation-induced exchange aberrations. DNA Repair 5(9-10): 1182-91.

Cornforth MN, Bailey SM, Goodwin EH 2002a. Dose responses for chromosome aberrations produced in noncycling primary human fibroblasts by alpha particles, and by gamma rays delivered at sublimiting low dose rates. Radiation Research 158(1): 43-53.

Cornforth MN, Greulich-Bode KM, Loucas BD, Arsuaga J, Vazquez M, Sachs RK, Bruckner M, Molls M, Hahnfeldt P, Hlatky L and others 2002b. Chromosomes are predominantly located randomly with respect to each other in interphase human cells. Journal of Cell Biology 159(2): 237-44.

Costes SV, Boissiere A, Ravani S, Romano R, Parvin B, Barcellos-Hoff MH 2006. Imaging features that discriminate between foci induced by high- and low-LET radiation in human fibroblasts. Radiation Research 165(5): 505-15.

Cram LS 2002. Flow Cytometry an Overview. Methods in Cell Science. Pp. 1-9.

Cram LS, Bell CS, Fawcett JJ 2002. Chromosome sorting and genomics. Methods in cell science: an official journal of the Society for *In Vitro* Biology 24(1-3): 1-3.

Criswell T, Klokov D, Beman M, Lavik J, Boothman DA 2003a. Repression of IR-inducible clusterin expression by the p53 tumor suppressor protein. Cancer Biology & Therapy 2(4): 372-80.

Criswell T, Leskov K, Miyamoto S, Luo GB, Boothman DA 2003b. Transcription factors activated in mammalian cells after clinically relevant doses of ionizing radiation. Oncogene 22(37): 5813-27.

Cucinotta EA 2008. Systems Biology and Understanding Radiation Sensitivity of Astronauts Radiation Research. Pp. 574-75.

Cucinotta FA, Durante M 2006. Cancer risk from exposure to galactic cosmic rays: implications for space exploration by human beings. Lancet Oncology 7(5): 431-35.

Curtis SB, Luebeck EG, Hazelton WD, Moolgavkar SH 2002. A new perspective of carcinogenesis from protracted high-LET radiation arises from the two-stage clonal expansion model. Advances in space research : the official journal of the Committee on Space Research (COSPAR) 30(4): 937-44.

Curtis SB, Hazelton WD, Luebeck EG, Moolgavkar SH 2004. From mechanisms to risk estimation - bridging the chasm. In: Cucinotta FRG ed. Space Life Sciences: Radiation Risk Assessment and Radiation Measurement in Low Earth Orbit. Pp. 1404-9.

Dainiak N, Berger P, Albanese J 2007. Relevance and feasibility of multi-parameter assessment for management of mass casualties from a radiological event. Experimental Hematology 35(4): 17-23.

Darakhshan F, Badie C, Moody J, Coster M, Finnon R, Finnon P, Edwards AA, Szluinska M, Skidmore CJ, Yoshida K and others 2006. Evidence for complex multigenic inheritance of radiation AML susceptibility in mice revealed using a surrogate phenotypic assay. Carcinogenesis 27(2): 311-18.

Dauer LT, Brooks AL, Hoel DG, Morgan WF, Stram D, Tran P 2010. Review and evaluation of updated research on the health effects associated with low-dose ionising radiation. Radiation Protection Dosimetry 140(2): 103-36.

Davis TW, Meyers M, Wilson-Van Patten C, Sharda N, C.R. Y, Kinsella TJ, Boothman D 2001. Transcriptional Responses to Damage Created by Ionizing Radiation: Moleculars Sensors. In: Nickoff JA, Hoekstra MF ed. DNA Damage and Repair, DNA Repair in Higher Eukaryotes. Totowa, Humana Press, Inc. Pp. 223-62.

Day TK, Hooker AM, Zeng G, Sykes PJ 2007a. Low dose X-radiation adaptive response in spleen and prostate of Atm knockout heterozygous mice. International Journal of Radiation Biology 83(8): 523-34.

Day TK, Zeng G, Hooker AM, Bhat M, Scott BR, Turner DR, Sykes PJ 2007b. Adaptive response for chromosomal inversions in pKZ1 mouse prostate induced by low doses of X radiation delivered after a high dose. Radiation Research 167(6): 682-92.

de Toledo SM, Azzam EI, Dahlberg WK, Gooding TB, Little JB 2000. ATM complexes with HDM2 and promotes its rapid phosphorylation in a p53-independent manner in normal and tumor human cells exposed to ionizing radiation. Oncogene 19(54): 6185-93.

de Toledo SM, Asaad N, Venkatachalam P, Li L, Howell RW, Spitz DR, Azzam EI 2006. Adaptive responses to low-dose/low-dose-rate gamma rays in normal human fibroblasts: The role of growth architecture and oxidative metabolism. Radiation Research 166(6): 849-57.

Deng X, Yin X, Allan R, Lu DD, Maurer CW, Haimovitz-Friedman A, Fuks Z, Shaham S, Kolesnick R 2008. Ceramide biogenesis is required for radiation-induced apoptosis in the germ line of C-elegans. Science 322(5898): 110-15.

Denko NC, Fornace AJ 2005. Editorial comment: Mutation research special issue on stress responses. Mutation Research-Fundamental and Molecular Mechanisms of Mutagenesis 569(1-2): 1-2.

Deppert W 2007. Introduction - Mutant p53: from guardian to fallen angel? Oncogene 26(15): 2142-44.

di Masi A, Antoccia A, Dimauro I, Argentino-Storino A, Mosiello A, Mango R, Novelli G, Tanzarella C 2006. Gene expression and apoptosis induction in p53-heterozygous irradiated mice. Mutation Research-Fundamental and Molecular Mechanisms of Mutagenesis 594(1-2): 49-62.

Ding LH, Shingyoji M, Chen FQ, Hwang JJ, Burma S, Lee C, Cheng JF, Chen DJ 2005. Gene expression profiles of normal human fibroblasts after exposure to ionizing radiation: A comparative study of low and high doses. Radiation Research 164(1): 17-26.

Dolinoy DC, Jirtle RL 2008. Environmental epigenomics in human health and disease. Environmental and Molecular Mutagenesis 49(1): 4-8.

Dolinoy DC, Weidman JR, Jirtle RL 2007a. Epigenetic gene regulation: Linking early developmental environment to adult disease. Reproductive Toxicology 23(3): 297-307.

Dolinoy DC, Das R, Weidman JR, Jirtle RL 2007b. Metastable epialleles, imprinting, and the fetal origins of adult diseases. Pediatric Research 61(5): 30R-37R.

Dong L-F, Swettenham E, Eliasson J, Wang X-F, Gold M, Medunic Y, Stantic M, Low P, Prochazka L, Witting PK and others 2007. Vitamin E analogues inhibit angiogenesis by selective induction of apoptosis in proliferating endothelial cells: The role of oxidative stress. Cancer Research 67(24): 11906-13.

Donoho G, Brenneman MA, Cui TX, Donoviel D, Vogel H, Goodwin EH, Chen DJ, Hasty P 2003. Deletion of Brca2 exon 27 causes hypersensitivity to DNA crosslinks, chromosomal instability, and reduced life span in mice. Genes Chromosomes & Cancer 36(4): 317-31.

Du G, Fischer BE, Voss K-O, Becker G, Taucher-Scholz G, Kraft G, Thiel G 2008. The absence of an early calcium response to heavy-ion radiation in mammalian cells. Radiation Research 170(3): 316-26.

Dugan LC, Bedford JS 2003. Are chromosomal instabilities induced by exposure of cultured normal human cells to low- or high-LET radiation? Radiation Research 159(3): 301-11.

Durant ST, Paffett KS, Shrivastav M, Timmins GS, Morgan WF, Nickoloff JA 2006. UV radiation induces delayed hyperrecombination associated with hypermutation in human cells. Molecular and Cellular Biology 26(16): 6047-55.

Durante M, Kronenberg A 2005. Ground-based research with heavy ions for space radiation protection. In: Hei TKMJ ed. Space Life Sciences: Ground-Based Iron-Ion Biology and Physics, Including Shielding. Pp. 180-84.

Durante M, George K, Cucinotta EA 2006. Chromosomes lacking telomeres are present in the progeny of human lymphocytes exposed to heavy ions. Radiation Research 165(1): 51-58.

Dutta J, Fan Y, Gupta N, Fan G, Gelinas C 2006. Current insights into the regulation of programmed cell death by NF-kappa B. Oncogene 25(51): 6800-16.

Dziegielewski J, Baulch JE, Goetz W, Coleman MC, Spitz DR, Murley JS, Grdina DJ, Morgan WF 2008. WR-1065, the active metabolite of amifostine, mitigates radiation-induced delayed genomic instability. Free Radical Biology and Medicine 45(12): 1674-81.

Easton DF, Pooley KA, Dunning AM, Pharoah PDP, Thompson D, Ballinger DG, Struewing JP, Morrison J, Field H, Luben R and others 2007. Genome-wide association study identifies novel breast cancer susceptibility loci. Nature 447(7148): 1087-U7.

Elmore E, Lao XY, Kapadia R, Redpath JL 2006. The effect of dose rate on radiation-induced neoplastic transformation *in vitro* by low doses of Low-LET radiation. Radiation Research 166(6): 832-38.

Elmore E, Lao XY, Kapadia R, Giedzinski E, Limoli C, Redpath JL 2008. Low doses of very low-dose-rate low-LET radiation suppress radiation-induced neoplastic transformation *in vitro* and induce an adaptive response. Radiation Research 169(3): 311-18.

Evans HH, Horng MF, Ricanati M, Diaz-Insua M, Jordan R, Schwartz JL 2003. Induction of genomic instability in TK6 human lymphoblasts exposed to (CS)-C-137 gamma radiation: Comparison to the induction by exposure to accelerated Fe-56 particles. Radiation Research 159(6): 737-47.

Ewan KB, Henshall-Powell RL, Ravani SA, Pajares MJ, Arteaga C, Warters R, Akhurst RJ, Barcellos-Hoff MH 2002. Transforming growth factor-beta 1 mediates cellular response to DNA damage in situ. Cancer Research 62(20): 5627-31.

Ewart-Toland A, Chan JM, Yuan JW, Balmain A, Ma J 2004. A gain of function TGFB1 polymorphism may be associated with late stage prostate cancer. Cancer Epidemiology Biomarkers & Prevention 13(5): 759-64.

Fakir H, Sachs RK, Stenerlow B, Hofmann W 2006. Clusters of DNA double-strand breaks induced by different doses of nitrogen ions for various LETs: Experimental measurements and theoretical analyses. Radiation Research 166(6): 917-27.

Fan M, Ahmed KM, Coleman MC, Spitz DR, Li JJ 2007. Nuclear factor-kappa B and manganese superoxide dismutase mediate adaptive radioresistance in low-dose irradiated mouse skin epithelial cells. Cancer Research 67(7): 3220-28.

Feinendegen LE 2005. Evidence for beneficial low level radiation effects and radiation hormesis. The British Journal of Radiology 78(925): 3-7.

Feinendegen LE, Paretzke HG, Neumann RD 2007. Damage propagation in complex biological systems following exposure to low doses of ionising radiation. ATOMS FOR PEACE 1(4): 336-54.

Feinendegen LE, Polycove M, Neumann RD 2007. Whole-body responses to low-level radiation exposure: New concepts in mammalian radiobiology. Exper Hematol 35: 37-46.

Feinendegen LE, Brooks AL, Morgan WF 2011. Biological consequences and health risks of low-level exposure to ionizing radiation: Commentary on the workshop. Health Phys. Health Physics 100(3): 247-59.

Fenech M, Holland N, Chang WP, Zeiger E, Bonassi S 1999. The HUman MicroNucleus Project - An international collaborative study on the use of the micronucleus technique for measuring DNA damage in humans. Mutation Research-Fundamental and Molecular Mechanisms of Mutagenesis 428(1-2): 271-83.

Filippova EM, Monteleone DC, Trunk JG, Sutherland BM, Quake SR, Sutherland JC 2003. Quantifying double-strand breaks and clustered damages in DNA by single-molecule laser fluorescence sizing. Biophysical Journal 84(2): 1281-90.

Flynn J, MacGregor D 2003. Commentary on hormesis and public risk communication: is there a basis for public discussions? Human & Experimental Toxicology 22(1): 31-34.

Folkard M, Vojnovic B, Prise KM, Michael BD 2002. The application of charged-particle microbeams in radiobiology. Nuclear Instruments & Methods in Physics Research Section B-Beam Interactions with Materials and Atoms 188: 49-54.

Folkard M, Schettino G, Vojnovic B, Gilchrist S, Michette AG, Pfauntsch SJ, Prise KM, Michael BD 2001a. A focused ultrasoft X-ray microbeam for targeting cells individually with submicrometer accuracy. Radiation Research 156(6): 796-804.

Folkard M, Prise KM, Vojnovic B, Gilchrist S, Schettino G, Belyakov OV, Ozols A, Michael BD 2001b. The impact of microbeams in radiation biology. Nuclear Instruments & Methods in Physics Research Section B-Beam Interactions with Materials and Atoms 181: 426-30.

Ford JR, Maslowski AJ, Redd RA, Braby LA 2005. Radiation Responses of Perfused Tracheal Tissue. Radiation Research 164(4): 487-92.

Fornace AJ, Amundson SA, Bittner M, Myers TG, Meltzer P, Weinsten JN, Trent J 1999. The complexity of radiation stress responses: Analysis by informatics and functional genomics approaches. Gene Expression 7(4-6): 387-400.

Fujimori A, Tachiiri S, Sonoda E, Thompson LH, Dhar PK, Hiraoka M, Takeda S, Zhang Y, Reth M, Takata M 2001. Rad52 partially substitutes for the Rad51 paralog XRCC3 in maintaining chromosomal integrity in vertebrate cells. Embo Journal 20(19): 5513-5520.

Geard CR, Jenkins-Baker G, Marino SA, Ponnaiya B 2002. Novel approaches with track segment alpha particles and cell co-cultures in studies of bystander effects. Radiation Protection Dosimetry 99(1-4): 233-36.

Georgakilas AG, Bennett PV, Sutherland BM 2002. High efficiency detection of bi-stranded abasic clusters in gamma-irradiated DNA by putrescine. Nucleic Acids Research 30(13): 2800-2808.

George K, Wu HL, Willingham V, Cucinotta FA 2002. Analysis of complex-type chromosome exchanges in astronauts' lymphocytes after space flight as a biomarker of High-LET exposure. Journal of Radiation Research 43: S129-S132.

Gerondakis S, Grumont R, Gugasyan R, Wong L, Isomura I, Ho W, Banerjee A 2006. Unravelling the complexities of the NF-kappa B signalling pathway using mouse knockout and transgenic models. Oncogene 25(51): 6781-99.

Gilbert ES 2009. The impact of dosimetry uncertainties on dose-response analyses. Health Phys. Health Physics 97(5): 487-92.

Gius D, Spitz DR 2006. Redox signaling in cancer biology. Antioxidants & Redox Signaling 8(7-8): 1249-52.

Glick AB, Perez-Lorenzo R, Mohammed J 2008. Context-dependent regulation of cutaneous immunological responses by TGF beta 1 and its role in skin carcinogenesis. Carcinogenesis 29(1): 9-14.

Glover D, Little JB, Lavin MF, Gueven N 2003. Low dose ionizing radiation-induced activation of connexin 43 expression. International Journal of Radiation Biology 79(12): 955-964.

Goans RE, Holloway EC, Berger ME, Ricks RC 2001. Early dose assessment in criticality accidents. Health Physics 81(4): 446-49.

Goldberg Z 2003. Clinical implications of radiation-induced genomic instability. Oncogene 22(45): 7011-17.

Goldberg Z, Lehnert BE 2002. Radiation-induced effects in unirradiated cells: A review and implications in cancer. International Journal of Oncology 21(2): 337-49.

Goldberg Z, Schwietert CW, Lehnert B, Stern R, Nami I 2004. Effects of low-dose ionizing radiation on gene expression in human skin biopsies. International Journal of Radiation Oncology Biology Physics 58(2): 567-74.

Grdina DJ, Murley JS, Kataoka Y, Calvin DP 2002. Differential activation of nuclear transcription factor kappa B, gene expression, and proteins by amifostine's free thiol in human microvascular endothelial and glioma cells. Seminars in Radiation Oncology 12(1): 103-11.

Greco O, Marples B, Wilson GD, Joiner MC, Scott SD 2005. Radiation-induced gene therapy. Radiation Research 163(6): 707-8.

Green LM, Bianski BM, Murray DK, Rightnar SS, Nelson GA 2005. Characterization of accelerated iron-ion-induced damage in gap junction-competent and -incompetent thyroid follicular cells. Radiation Research 163(2): 172-82.

Green LM, Tran DT, Murray DK, Rightnar SS, Todd S, Nelson GA 2002. Response of thyroid follicular cells to gamma irradiation compared to proton irradiation: II. The role of connexin 32. Radiation Research 158(4): 475-85.

Gridley DS, Coutrakon GB, Rizvi A, Bayeta EJM, Luo-Owen X, Makinde AY, Baqai F, Koss P, Slater JM, Pecaut MJ 2008. Low-dose photons modify liver response to simulated solar particle event protons. Radiation Research 169(3): 280-87.

Grigorova M, Staines JM, Ozdag H, Caldas C, Edwards PAW 2004. Possible causes of chromosome instability: comparison of chromosomal abnormalities in cancer cell lines with mutations in BRCA1, BRCA2, CHK2 and BUB1. Cytogenetic and Genome Research 104(1-4): 333-40.

Grishko VI, Rachek LI, Spitz DR, Wilson GL, LeDoux SP 2005. Contribution of mitochondrial DNA repair to cell resistance from oxidative stress. Journal of Biological Chemistry 280(10): 8901-05.

Groesser T, Chun E, Rydberg B 2007. Relative biological effectiveness of high-energy iron ions for micronucleus formation at low doses. Radiation Research 168(6): 675-82.

Groesser T, Cooper B, Rydberg B 2008. Lack of Bystander Effects from High-LET Radiation for Early Cytogenetic End Points. Radiation Research 170(6): 794-802.

Guo GZ, Wang TL, Gao Q, Tamae D, Wong P, Chen T, Chen WC, Shively JE, Wong JYC, Li JJ 2004. Expression of ErbB2 enhances radiation-induced NF-kappa B activation. Oncogene 23(2): 535-45.

Guo GZ, Yan-Sanders Y, Lyn-Cook BD, Wang TL, Tamae D, Ogi J, Khaletskiy A, Li ZK, Weydert C, Longmate JA and others 2003. Manganese superoxide dismutase-mediated gene expression in radiation-induced adaptive responses. Molecular and Cellular Biology 23(7): 2362-78.

Guo HR, Chen CH, Ho SY, Ho YS, Chen RJ, Wang YJ 2006. Staurosporine modulates radiosensitivity and radiation-induced apoptosis in U937 cells. International Journal of Radiation Biology 82(2): 97-109.

Hada M, Sutherland BM 2006. Spectrum of complex DNA damages depends on the incident radiation. Radiation Research 165(2): 223-30.

Hada M, Cucinotta FA, Gonda SR, Wu H 2007. mBAND analysis of chromosomal aberrations in human epithelial cells exposed to low- and high-LET radiation. Radiation Research 168(1): 98-105.

Hadjidekova VB, Bulanova M, Bonassi S, Neri M 2003. Micronucleus frequency is increased in peripheral blood lymphocytes of nuclear power plant workers. Radiation Research 160(6): 684-90.

Hakim TS, Sugimori K, Camporesi EM, Anderson G 1996. Half-life of nitric oxide in aqueous solutions with and without haemoglobin. Physiological Measurement 17(4): 267-77.

Hall EJ 2000a. Radiobiology for the radiologist. Philadelphia, Lippincott Williams & Wilkins.

Hall EJ 2000b. A radiation biologist looks to the future. International Journal of Radiation Oncology Biology Physics 46(1): 1-2.

Hall EJ 2003. The bystander effect. Health Physics 85(1): 31-35.

Hall EJ 2006. The Bystander Effect: Paradigm Shifts in Interpreting Radiation Effects. Advances in Medical Physics. Pp. 260-64.

Hall EJ, Hei TK 2003. Genomic instability and bystander effects induced by high-LET radiation. Oncogene 22(45): 7034-42.

Hall EJ, Giaccia A 2006. Radiobiology for the Radiologist. 6th ed. Philadelphia, Lippincott Williams & Wilkins.

Hall EJ, Brenner DJ, Worgul B, Smilenov L 2005. Genetic susceptibility to radiation. In: Hei TKMJ ed. Space Life Sciences: Ground-Based Iron-Ion Biology and Physics, Including Shielding. Pp. 249-53.

Hamasaki K, Imai K, Nakachi K, Takahashi N, Kodama Y, Kusunoki Y 2007. Short-term culture and gamma H2AX flow cytometry determine differences in individual radiosensitivity in human peripheral T lymphocytes. Environmental and Molecular Mutagenesis 48(1): 38-47.

Han W, Zhu L, Jiang E, Wang J, Chen S, Bao L, Zhao Y, Xu A, Yu ZL, Wu LJ 2007. Elevatedsodiumchlorideconcentrations enhance the bystander effects induced by low dose alpha-particle irradiation. Mutation Research. Pp. 124-31.

Hanahan D, Weinberg RA 2000. The hallmarks of cancer. Cell 100(1): 57-70.

Hande MP, Azizova TV, Geard CR, Burak LE, Mitchell CR, Khokhryakov VF, Vasilenko EK, Brenner DJ 2003. Past exposure to densely ionizing radiation leaves a unique permanent signature in the genome. American Journal of Human Genetics 72(5): 1162-70.

Hanin L, Hyrien O, Bedford J, Yakovlev A 2006. A comprehensive stochastic model of irradiated cell populations in culture. Journal of Theoretical Biology 239(4): 401-16.

Harney J, Short SC, Shah N, Joiner M, Saunders MI 2004a. Low dose hyper-radiosensitivity in metastatic tumors. International Journal of Radiation Oncology Biology Physics 59(4): 1190-95.

Harney J, Shah N, Short S, Daley F, Groom N, Wilson GD, Joiner MC, Saunders MI 2004b. The evaluation of low dose hyper-radiosensitivity in normal human skin. Radiotherapy and Oncology 70(3): 319-29.

Hazelton W, Curtis SB, Moolgavkar SH 2006. Analysis of Radiation Effects Using a Combined Cell Cycle and Multistage Carcinogenesis Model. Advances in Space Research. Pp. 1809-12.

Hei TK, Zhou H, Ivanov VN, Hong M, Lieberman HB, Brenner DJ, Amundson SA, Geard CR 2008. Mechanism of radiation-induced bystander effects: a unifying model. Journal of Pharmacy and Pharmacology 60(8): 943-50.

Hendricks CA, Engelward BP 2004. "Recombomice": The past, present, and future of recombination-detection in mice. DNA Repair 3(10): 1255-61.

Hill MA, Stevens DL, Kadhim M, Blake-James M, Mill AJ, Goodhead DT 2006. Experimental techniques for studying bystander effects *in vitro* by high and low-let ionising radiation. Radiation Protection Dosimetry 122(1-4): 260-65.

Hinz JM, Nham PB, Salazar EP, Thompson LH 2006. The Fanconi anemia pathway limits the severity of mutagenesis. DNA Repair 5(8): 875-84.

Hinz JM, Nham PB, Urbin SS, Jones IM, Thompson LH 2007. Disparate contributions of the Fanconi anemia pathway and homologous recombination in preventing spontaneous mutagenesis. Nucleic Acids Research 35(11): 3733-40.

Hirano S, Yamamoto K, Ishiai M, Yamazoe M, Seki M, Matsushita N, Ohzeki M, Yamashita YM, Arakawa H, Buerstedde JM and others 2005. Functional relationships of FANCC to homologous recombination, translesion synthesis, and BLM. Embo Journal 24(2): 418-27.

Hlatky L, Sachs RK, Vazquez M, Cornforth MN 2002. Radiation-induced chromosome aberrations: insights gained from biophysical modeling. Bioessays 24(8): 714-23.

Hoel DG 1987a. Cancer Risk Models for Ionizing-Radiation. Environmental Health Perspectives 76: 121-24.

Hoel DG 1987b. Radiation Risk-Estimation Models. Environmental Health Perspectives 75: 105-7.

Hoel DG, Li P 1998. Threshold models in radiation carcinogenesis. Health Physics 75(3): 241-50.

Hofman-Huther H, Peuckert H, Ritter S, Virsik-Koepp P 2006. Chromosomal instability and delayed apoptosis in long-term T-lymphocyte cultures irradiated with carbon ions and X rays. Radiation Research 166(6): 858-69.

Hollander MC, Sheikh MS, Bulavin DV, Lundgren K, Augeri-Henmueller L, Shehee R, Molinaro TA, Kim KE, Tolosa E, Ashwell JD and others 1999. Genomic instability in Gadd45a-deficient mice. Nature Genetics 23(2): 176-84.

Holley WR, Mian IS, Park SJ, Rydberg B, Chatterjee A 2002. A model for interphase chromosomes and evaluation of radiation-induced aberrations. Radiation Research 158(5): 568-80.

Honma M, Momose M, Tanabe H, Sakamoto H, Yu YJ, Little JB, Sofuni T, Hayashi W 2000. Requirement of wild-type p53 protein for maintenance of chromosomal integrity. Molecular Carcinogenesis 28(4): 203-14.

Hooker AM, Morley AA, Tilley WD, Sykes PJ 2004a. Cancer-associated genes can affect somatic intrachromosomal recombination early in carcinogenesis. Mutation Research-Fundamental and Molecular Mechanisms of Mutagenesis 550(1-2): 1-10.

Hooker AM, Bhat M, Day TK, Lane JM, Swinburne SJ, Morley AA, Sykes PJ 2004b. The linear no-threshold model does not hold for low-dose ionizing radiation. Radiation Research 162(4): 447-52.

Horn HF, Vousden KH 2007. Coping with stress: multiple ways to activate p53. Oncogene 26(9): 1306-16.

Hu BR, Wu LJ, Han W, Zhang LL, Chen SP, Xu A, Hei TK, Yu ZL 2006. The time and spatial effects of bystander response in mammalian cells induced by low dose radiation. Carcinogenesis 27(2): 245-51.

Huang L, Snyder AR, Morgan WF 2003. Radiation-induced genomic instability and its implications for radiation carcinogenesis. Oncogene 22(37): 5848-54.

Huang L, Kim PM, Nickoloff JA, Morgan WF 2007. Targeted and nontargeted effects of low-dose ionizing radiation on delayed genomic instability in human cells. Cancer Research 67(3): 1099-1104.

Huang L, Grim S, Smith LE, Kim PN, Nickoloff JA, Goloubeva OG, Morgan WF 2004. Ionizing radiation induces delayed hyperrecombination in mammalian cells. Molecular and Cellular Biology 24(11): 5060-68.

Huang TT, Wuerzberger-Davis SM, Seufzer BJ, Shumway SD, Kurama T, Boothman DA, Miyamoto S 2000. NF-kappa B activation by camptothecin - A linkage between nuclear DNA damage and cytoplasmic signaling events. Journal of Biological Chemistry 275(13): 9501-9.

Hunter AJ, Hendrikse AS, Renan MJ 2007. Can radiation-induced apoptosis be modulated by inhibitors of energy metabolism? International Journal of Radiation Biology 83(2): 105-14.

Huo LH, Nagasawa H, Little JB 2001. HPRT mutants induced in bystander cells by very low fluences of alpha particles result primarily from point mutations. Radiation Research 156(5): 521-25.

Hwang SL, Guo HR, Hsieh WA, Hwang JS, Lee SD, Tang JL, Chen CC, Chang TC, Wang JD, Chang WP 2006. Cancer risks in a population with prolonged low dose-rate gamma-radiation exposure in radiocontaminated buildings, 1983-2002. International Journal of Radiation Biology 82(12): 849-58.

ICRU 2011. Quantification and reporting of low-dose and other heterogeneous exposures. Oxford, Oxford University Press.

Ishizaki K, Hayashi Y, Nakamura H, Yasui Y, Komatsu K, Tachibana A 2004. No induction of p53 phosphorylation and few focus formation of phosphorylated H2AX suggest efficient repair of DNA damage during chronic low-dose-rate irradiation in human cells. Journal of Radiation Research 45(4): 52125.

Jang D-L, Guo M, Wang D 2007. Proteomic and biochemical studies of calcium- and phosphorylation-dependent calmodulin complexes in mammalian cells. Journal of Proteome Research 6(9): 3718-28.

Jaworowski Z 2008. The Paradigm that Failed. International Journal of Low Radiation Pp. 151-55.

Jeggo PA 1998. DNA breakage and repair. Advances in Genetics, Vol 38 38: 185-218.

Jenkins-Smith HC, Silva CL, Murray C 2009. Beliefs about Radiation: Scientists, the Public and Public Policy. Health Physics 97(5): 519-27.

Jin S, Daly DS, Springer DL, Miller JH 2008. The effects of shared peptides on protein quantitation in label-free proteomics by LC/MS/MS. Journal of Proteome Research 7(1): 164-69.

Jirtle RL, Skinner MK 2007. Environmental epigenomics and disease susceptibility. Nature Reviews Genetics 8(4): 253-62.

Jobling ME, Mott JD, Finnegan MT, Jurukovski V, Erickson AC, Walian PJ, Taylor SE, Ledbetter S, Lawrence CM, Rifkin DB and others 2006. Isoform-specific activation of latent transforming growth factor beta (LTGF-beta) by reactive oxygen species. Radiation Research 166(6): 839-48.

Joiner M 2004. New frontiers or red herrings in radiation oncology? The Lancet Oncology 5(1).

Joiner MC, Johns H 1988. Renal damage in the mouse - the response to very small doeses per fraction. Radiation Research 114(2): 385-98.

Joiner MC, Denekamp J, Maughan RL 1986. The use of top-up experiments to investigate the effect of very small doses per fraction in mouse skin. International Journal of Radiation Biology 49(4): 565-80.

Joiner MC, Marples B, Lambin P, Short SC, Turesson I 2001. Low-dose hypersensitivity: Current status and possible mechanisms. International Journal of Radiation Oncology Biology Physics 49(2): 379-89.

Jostes RF, Fleck EW, Morgan TL, Stiegler GL, Cross FT 1994. Southern blot and polymerase chain-reaction exon analyses of HPRT(-) mutations induced by radon and radon progeny. Radiation Research 137(3): 371-79.

Kadhim MA 2003. Role of genetic background in induced instability. Oncogene 22(45): 6994-99.

Kadhim MA, Moore SR, Goodwin EH 2004. Interrelationships amongst radiation-induced genomic instability, bystander effects, and the adaptive response. Mutation Research-Fundamental and Molecular Mechanisms of Mutagenesis 568(1): 21-32.

Kadhim MA, Hill MA, Moore SR 2006. Genomic instability and the role of radiation quality. Radiation Protection Dosimetry 122(1-4): 221-27.

Kadhim MA, Macdonald DA, Goodhead DT, Lorimore SA, Marsden SJ, Wright EG 1992. Transmission of Chromosomal Instability after Plutonium Alpha-particle Irradiation. Nature 355(6362): 738-40.

Kadhim MA, Lorimore SA, Townsend KMS, Goodhead DT, Buckle VJ, Wright EG 1995. Radiation-Induced Genomic Instability-Delayed Cytogenetic Aberrations and Apoptosis in Primary Human Bone-Marrow Cells. International Journal of Radiation Biology 67(3): 287-93.

Kagawa S, He C, Gu J, Koch P, Rha SJ, Roth JA, Curley SA, Stephens LC, Fang BL 2001. Antitumor activity and bystander effects of the tumor necrosis factor-related apoptosis-inducing ligand (TRAIL) gene. Cancer Research 61(8): 3330-38.

Kahn MA, Hill RP, Van Dyke J 1998. Partial volume rat lung irradiation: an evaluation of early DNA damage. International Journal of Radiation Oncology Biology and Physics 40: 467-76.

Kalka K, Ahmad N, Criswell T, Boothman D, Mukhtar H 2000. Up-regulation of clusterin during phthalocyanine 4 photodynamic therapy-mediated apoptosis of tumor cells and ablation of mouse skin tumors. Cancer Research 60(21): 5984-87.

Kaplan MI, Morgan WF 1998. The nucleus is the target for radiation-induced chromosomal instability. Radiation Research 150(4): 382-90.

Kataoka Y, Murley JS, Khodarev NN, Weichselbaum RR, Grdina DJ 2002. Activation of the nuclear transcription factor kappa B (NF kappa B) and differential gene expression in U87 glioma cells after exposure to the cytoprotector amifostine. International Journal of Radiation Oncology Biology Physics 53(1): 180-89.

Kataoka Y, Bindokas VP, Duggan RC, Murley JS, Grdina DJ 2006. Flow cytometric analysis of phosphorylated histone H2AX following exposure to ionizing radiation in human microvascular endothelial cells. Journal of Radiation Research 47(3-4): 245-57.

Kathren RL 1996. Pathway to a paradigm: The linear nonthreshold dose-response model in historical context: The American Academy of Health Physics 1995 Radiology Centennial Hartman Oration. Health Physics 70(5): 621-35.

Kato TA, Okayasu R, Bedford JS 2008. Comparison of the induction and disappearance of DNA double strand breaks and gamma-H2AX foci after irradiation of chromosomes in G1-phase or in condensed metaphase cells. Mutation Research-Fundamental and Molecular Mechanisms of Mutagenesis 639(1-2): 108-12.

Kato TA, Nagasawa H, Weil MM, Little JB, Bedford JS 2006. Levels of gamma-H2AX foci after low-dose-rate irradiation reveal a DNA DSB rejoining defect in cells from human ATM heterozygotes in two AT families and in another apparently normal individual. Radiation Research 166(3): 443-53.

Kato TA, Wilson PF, Nagasaw H, Peng Y, Weil MM, Bedford JS, Little JB 2009. Variations in radiosensitivity among individuals: A potential impact on risk assessment? Health Phys. Health Physics 97(5): 470-80.

Kawata T, Gotoh E, Durante M, Wu H, George K, Furusawa Y, Cucinotta FA 2000. High-LET radiation-induced aberrations in prematurely condensed G_2 chromosomes of human fibroblasts. International Journal of Radiation Biology 76(7): 929-37.

Kegelmeyer LM, Tomascik-Cheeseman L, Burnett MS, van Hummelen P, Wyrobek AJ 2001. A groundtruth approach to accurate quantitation of fluorescence microarrays. In: Bittner MLCYDDANDER ed. Microarrays: Optical Technologies and Informatics. Pp. 35-45.

Kellerer AM, Rossi HH 1978. Generalized formulation of dual radiation action. Radiation Research 75(3): 471-88.

Kennedy AR, Zhou Z, Donahue JJ, Ware JH 2006. Protection against adverse biological effects induced by space radiation by the Bowman-Birk inhibitor and antioxidants. Radiation Research 166(2): 327-32.

Khan MA, Hill RP, Van Dyk J 1998. Partial volume rat lung irradiation: An evaluation of early DNA damage. International Journal of Radiation Oncology Biology Physics 40(2): 467-76.

Khan N, Afaq F, Mukhtar H 2007. Apoptosis by dietary factors: the suicide solution for delaying cancer growth. Carcinogenesis 28(2): 233-39.

Kim GJ, Chandrasekaran K, Morgan WF 2006. Mitochondrial dysfunction, persistently elevated levels of reactive oxygen species and radiation-induced genomic instability: a review. Mutagenesis 21(6): 361-67.

Kim J, Fiskum G, Morgan WF 2000. A role for mitochondrial dysfunction in perpetuating radiation-induced genomic instability. Cancer Research. Pp. 10377-83.

Kimmel RR, Agnani S, Yang Y, Jordan R, Schwartz JL 2008. DNA copy-number instability in low-dose gamma-irradiated TK6 lymphoblastoid clones. Radiation Research 169(3): 259-69.

King MC, Marks JH, Mandell JB, New York Breast Canc Study G 2003. Breast and ovarian cancer risks due to inherited mutations in BRCA1 and BRCA2. Science 302(5645): 643-46.

Kirshner J, Jobling MF, Pajares MJ, Ravani SA, Glick AB, Lavin MJ, Koslov S, Shiloh Y, Barcellos-Hoff MH 2006. Inhibition of transforming growth factor-beta 1 signaling attenuates ataxia telanglectasia mutated activity in response to genotoxic stress. Cancer Research 66(22): 10861-+.

Kleiman NJ, David J, Elliston CD, Hopkihs KM, Smilenov LB, Brenner DJ, Worgul BV, Hall EJ, Lieberman HB 2007. Mrad9 and atm haploinsufficiency enhance spontaneous and X-ray-induced cataractogenesis in mice. Radiation Research 168(5): 567-73.

Klokov D, Criswell T, Leskov KS, Araki S, Mayo L, Boothman DA 2004. IR-inducible clusterin gene expression: a protein with potential roles in ionizing radiation-induced adaptive responses, genomic instability, and bystander effects. Mutation Research-Fundamental and Molecular Mechanisms of Mutagenesis 568(1): 97-110.

Klokov D, Criswell T, Sampath L, Leskov KS, Frinkley K, Araki S, Beman M, Wilson DL, Boothman DA 2003. Clusterin: a protein with multiple functions as a potential ionizing radiation exposure marker. In: Shibata YYSWMTM ed. Radiation and Humankind. Pp. 219-32.

Ko SJ, Liao XY, Molloi S, Elmore E, Redpath JL 2004. Neoplastic transformation *in vitro* after exposure to low doses of mammographic-energy X rays: Quantitative and mechanistic aspects. Radiation Research 162(6): 646-54.

Koch-Paiz CA, Momenan R, Amundson SA, Lamoreaux E, Fornace AJ 2000. Estimation of relative mRNA content by filter hybridization to a polyuridylic probe. Biotechniques 29(4): 706-+.

Kocher DC, Apostoaei AI, Henshaw RW, Hoffman FO, Schubauer-Berigan MK, Stancescu DO, Thomas BA, Trabalka JR, Gilbert ES, Land CE 2008. Interactive RadioEpidemiological Program (IREP): A web-based tool for estimating probability of causation/assigned share of radiogenic cancers. Health Physics 95(1): 119-47.

Konopacka M, Rzeszowska-Wolny J 2006. The bystander effect-induced formation of micronucleated cells is inhibited by antioxidants, but the parallel induction of apoptosis and loss of viability are not affected. Mutation Research-Fundamental and Molecular Mechanisms of Mutagenesis 593(1-2): 32-38.

Kovalchuk O, Baulch JE 2008. Epigenetic changes and nontargeted radiation effects - Is there a link? Environmental and Molecular Mutagenesis 49(1): 16-25.

Kovalchuk O, Hendricks CA, Cassie S, Engelward AJ, Engelward BP 2004. *In vivo* recombination after chronic damage exposure falls to below spontaneous levels in "recombomice". Molecular Cancer Research 2(10): 567-73.

Krause M, Prager J, Wohlfarth J, Hessel F, Dorner D, Haase M, Joiner MC, Baumann M 2005a. Ultrafractionation does not Improve the Results of Radiotherapy in Radioresistant Murine DDL1 Lymphoma. Strahlentherapie und Onkologie. 181(8): 540.

Krause M, Wohlfarth J, Georgi B, Pimentel N, Dorner D, Zips D, Eicheler W, Hessel F, Short S, Joiner M and others 2005b. Low-dose hyperradiosensitivity of human glioblastoma cell lines *in vitro* does not translate into improved outcome of ultrafractionated radiotherapy *in vivo*. International Journal of Radiation Biology 81(10): 751-58.

Krueger SA, Joiner MC, Weinfeld M, Piasentin E, Marples B 2007a. Role of apoptosis in low-dose hyper-radiosensitivity. Radiation Research 167(3): 260-67.

Krueger SA, Collis SJ, Joiner MC, Wilson GD, Marples B 2007b. Transition in survival from low-dose hyper-radiosensitivity to increased radioresistance is independent of activation of atm ser1981 activity. International Journal of Radiation Oncology Biology Physics 69(4): 1262-71.

Kryscio A, Muller WUU, Wojcik A, Kotschy N, Grobelny S, Streffer C 2001. A cytogenetic analysis of the long-term effect of uranium mining on peripheral lymphocytes using the micronucleus-centromere assay. International Journal of Radiation Biology 77(11): 1087-93.

Kuhne M, Riballo E, Rief N, Rothkamm K, Jeggo PA, Lobrich M 2004. A double-strand break repair defect in ATM-deficient cells contributes to radiosensitivity. Cancer Research 64(2): 500-8.

Kuo WL, Das D, Ziyad S, Bhattacharya S, Gibb WJ, Heiser LM, Sadanandam A, Fontenay GV, Wang NJ, Bayani N and others 2009. A systems analysis of the chemosensitivity of breast cancer cells to the polyamine analogue PG-11047. BMC Med. BMC Medicine 7.

Kurpinski K, Jang D-J, Bhattacharya S, Rydberg B, Chu J, So J, Wyrobek A, Li S, Wang D 2009. Differential Effects of X-Rays and High-Energy Fe-56 Ions on Human Mesenchymal Stem Cells. International Journal of Radiation Oncology Biology Physics 73(3): 869-77.

Laiakis EC, Morgan WF 2005. Communicating the Non-targeted Effects of Radiation from Irradiated to Non-irradiated Cells. Acta Med Nagasaki. Pp. 79-80.

Laiakis EC, Baulch JE, Morgan WF 2008. Interleukin 8 exhibits a pro-mitogenic and pro-survival role in radiation induced genomically unstable cells. Mutation Research-Fundamental and Molecular Mechanisms of Mutagenesis 640(1-2): 74-81.

Lall R, Ganapathy S, Yang M, Xiao S, Xu T, Su H, Shadfan M, Asara JM, Ben-Sahra I, Manning BD, Little JB, Yuan Z-M. 2014. Low-dose radiation exposure induces a HIF-1-mediated adaptive and protective metabolic response. Cell Death and Differ 4: 1-9.

Lamerdin JE, Yamada NA, George JW, Souza B, Christian AT, Jones NJ, Thompson LH 2004. Characterization of the hamster FancG/Xrcc9 gene and mutations in CHOUV40 and NM3. Mutagenesis 19(3): 237-44.

Land CE 2009. Low-Dose Extrapolation of Radiation Health Risks: Some Implications of Uncertainty for Radiation Protection at Low Doses. Health Physics 97(5): 407-15.

Lavin MF 1999. ATM: the product of the gene mutated in ataxia-telangiectasia. International Journal of Biochemistry & Cell Biology 31(7): 735-40.

Lavin MF, Birrell G, Chen P, Kozlov S, Scott S, Gueven N 2005. ATM signaling and genornic stability in response to DNA damage. Mutation Research-Fundamental and Molecular Mechanisms of Mutagenesis 569(1-2): 123-32.

Lea DE 1955. The target theory. In: Actions of radiations on living cells. Cambridge Eng., Canbridge University Press.

Leatherbarrow EL, Harper JV, Cucinotta FA, O'Neill P 2006. Induction and quantification of gamma-H2AX foci following low and high LET-irradiation. International Journal of Radiation Biology 82(2): 111-18.

Lee S, Moore JK, Haber JE, Greider C 1999. RAD50 and RAD51 define two independent pathways that collaborate to maintain telomeres in the absence of telomerase. Genetics 152(1): 27-40.

Lee S, Chen JJ, Zhou GL, Shi RZ, Bouffard GG, Kocherginsky M, Ge XJ, Sun M, Jayathilaka N, Kim YC and others 2006. Gene expression profiles in acute myeloid leukemia with common translocations using SAGE. Proceedings of the National Academy of Sciences of the United States of America 103(4): 1030-35.

Lee SE, Bressan DA, Petrini JHJ, Haber JE 2002. Complementation between N-terminal Saccharomyces cerevisiae mre11 alleles in DNA repair and telomere length maintenance. DNA Repair 1(1): 27-40.

Leenhouts H, Chadwick K 1994. A Two-Mutation Model of Radiation Carcinogenesis: Application to Lung Tumours in Rodents and Implications for Risk Evaluation. Journal of Radiological Protection. Pp. 115-30.

Lehnert BE, Goodwin EH 1997. Extracellular factor(s) following exposure to alpha particles can cause sister chromatid exchanges in normal human cells. Cancer Research 57(11): 2164-71.

Lengauer C, Kinzler KW, Vogelstein B 1998. Genetic instabilities in human cancers. Nature 396(6712): 643-49.
Leonard BE 2007a. Adaptive response: Part II. Further modeling for dose rate and time influences. International Journal of Radiation Biology 83(6): 395-408.
Leonard BE 2007b. Adaptive response and human benefit: Part I. A microdosimetry dose-dependent model. International Journal of Radiation Biology 83(2): 115-31.
Leonard BE 2008. Common sense about the linear no-threshold controversy - Give the general public a break. Radiation Research 169(2): 245-46.
Leskov K, Antonio S, Criswell T, Yang CR, Kinsella TJ, Boothman DA 2001a. Radiation-inducible clusterin (CLU): A molecular switch between life and death. Radiation Research 156(4): 441-441.
Leskov KS, Klokov DY, Li J, Kinsella TJ, Boothman DA 2003. Synthesis and functional analyses of nuclear clusterin, a cell death protein. Journal of Biological Chemistry 278(13): 11590-600.
Leskov KS, Criswell T, Antonio S, Li J, Yang CR, Kinsella TJ, Boothman DA 2001b. When X-ray-inducible proteins meet DNA double strand break repair. Seminars in Radiation Oncology 11(4): 352-72.
Levy D, Vazquez M, Cornforth M, Loucas B, Sachs RK, Arsuaga J 2004. Comparing DNA damage-processing pathways by computer analysis of chromosome painting data. Journal of Computational Biology 11(4): 626-41.
Levy D, Reeder C, Loucas B, Hlatky L, Chen A, Cornforth M, Sachs R 2007. Interpreting chromosome aberration spectra. Journal of Computational Biology 14(2): 144-55.
Li CY, Little JB, Hu K, Zhang W, Zhang L, Dewhirst MW, Huang Q 2001. Persistent genetic instability in cancer cells induced by non-DNA-damaging stress exposures. Cancer Research 61(2): 428-32.
Li H, Liu N, Rajendran GK, Gernon TJ, Rockhill JK, Schwartz JL, Gu Y 2008a. A role for endogenous and radiation-induced DNA double-strand breaks in p53-dependent apoptosis during cortical neurogenesis. Radiation Research 169(5): 513-22.
Li Y, Lu J, Cohen D, Prochownik EV 2008b. Transformation, genomic instability and senescence mediated by platelet/megakaryocyte glycoprotein Ib alpha. Oncogene 27(11): 1599-1609.
Liang L, Mendonca MS, Deng L, Nguyen SC, Shao C, Tischfield JA 2007. Reduced apoptosis and increased deletion mutations at Aprt locus *in vivo* in mice exposed to repeated ionizing radiation. Cancer Research 67(5): 1910-17.
Lim YP, Lim TT, Chan YL, Song ACM, Yeo BH, Vojtesek B, Coomber D, Rajagopal G, Lane D 2007. The p53 knowledgebase: an integrated information resource for p53 research. Oncogene 26(11): 1517-21.
Limoli CL, Kaplan MI, Giedzinski E, Morgan WF 2001a. Attenuation of radiation-induced genomic instability by free radical scavengers and cellular proliferation. Free Radical Biology and Medicine 31(1): 10-19.
Limoli CL, Corcoran JJ, Milligan JR, Ward JF, Morgan WF 1999. Critical target and dose and dose-rate responses for the induction of chromosomal instability by ionizing radiation. Radiation Research 151(6): 677-85.
Limoli CL, Ponnaiya B, Corcoran JJ, Giedzinski E, Morgan WF 2000a. Chromosomal instability induced by heavy ion irradiation. International Journal of Radiation Biology 76(12): 1599-1606.

Limoli CL, Corcoran JJ, Jordan R, Morgan WF, Schwartz JL 2001b. A role for chromosomal instability in the development of and selection for radioresistant cell variants. British Journal of Cancer 84(4): 489-92.

Limoli CL, Giedzinski E, Morgan WF, Swarts SG, Jones GDD, Hyun W 2003. Persistent oxidative stress in chromosomally unstable cells. Cancer Research 63(12): 3107-11.

Limoli CL, Giedzinski E, Rola R, Otsuka S, Palmer TD, Fike JR 2004. Radiation response of neural precursor cells: Linking cellular sensitivity to cell cycle checkpoints, apoptosis and oxidative stress. Radiation Research 161(1): 17-27.

Limoli CL, Hartmann A, Shephard L, Yang CR, Boothman DA, Bartholomew J, Morgan WF 1998. Apoptosis, reproductive failure, and oxidative stress in Chinese hamster ovary cells with compromised genomic integrity. Cancer Research 58(16): 3712-18.

Limoli CL, Ponnaiya B, Corcoran JJ, Giedzinski E, Kaplan MI, Hartmann A, Morgan WF 2000b. Genomic instability induced by high and low let ionizing radiation. In: Slenzka KVMCFA ed. Life Sciences: Microgravity and Space Radiation Effects. Pp. 2107-2117.

Little JB 1999. Induction of genetic instability by ionizing radiation. Comptes Rendus De L Academie Des Sciences Serie Iii-Sciences De La Vie-Life Sciences 322(2-3): 127-34.

Little JB 2003. Genomic instability and radiation. Journal of Radiological Protection 23(2): 173-81.

Little JB, Azzam EI, de Toledo SM, Nagasawa H 2002a. Bystander effects: Intercellular transmission of radiation damage signals. Radiation Protection Dosimetry 99(1-4): 159-62.

Little JB, Nagasawa H, Dahlberg WK, Zdzienicka MZ, Burma S, Chen DJ 2002b. Differing responses of nijmegen breakage syndrome and ataxia telangiectasia cells to ionizing radiation. Radiation Research 158(3): 319-26.

Littlefield LG, McFee AF, Salomaa SI, Tucker JD, Inskip PD, Sayer AM, Lindholm C, Makinen S, Mustonen R, Sorensen K and others 1998. Do recorded doses overestimate true doses received by Chernobyl cleanup workers? Results of cytogenetic analyses of Estonian workers by fluorescence in situ hybridization. Radiation Research 150(2): 237-49.

Liu N, Lamerdin JE, Tebbs RS, Schild D, Tucker JD, Shen MR, Brookman KW, Siciliano MJ, Walter CA, Fan WF and others 1998. XRCC2 and XRCC3, new human Rad51-family members, promote chromosome stability and protect against DNA cross-links and other damages. Molecular Cell 1(6): 783-93.

Locke PA 2009. Incorporating information from the U.S. department of energy low-dose program into regulatory decision-making: Three policy integration challenges. Health Physics 97(5): 510-15.

Lopez-Ferrer D, Heibeck TH, Petritis K, Hixson KK, Qian W, Monroe ME, Mayampurath A, Moore RJ, Belov ME, Camp DG, II and others 2008. Rapid sample processing for LC-MS-based quantitative proteomics using high intensity focused ultrasound. Journal of Proteome Research 7(9): 3860-67.

Lorimore SA, Wright EG 2003. Radiation-induced genomic instability and bystander effects: related inflammatory-type responses to radiation-induced stress and injury? A review. International Journal of Radiation Biology 79(1): 15-25.

Lorimore SA, Coates PJ, Wright EG 2003. Radiation-induced genomic instability and bystander effects: inter-related nontargeted effects of exposure to ionizing radiation. Oncogene 22(45): 7058-69.

Lou J, He J, Zheng W, Jin L, Chen Z, Chen S, Lin Y, Xu S 2007. Investigating the genetic instability in the peripheral lymphocytes of 36 untreated lung cancer patients with comet assay and micronucleus assay. Mutation Research-Fundamental and Molecular Mechanisms of Mutagenesis 617(1-2): 104-10.

Loucas BD, Cornforth MN 2001. Complex chromosome exchanges induced by gamma rays in human lymphocytes: An mFISH study. Radiation Research 155(5): 660-71.

Loucas BD, Eberle RL, Durante M, Cornforth MN 2004a. Complex chromatid-isochromatid exchanges following irradiation with heavy ions? Cytogenetic and Genome Research 104(1-4): 206-10.

Loucas BD, Eberle R, Bailey SM, Cornforth MN 2004b. Influence of dose rate on the induction of simple and complex chromosome exchanges by gamma rays. Radiation Research 162(4): 339-49.

Lowe XR, Bhattacharya S, Marchetti F, Wyrobek AJ 2009. Early Brain Response to Low-Dose Radiation Exposure Involves Molecular Networks and Pathways Associated with Cognitive Functions, Advanced Aging and Alzheimer's Disease. Radiation Research 171(1): 53-65.

Lu X, Yang C, Hill R, Yin C, Hollander MC, Fornace AJ, Jr., Van Dyke T 2008. Inactivation of gadd45a sensitizes epithelial cancer cells to ionizing radiation *in vivo* resulting in prolonged survival. Cancer Research 68(10): 3579-83.

Luckey T 1991. Radiation Hormesis. CRC Press, Boca Raton.

Luebeck EG, Curtis SB, Hazelton WD, Moolgavkar SH 1999. A biologically based model for radon induced malignant lung tumors, indoor radon exposure and its health consequences. Kodansha Scientific Ltd.: 143-53.

Lynch DJ, Wilson WE, Batdorf MT, Resat MBS, Kimmel GA, Miller JH 2005. Monte Carlo simulation of the spatial distribution of energy deposition for an electron microbeam. Radiation Research 163(4): 468-72.

Lyng FM, Seymour CB, Mothersill C 2000. Production of a signal by irradiated cells which leads to a response in unirradiated cells characteristic of initiation of apoptosis. British Journal of Cancer 83(9): 1223-30.

Lyng FM, Seymour CB, Mothersill C 2002a. Initiation of apoptosis in cells exposed to medium from the progeny of irradiated cells: A possible mechanism for bystander-induced genomic instability? Radiation Research 157(4): 365-70.

Lyng FM, Seymour CB, Mothersill C 2002b. Early events in the apoptotic cascade initiated in cells treated with medium from the progeny of irradiated cells. Radiation Protection Dosimetry 99(1-4): 169-72.

Lyng FM, Maguire P, McClean B, Seymour C, Mothersill C 2006. The involvement of calcium and MAP kinase signaling pathways in the production of radiation-induced bystander effects. Radiation Research 165(4): 400-9.

MacGregor DG, Slovic P, Malmfors T 1999. "How exposed is exposed enough?" Lay inferences about chemical exposure. Risk Analysis: an Official Publication of the Society for Risk Analysis 19(4): 649-59.

Maguire P, Mothersill C, McClean B, Seymour C, Lyng FM 2007. Modulation of radiation responses by pre-exposure to irradiated cell conditioned medium. Radiation Research 167(4): 485-92.

Mainardi E, Donahue RJ, Wilson WE, Blakely EA 2004. Comparison of microdosimetric Simulations using PENELOPE and PITS for a 25 keV electron microbeam in water. Radiation Research 162(3): 326-31.

Mao JH, Wu D, DelRosario R, Castellanos A, Balmain A, Perez-Losada J 2008. Atm heterozygosity does not increase tumor susceptibility to ionizing radiation alone or in a p53 heterozygous background. Oncogene 27(51): 6596-6600.

Mao JH, Li JZ, Jiang T, Li Q, Wu D, Perez-Losada J, DelRosario R, Peterson L, Balmain A, Cai WW 2005. Genomic instability in radiation-induced mouse lymphoma from p53 heterozygous mice. Oncogene 24(53): 7924-34.

Marchetti F, Coleman MA, Jones IM, Wyrobek AJ 2006. Candidate protein biodosimeters of human exposure to ionizing radiation. International Journal of Radiation Biology 82(9): 605-39.

Marder BA, Morgan WF 1993. Delayed chromosomal instability induced by DNA-damage. Molecular and Cellular Biology 13(11): 6667-77.

Marples B, Joiner MC 1993. The response of Chinese-hamster V79 cells to low radiation-doses - evidence of enhanced sensitivity of the whole cell population. Radiation Research 133(1): 41-51.

Marples B, Joiner MC 2000. Modification of survival by DNA repair modifiers: a probable explanation for the phenomenon of increased radioresistance. International Journal of Radiation Biology 76(3): 305-12.

Marples B, Collis SJ 2008. Low-dose hyper-radiosensitivity: Past, present, and future. International Journal of Radiation Oncology Biology Physics 70(5): 1310-18.

Marples B, Joiner MC, Skov KA 1994. The effect of oxygen on low-dose hypersensitivity and increased radioresistance in Chinese hamster V79-379A cells. Radiation Research 138(1): S17-S20.

Marples B, Wouters BG, Joiner MC 2003. An association between the radiation-induced arrest of G_2-phase cells and low-dose hyper-radiosensitivity: a plausible underlying mechanism? Radiation Research 160(1): 38-45.

Marples B, Lambin P, Skov KA, Joiner MC 1997. Low dose hyper-radiosensitivity and increased radioresistance in mammalian cells. International Journal of Radiation Biology. Pp. 721-35.

Marples B, Cann NE, Mitchell CR, Johnston PJ, Joiner MC 2002. Evidence for the involvement of DNA-dependent protein kinase in the phenomena of low dose hyper-radiosensitivity and increased radioresistance. International Journal of Radiation Biology 78(12): 1139-47.

Marples B, Wouters BG, Collis SJ, Chalmers AJ, Joiner MC 2004. Low-dose hyper-radiosensitivity: A consequence of ineffective cell cycle arrest of radiation-damaged G(2)-phase cells. Radiation Research 161(3): 247-55.

Matsumoto H, Takahashi A, Ohnishi T 2007. Nitric oxide radicals choreograph a radio-adaptive response. Cancer Research 67(18): 8574-79.

Maxwell CA, Fleisch MC, Costes SV, Erickson AC, Boissiere A, Gupta R, Ravani SA, Parvin B, Barcellos-Hoff MH 2008. Targeted and nontargeted effects of ionizing radiation that impact genomic instability. Cancer Research 68(20): 8304-11.

Medina D, Ullrich R, Meyn R, Wiseman R, Donehower L 2002. Environmental carcinogens and p53 tumor-suppressor gene interactions in a transgenic mouse model for mammary carcinogenesis. Environmental and Molecular Mutagenesis 39(2-3): 178-83.

Medvedeva N, Ford J, Braby L 2004. Changes in micronucleus frequency resulting from preirradiation of cell culture surfaces. Radiation Research 162(6): 660-66.

Mendonca MS, Howard KL, Farrington DL, Desmond LA, Temples TM, Mayhugh BM, Pink JJ, Boothman DA 1999. Delayed apoptotic responses associated with radiation-induced neoplastic transformation of human hybrid cells. Cancer Research 59(16): 3972-79.

Mendonca MS, Chin-Sinex H, Gomez-Millan J, Datzman N, Hardacre M, Comerford K, Nakshatri H, Nye M, Benjamin L, Mehta S and others 2007. Parthenolide sensitizes cells to X-ray-induced inhibition of NF-kappa B and split-dose cell killing through repair. Radiation Research 168(6): 689-97.

Meng A, Yu T, Chen G, Brown SA, Wang Y, Thompson JS, Zhou D 2003. Cellular origin of ionizing radiation-induced NF-kappa B activation *in vivo* and role of NF-kappa B in ionizing radiation-induced lymphocyte apoptosis. International Journal of Radiation Biology 79(11): 849-61.

Mettler FA 2011. Communication of radiation benefits and risks in decision making: Communication on children's imaging and computed tomography. Health Phys. Health Physics 101(5): 589-90.

Meyn RE, Milas L, Ang KK 2009. The role of apoptosis in radiation oncology. International Journal of Radiation Biology 85(2): 107-15.

Michael BD, Held KD, Schettino G, Folkard M, Prise KM, Vojnovic B 2001. Charged-particle and focused soft X-ray microbeams for investigating individual and collective radiation responses of cells. Radiation Research 156(4): 439-40.

Miller JH, Zheng F 2004. Large-scale simulations of cellular signaling processes. Parallel Computing 30(9-10): 1137-49.

Miller JH, Wilson WE, Lynch DJ, Sowa Resat MB, Trease HE 2001. Computational dosimetry for electron microbeams: Monte Carlo track simulation combined with confocal microscopy. Radiation Research 156(4): 438-39.

Miller JH, Jin S, Morgan WF, Yang A, Wan Y, Aypar U, Peters JS, Springer DL 2008. Profiling mitochondrial proteins in radiation-induced genome-unstable cell lines with persistent oxidative stress by mass spectrometry. Radiation Research 169(6): 700-6.

Miller KA, Hinz JM, Yamada NA, Thompson LH, Albala JS 2005. Nuclear localization of Rad51B is independent of Rad51C and BRCA2. Mutagenesis 20(1): 57-63.

Miller RC, Randers-Pehrson G, Geard CR, Hall EJ, Brenner DJ 1999. The oncogenic transforming potential of the passage of single alpha particles through mammalian cell nuclei. Proceedings of the National Academy of Sciences of the United States of America 96(1): 19-22.

Miller RC, Richards M, Brenner DJ, Hall EJ, Jostes R, Hui TE, Brooks AL 1996. The biological effectiveness of radon-progeny alpha particles .5. Comparison of oncogenic transformation by accelerator-produced monoenergetic alpha particles and by polyenergetic alpha particles from radon progeny. Radiation Research 146(1): 75-80.

Mitchel R, Azzam EI, De Toledo SM 1997. Adaption to Ionizing Radiation in Mammalian Cells. Stress-Inducible Processes in Higher Eukaryotes. Pp. 221-43.

Mitchel REJ 2006. Adaptive Response *In Vitro* and *In Vivo* and it's Impact on Low Dose Radiation. In: Wolbarst A, Zamenhof R, Hendee W ed. Biological Effects of Low Doses of Ionizing Radiation. Madison, Medical Physics Publishing. Pp. 264-70.

Mitchell CR, Joiner MC 2002. Effect of subsequent acute-dose irradiation on cell survival *in vitro* following low dose-rate exposures. International Journal of Radiation Biology 78(11): 981-90.

Mitchell CR, Folkard M, Joiner MC 2002. Effects of exposure to low-dose-rate Co-60 gamma rays on human tumor cells *in vitro*. Radiation Research 158(3): 311-18.

Mitchell CR, Azizova TV, Hande MP, Burak LE, Tsakok JM, Khokhryakov VF, Geard CR, Brenner DJ 2004a. Stable intrachromosomal biomarkers of past exposure to densely ionizing radiation in several chromosomes of exposed individuals. Radiation Research 162(3): 257-63.

Mitchell SA, Randers-Pehrson G, Brenner DJ, Hall EJ 2004b. The bystander response in C3H 10T1/2 cells: The influence of cell-to-cell contact. Radiation Research 161(4): 397-401.

Mitchell SA, Marino SA, Brenner DJ, Hall EJ 2004c. Bystander effect and adaptive response in C3H 10T(1)/(2) cells. International Journal of Radiation Biology 80(7): 465-72.

Moolgavkar SH 1983. Model for human carcinogenesis - action of environmental agents. Environmental Health Perspectives 50(Apr): 285-91.

Moolgavkar SH, Knudson AG 1981. Mutation and cancer - a model for human carcinogenesis. Journal of the National Cancer Institute 66(6): 1037-52.

Moolgavkar SH, Cross FT, Luebeck G, Dagle GE 1990. A 2-mutation model for radon-induced lung thmors in rats. Radiation Research 121(1): 28-37.

Moore SR, Marsden S, Macdonald D, Mitchell S, Folkard M, Michael B, Goodhead DT, Prise KM, Kadhim MA 2005. Genomic instability in human lymphocytes irradiated with individual charged particles: Involvement of tumor necrosis factor a in irradiated cells but not bystander cells. Radiation Research 163(2): 183-90.

Morgan WF 2002. Genomic instability and bystander effects: a paradigm shift in radiation biology? Military Medicine 167(2): 44-45.

Morgan WF 2003a. Is there a common mechanism underlying genomic instability, bystander effects and other nontargeted effects of exposure to ionizing radiation? Oncogene 22(45): 7094-99.

Morgan WF 2003b. Non-targeted and delayed effects of exposure to ionizing radiation: II. Radiation-induced genomic instability and bystander effects *in vivo*, clastogenic factors and transgenerational effects. Radiation Research 159(5): 581-96.

Morgan WF 2003c. Non-targeted and delayed effects of exposure to ionizing radiation: I. Radiation-induced genomic instability and bystander effects *In Vitro*. Radiation Research 159(5): 567-80.

Morgan WF 2006. Will radiation-induced bystander effects or adaptive responses impact on the shape of the dose response relationships at low doses of ionizing radiation?, Dose-Response an International Journal. Pp. 257-62.

Morgan WF, Sowa MB 2005. Effects of ionizing radiation in nonirradiated cells. Proceedings of the National Academy of Sciences of the United States of America 102(40): 14127-28.

Morgan WF, Sowa MB 2007. Non-targeted bystander effects induced by ionizing radiation. Mutation Research-Fundamental and Molecular Mechanisms of Mutagenesis 616(1-2): 159-64.

Morgan WF, Sowa MB 2009. Non-targeted effects of ionizing radiation: Implications for risk assessment and the radiation dose response profile. Health Physics 97(5): 426-432.

Morgan WF, Day JP, Kaplan MI, McGhee EM, Limoli CL 1996. Genomic instability induced by ionizing radiation. Radiation Research 146(3): 247-58.

Morgan WF, Hartmann A, Limoli CL, Nagar S, Ponnaiya B 2002. Bystander effects in radiation-induced genomic instability. Mutation Research-Fundamental and Molecular Mechanisms of Mutagenesis 504(1-2): 91-100.

Mothersill C, Seymour C 1997. Lethal mutations and genomic instability. International Journal of Radiation Biology 71(6): 751-58.

Mothersill C, Seymour C 2001. Radiation-induced bystander effects: Past history and future directions. Radiation Research 155(6): 759-67.

Mothersill C, Seymour RJ, Seymour CB 2006. Increased radiosensitivity in cells of two human cell lines treated with bystander medium from irradiated repair-deficient cells. Radiation Research 165(1): 26-34.

Mothersill C, Stamato TD, Perez ML, Cummins R, Mooney R, Seymour CB 2000. Involvement of energy metabolism in the production of 'bystander effects' by radiation. British Journal of Cancer 82(10): 1740-46.

Mothersill C, Lyng F, Seymour C, Maguire P, Lorimore S, Wright E 2005. Genetic factors influencing bystander signaling in murine bladder epithelium after low-dose irradiation *in vivo*. Radiation Research 163(4): 391-99.

Mukaida N, Kodama S, Suzuki K, Oshimura M, Watanabe M 2007. Transmission of genomic instability from a single irradiated human chromosome to the progeny of unirradiated Cells. Radiation Research 167(6): 675-81.

Muller WU, Kryscio A, Streffer C 2004. Micronuclei in lymphocytes of uranium miners of the former Wismut SDAG. Cytogenetic and Genome Research 104(1-4): 295-98.

Murley JS, Kataoka Y, Weydert CJ, Oberley LW, Grdina DJ 2002. Delayed cytoprotection after enhancement of Sod2 (MnSOD) gene expression in SA-NH mouse sarcoma cells exposed to WR-1065, the active metabolite of amifostine. Radiation Research 158(1): 101-9.

Murley JS, Kataoka Y, Cao DC, Li JJ, Oberley LW, Grdina DJ 2004. Delayed radioprotection by NF kappa B-mediated induction of Sod2 (MnSOD) in SA-NH tumor cells after exposure to clinically used thiol-containing drugs. Radiation Research 162(5): 536-46.

Murley JS, Kataoka Y, Baker KL, Diamond AM, Morgan WE, Grdina DJ 2007. Manganese superoxide dismutase (SOD2)-mediated delayed radioprotection induced by the free thiol form of amifostine and tumor necrosis factor alpha. Radiation Research 167(4): 465-74.

Murley JS, Nantajit D, Baker KL, Kataoka Y, Li JJ, Grdina DJ 2008. Maintenance of manganese superoxide dismutase (SOD2)-mediated delayed radioprotection induced by repeated administration of the free thiol form of amifostine. Radiation Research 169(5): 495-505.

Nagar S, Morgan WF 2005. The death-inducing effect and genomic instability. Radiation Research 163(3): 316-23.

Nagar S, Smith LE, Morgan WF 2003a. Characterization of a novel epigenetic effect of ionizing radiation: The death-inducing effect. Cancer Research 63(2): 324-28.

Nagar S, Smith LE, Morgan WF 2003b. Mechanisms of cell death associated with death-inducing factors from genomically unstable cell lines. Mutagenesis 18(6): 549-60.

Nagar S, Smith LE, Morgan WF 2005. Variation in apoptosis profiles in radiation-induced genomically unstable cell lines. Radiation Research 163(3): 324-31.

Nagasawa H, Little JB 1992. Induction of sister chromatid exchanges by extremely low doses of alpha particles. Cancer Research 52(22): 6394-96.

Nagasawa H, Cremesti A, Kolesnick R, Fuks Z, Little JB 2002. Involvement of membrane signaling in the bystander effect in irradiated cells. Cancer Research 62(9): 2531-34.

Nagasawa H, Wilson PE, Chen DJ, Thompson LH, Bedford JS, Little JB 2008. Low doses of alpha particles do not induce sister chromatid exchanges in bystander Chinese hamster cells defective in homologous recombination. DNA Repair 7(3): 515-22.

Nagasawa H, Peng Y, Wilson PF, Lio YC, Chen DJ, Bedford JS, Little JB 2005. Role of homologous recombination in the alpha-particle-induced bystander effect for sister chromatid exchanges and chromosomal aberrations. Radiation Research 164(2): 141-47.

Nakamura A, Sedelnikova OA, Redon C, Pilch DR, Sinogeeva NI, Shroff R, Lichten M, Bonner WM 2006. Techniques for gamma-H2AX detection. In: Campbell JLMP ed. DNA Repair, Pt B. Pp. 236-+.

Natarajan M, Gibbons CF, Mohan S, Moore S, Kadhim MA 2007. Oxidative stress signalling: a potential mediator of tumour necrosis factor alpha-induced genomic instability in primary vascular endothelial cells. British Journal of Radiology 80: S13-S22.

NCRP 2006. Cesium-137 in the environment: Radioecology and approaches to assessment and management, NCRP Report No. 154. National Council on Radiation Protection and Measurements, Bethesda, MD, USA.

NCRP 2010. Potential impact of individual genetic susceptibility and previous radiation exposure on radiation risk for astronauts. Bethesda, Md, National Council on Radiation Protection and Measurements.

Nelson CM, Bissell MJ 2005. Modeling dynamic reciprocity: Engineering three-dimensional culture models of breast architecture, function, and neoplastic transformation. Seminars in Cancer Biology 15(5): 342-52.

Nelson JM, Brooks AL, Metting NF, Khan MA, Buschbom RL, Duncan A, Miick R, Braby LA 1996. Clastogenic effects of defined numbers of 3.2 MeV alpha particles on individual CHO-K1 cells. Radiation Research 145(5): 568-74.

Nikjoo H, Bolton CE, Watanabe R, Terrissol M, O'Neill P, Goodhead DT 2002. Modelling of DNA damage induced by energetic electrons (100 eV to 100 keV). Radiation Protection Dosimetry 99(1-4): 77-80.

NRC 1999. Health effects of exposure to radon http://search.ebscohost.com/login.aspx?direct=true&scope=site&db=nlebk&db=nlabk&AN=963.

NRC 2005. Health Risks from Exposure to Low Levels of Ionizing Radiation.

NRC 2006. Health risks from exposure to low levels of ionizing radiation : BEIR VII Phase 2. Washington, D.C., National Academies Press.

Nugent CI, Bosco G, Ross LO, Evans SK, Salinger AP, Moore JK, Haber JE, Lundblad V 1998. Telomere maintenance is dependent on activities required for end repair of double-strand breaks. Current Biology: CB 8(11): 657-60.

Odegaard E, Yang CR, Boothman DA 1998. DNA-dependent protein kinase does not play a role in adaptive survival responses to ionizing radiation. Environmental Health Perspectives 106: 301-5.

Okazaki R, Ootsuyama A, Norimura T 2007. TP53 and TP53-related genes associated with protection from apoptosis in the radioadaptive response. Radiation Research 167(1): 51-57.

Okladnikova ND, Scott BR, Tokarskaya ZB, Zhuntova GV, Khokhryakov VF, Syrchikov VA, Grigoryeva ES 2005. Chromosomal aberrations in lymphocytes of peripheral blood among Mayak facility workers who inhaled insoluble forms of (PU)-P-239. Radiation Protection Dosimetry 113(1): 3-13.

Olive PL 1998. The role of DNA single- and double-strand breaks in cell killing by ionizing radiation. Radiation Research 150(5): S42-S51.

Olivieri G, Bodycote J, Wolff S 1984. Adaptive response of human-lymphocytes to low concentrations of radioactive thymidine. Science 223(4636): 594-97.

Ozeki M, Tamae D, Hou DX, Wang TL, Lebon T, Spitz DR, Li JJ 2004. Response of cyclin B1 to ionizing radiation: Regulation by NF-kappa B and mitochondrial antioxidant enzyme MnSOD. Anticancer Research 24(5A): 2657-63.

Paap B, Wilson DM, III, Sutherland BM 2008. Human abasic endonuclease action on multilesion abasic clusters: implications for radiation-induced biological damage. Nucleic Acids Research 36(8): 2717-27.

Park CC, Bissell MJ, Barcellos-Hoff MH 2000. The influence of the microenvironment on the malignant phenotype. Molecular Medicine Today 6(8): 324-29.

Park HJ, Ahn KJ, Ahn SD, Choi E, Lee SW, Williams B, Kim EJ, Griffin R, Bey EA, Bornmann WG and others 2005. Susceptibility of cancer cells to beta-lapachone is enhanced by ionizing radiation. International Journal of Radiation Oncology Biology Physics 61(1): 212-19.

Pati D, Haddad BR, Haegele A, Thompson H, Kittrell FS, Shepard A, Montagna C, Zhang NG, Ge GQ, Otta SK and others 2004. Hormone-induced chromosomal instability in p53-null mammary epithelium. Cancer Research 64(16): 5608-16.

Patterson A, Lei H, Eichler G, Krauz K, Weinstein J, Fornace A, Gonzalez F, Idle J 2008. UPLC-ESI-TOFMS-Based Metabolomics and Gene Expression Dynamics Inspector Self-Organizing Metabolomic Maps as Tools for Understanding the Cellular Response to Ionizing Radiation. Analytical Chemistry. Pp. 665-74.

Paul S, Gros L, Laval J, Sutherland BM 2006. Expression of the E. coli fpg protein in CHO cells lowers endogenous oxypurine clustered damage levels and decreases accumulation of endogenous Hprt mutations. Environmental and Molecular Mutagenesis 47(5): 311-9.

Perera SA, Maser RS, Xia H, McNamara K, Protopopov A, Chen L, Hezel AF, Kim CF, Bronson RT, Castrillon DH and others 2008. Telomere dysfunction promotes genome instability and metastatic potential in a K-ras p53 mouse model of lung cancer. Carcinogenesis 29(4): 747-53.

Perez-Losada J, Mao JH, Balmain A 2005. Control of genomic instability and epithelial tumor development by the p53-Fbxw7/Cdc4 pathwayle. Cancer Research 65(15): 6488-92.

Persaud R, Zhou H, Hei TK, Hall EJ 2007. Demonstration of a radiation-induced bystander effect for low dose low LET beta-particles. Radiation and Environmental Biophysics 46(4): 395-400.

Petrini JHJ 2007. A touching response to damage. Science 316(5828): 1138-39.

Pichiorri F, Ishii H, Okumura H, Trapass F, Wang Y, Huebner K 2008. Molecular parameters of genome instability: Roles of fragile genes at common fragile sites. Journal of Cellular Biochemistry 104(5): 1525-33.

Pierce AJ, Johnson RD, Thompson LH, Jasin M 1999. XRCC3 promotes homology-directed repair of DNA damage in mammalian cells. Genes & Development 13(20): 2633-2638.

Pierce DA 2003. Mechanistic models for radiation carcinogenesis and the atomic bomb survivor data. Radiation Research 160(6): 718-23.

Pinto M, Prise KM, Michael BD 2005. Evidence for complexity at the nanometer scale of radiation-induced DNA DSBs as a determinant of rejoining kinetics. Radiation Research 164(1): 73-85.

Plan Y, Hlatky L, Hahnfeldt P, Sachs R, Loucas B, Cornforth M 2005. Full-color painting reveals an excess of radiation-induced dicentrics involving homologous chromosomes. International Journal of Radiation Biology 81(8): 613-20.

Pluth JM, Yamazaki V, Cooper BA, Rydberg BE, Kirchgessner CU, Cooper PK 2008. DNA double-strand break and chromosomal rejoining defects with misrejoining in Nijmegen breakage syndrome cells. DNA Repair 7(1): 108-18.

Ponnaiya B, Cornforth MN, Ullrich RL 1997a. Radiation-induced chromosomal instability in BALB/c and C57BL/6 mice: The difference is as clear as black and white. Radiation Research 147(2): 121-25.

Ponnaiya B, Cornforth MN, Ullrich RL 1997b. Induction of chromosomal instability in human mammary cells by neutrons and gamma rays. Radiation Research 147(3): 288-94.

Ponnaiya B, Jenkins-Baker G, Randers-Pherson G, Geard CR 2007. Quantifying a bystander response following microbearn irradiation using single-cell RT-PCR analyses. Experimental Hematology 35(4): 64-68.

Ponnaiya B, Jenkins-Baker G, Brenner DJ, Hall EJ, Randers-Pehrson G, Geard CR 2004. Biological responses in known bystander cells relative to known microbeam-irradiated cells. Radiation Research 162(4): 426-32.

Ponomarev AL, Cucinotta FA, Sachs RK, Brenner DJ 2001a. Monte Carlo predictions of DNA fragment-size distributions for large sizes after HZE particle irradiation. Physica Medica 17: 153-56.

Ponomarev AL, Cucinotta FA, Sachs RK, Brenner DJ, Peterson LE 2001b. Extrapolation of the DNA fragment-size distribution after high-dose irradiation to predict effects at low doses. Radiation Research 156(5): 594-97.

Ponomarev AL, Belli M, Hahnfeldt PJ, Hlatky L, Sachs RK, Cucinotta FA 2006. A robust procedure for removing background damage in assays of radiation-induced DNA fragment distributions. Radiation Research 166(6): 908-16.

Portess DI, Bauer G, Hill MA, O'Neill P 2007. Low-dose irradiation of nontransformed cells stimulates the selective removal of precancerous cells via intercellular induction of apoptosis. Cancer Research 67(3): 1246-53.

Poston SJW, Ford JR 2009. How do we combine science and regulations for decision making following a terrorist incident involving radioactive materials? Health Phys. Health Physics 97(5): 537-41.

Preston RJ 2003. The LNT model is the best we can do - today. Journal of Radiological Protection 23(3): 263-68.

Prise KM, Folkard M, Michael BD 2003. Bystander responses induced by low LET radiation. Oncogene 22(45): 7043-49.

Prise KM, Folkard M, Michael BD 2006a. Radiation-induced bystander and adaptive responses in cell and tissue models. Dose-Response: a Publication of the International Hormesis Society 4(4): 263-76.

Prise KM, Belyakov OV, Folkard M, Michael BD 1998. Studies of bystander effects in human fibroblasts using a charged particle microbeam. International Journal of Radiation Biology 74(6): 793-98.

Prise KM, Pinto M, Newman HC, Michael BD 2001. A review of studies of ionizing radiation-induced double-strand break clustering. Radiation Research 156(5): 572-76.

Prise KM, Folkard M, Kuosaite V, Tartier L, Zyuzikov N, Shao CL 2006b. What role for DNA damage and repair in the bystander response? Mutation Research-Fundamental and Molecular Mechanisms of Mutagenesis 597(1-2): 1-4.

Prise KM, Belyakov OV, Folkard M, Ozols A, Schettino G, Vojnovic B, Michael BD 2002. Investigating the cellular effects of isolated radiation tracks using microbeam techniques. In: Ijiri KSKKA ed. Space Life Sciences: Biological Research and Space Radiation. Pp. 871-76.

Purchase I, Slovic P 1999. PERSPECTIVE: Quantitative Risk Assessment Breeds Fear. Human and Ecological Risk Assessment: 445-53.

Puskin JS 2009. Perspective on the use of LNT for Radiation Protection and Risk Assessment by the US Environmental Protection Agency. Dose-Response 7(4): 284-91.

Pyke EL, Stevens DL, Hill MA 2006. Keeping up with the neighbours - Measuring the bystander response. Radiation Protection Dosimetry 122(1-4): 266-70.

Radivoyevitch T, Kozubek S, Sachs RK 2001. Biologically based risk estimation for radiation-induced CML - Inferences from BCR and ABL geometric distributions. Radiation and Environmental Biophysics 40(1): 1-9.

Radivoyevitch T, Kozubek S, Sachs RK 2002. The risk of chronic myeloid leukemia: Can the dose-response curve be U-shaped? Radiation Research 157(1): 106-9.

Rainaldi G, Romano R, Indovina P, Ferrante A, Motta A, Indovina PL, Santini MT 2008. Metabolomics using H-1-NMR of apoptosis and necrosis in HL60 leukemia cells: Differences between the two types of cell death and independence from the stimulus of apoptosis used. Radiation Research 169(2): 170-80.

Rainey MD, Charlton ME, Stanton RV, Kastan MB 2008. Transient inhibition of ATM kinase is sufficient to enhance cellular sensitivity to ionizing radiation. Cancer Research 68(18): 7466-74.

Raman S, Maxwell CA, Barcellos-Hoff MH, Parvin B 2007. Geometric approach to segmentation and protein localization in cell culture assays. Journal of Microscopy-Oxford 225(1): 22-30.

Randers-Pehrson G, Geard CR, Johnson G, Elliston CD, Brenner DJ 2001. The Columbia University single-ion microbeam. Radiation Research 156(2): 210-14.

Rao RS, Visuri SR, McBride MT, Albala JS, Matthews DL, Coleman MA 2004. Comparison of multiplexed techniques for detection of bacterial and viral proteins. Journal of Proteome Research 3(4): 736-42.

Rassool FV, Gaymes TJ, Omidvar N, Brady N, Beurlet S, Pla M, Reboul M, Lea N, Chomienne C, Thomas NSB and others 2007. Reactive oxygen species, DNA damage, and error-prone repair: A model for genomic instability with progression in myeloid leukemia? Cancer Research 67(18): 8762-71.

Ray S, Chatterjee A 2007. Influence of glutathione on the induction of chromosome aberrations, delay in cell cycle kinetics and cell cycle regulator proteins in irradiated mouse bone marrow cells. International Journal of Radiation Biology 83(5): 347-54.

Redpath JL 2004. Radiation-induced neoplastic transformation *in vitro*: Evidence for a protective effect at low doses of low LET radiation. Cancer and Metastasis Reviews 23(3-4): 333-39.

Redpath JL 2005. Nonlinear response for neoplastic transformation following low doses of low LET radiation. Nonlinearity in Biology, Toxicology, Medicine 3(1): 113-24.

Redpath JL 2006a. Health risks of low photon energy imaging. Radiation Protection Dosimetry 122(1-4): 528-33.

Redpath JL 2006b. Suppression of Neoplastic Transformation *In Vitro* by Low Doses of Low LET Radiation. Dose Response. Pp. 302-8.

Redpath JL 2007. *In vitro* radiation-induced neoplastic transformation: Suppressive effects at low doses. Radiation Research 167(3): 345-46.

Redpath JL, Elmore E 2007. Radiation-Induced Neoplastic Transformation *in Vitro*, Hormesis and Risk Assessment. Dose Response An International Journal. Pp. 123-30.

Redpath JL, Sun C, Colman M, Stanbridge EJ 1987. Neoplastic transformation of human hybrid-cells by gamma radiation - a quantitative assay. Radiation Research 110(3): 468-72.

Redpath JL, Liang D, Taylor TH, Christie C, Elmore E 2001. The shape of the dose-response curve for radiation-induced neoplastic transformation *in vitro*: Evidence for an adaptive response against neoplastic transformation at low doses of low-LET radiation. Radiation Research 156(6): 700-7.

Redpath JL, Lu Q, Lao X, Molloi S, Elmore E 2003. Low doses of diagnostic energy X-rays protect against neoplastic transformation *in vitro*. International Journal of Radiation Biology 79(4): 235-40.

Resat MBS, Morgan WF 2004a. Radiation-induced genomic instability: A role for secreted soluble factors in communicating the radiation response to non-irradiated cells. Journal of Cellular Biochemistry 92(5): 1013-19.

Resat MS, Morgan WF 2004b. Microbeam developments and applications: A low linear energy transfer perspective. Cancer and Metastasis Reviews 23(3-4): 323-31.

Ridnour LA, Sim JE, Choi J, Dickinson DA, Forman HJ, Ahmad IM, Coleman MC, Hunt CR, Goswami PC, Spitz DR 2005. Nitric oxide-induced resistance to hydrogen peroxide stress is a glutamate cysteine ligase activity-dependent process. Free Radical Biology and Medicine 38(10): 1361-71.

Rithidech K, Honikel L, Whorton E 2007a. mFISH analysis of chromosomal damage in bone marrow cells collected from CBA/CaJ mice following whole body exposure to heavy ions (^{56}Fe ions). Radiation and Environmental Biophysics 46(2): 137-45.

Rithidech K, Dunn JJ, Bond VP, Gordon CR, Cronkite EP 1999. Characterization of genetic instability in radiation- and benzene-induced murine acute leukemia. Mutation Research-Fundamental and Molecular Mechanisms of Mutagenesis 428(1-2): 33-39.

Rithidech KN, Honikel L, Simon SR 2007b. Radiation leukemogenesis: A proteomic approach. Experimental Hematology 35(4): 117-24.

Rithidech KN, Tungjai M, Arbab E, Simon SR 2005. Activation of NF-kappa B in bone marrow cells of BALB/cJ mice following exposure *in vivo* to low doses of Cs-137 gamma-rays. Radiation and Environmental Biophysics 44(2): 139-43.

Robson TA, Lohrer H, Bailie JR, Hirst DG, Joiner MC, Arrand JE 1997. Gene regulation by low-dose ionizing radiation in a normal human lung epithelial cell line. Biochemical Society Transactions 25(1): 335-42.

Rocke DM, Goldberg Z, Schweitert C, Santana A 2005. A method for detection of differential gene expression in the presence of inter-individual variability in response. Bioinformatics 21(21): 3990-92.

Rodemann HP, Dittmann K, Toulany M 2007. Radiation-induced EGFR-signaling and control of DNA-damage repair. International Journal of Radiation Biology 83(11-12): 781-791.

Rodgers BE, Baker RJ 2000. Frequencies of micronuclei in bank voles from zones of high radiation at Chornobyl, Ukraine. Environmental Toxicology and Chemistry 19(6): 1644-48.

Rodgers BE, Chesser RK, Wickliffe JK, Phillips CJ, Baker RJ 2001. Subchronic exposure of BALB/c and C57BL/6 strains of Mus musculus to the radioactive environment of the Chornobyl, Ukraine exclusion zone. Environmental Toxicology and Chemistry 20(12): 2830-35.

Rogakou EP, Pilch DR, Orr AH, Ivanova VS, Bonner WM 1998. DNA double-stranded breaks induce histone H2AX phosphorylation on serine 139. Journal of Biological Chemistry 273(10): 5858-68.

Romney CA, Paulauskis JD, Nagasawa H, Little JB 2001. Multiple manifestations of X-ray-induced genomic instability in Chinese hamster ovary (CHO) cells. Molecular Carcinogenesis 32(3): 118-27.

Rothkamm K, Lobrich M 2003. Evidence for a lack of DNA double-strand break repair in human cells exposed to very low x-ray doses. Proceedings of the National Academy of Sciences of the United States of America 100(9): 5057-62.

Ryan LA, Wilkins RC, McFarlane NM, Sung MM, McNamee JP, Boreham DR 2006. Relative biological effectiveness of 280 keV neutrons for apoptosis in human lymphocytes. Health Physics 91(1): 68-75.

Rydberg B, Lobrich M, Cooper PK 1994. DNA double-strand breaks induced by high-energy neon and iron ions in human fibroblasts .1. Pulsed-field gel-electrophoresis method. Radiation Research 139(2): 133-41.

Rydberg B, Cooper B, Cooper PK, Holley WR, Chatterjee A 2005. Dose-dependent misrejoining of radiation-induced DNA double-strand breaks in human fibroblasts: Experimental and theoretical study for high- and low-LET radiation. Radiation Research 163(5): 526-34.

Sachs RK, Levy D, Hahnfeldt P, Hlatky L 2004. Quantitative analysis of radiation-induced chromosome aberrations. Cytogenetic and Genome Research 104(1-4): 142-48.

Sachs RK, Chan M, Hlatky L, Hahnfeldt P 2005. Modeling intercellular interactions during carcinogenesis. Radiation Research 164(3): 324-31.

Sachs RK, Arsuaga J, Vazquez M, Hlatky L, Hahnfeldt P 2002. Using graph theory to describe and model chromosome aberrations. Radiation Research 158(5): 556-67.

Sachs RK, Levy D, Chen AM, Simpson PJ, Cornforth MN, Ingerman EA, Hahnfeld P, Hlatky LR 2000. Random breakage and reunion chromosome aberration formation model; an interaction-distance version based on chromatin geometry. International Journal of Radiation Biology 76(12): 1579-88.

Sahijdak WM, Yang CR, Zuckerman JS, Meyers M, Boothman DA 1994. Alterations in transcription factor-binding in radioresistant human-melanoma cells after ionizing-radiation. Radiation Research 138(1): S47-S51.

Sakai K, Hoshi Y, Nomura T, Oda T, Iwasaki T, Fujita K, Yamada T, Tanooka H 2003. Suppression of carcinogenic processes in mice by low dose rate gamma-irradiation. International Journal of Low Radiation. Pp. 142-46.

Sanders CL, Scott BR 2008. Smoking and hormesis as confounding factors in radiation pulmonary carcinogenesis. Dose-Response 6(1): 53-79.

Sandfort V, Koch U, Cordes N 2007. Cell adhesion-mediated radioresistance revisited. International Journal of Radiation Biology 83(11-12): 727-32.

Sawant SG, Randers-Pehrson G, Metting NF, Hall EJ 2001a. Adaptive response and the bystander effect induced by radiation in C3H 10T(1)/(2) cells in culture. Radiation Research 156(2): 177-80.

Sawant SG, Randers-Pehrson G, Geard CR, Brenner DJ, Hall EJ 2001b. The bystander effect in radiation oncogenesis: I. Transformation in C3H 10T(1)/(2) cells *in vitro* can be initiated in the unirradiated neighbors of irradiated cells. Radiation Research 155(3): 397-401.

Schafer J, Bachtler J, Engling A, Little JB, Weber KJ, Wenz F 2002. Suppression of apoptosis and clonogenic survival in irradiated human lymphoblasts with different TP53 status. Radiation Research 158(6): 699-706.

Schettino G, Folkard M, Michael BD, Prise KM 2005. Low-dose binary behavior of bystander cell killing after microbeam irradiation of a single cell with focused C-K X rays. Radiation Research 163(3): 332-36.

Schettino G, Folkard M, Michette AG, Prise KM, Vojnovic B, Michael B 2000. A Focused Soft X-ray Microbeam for Investigating the Radiation Responses of Individual Cells. In: Prade H ed. Workshop on X-rays from Electron Beams. Pp. 229-46.

Schettino G, Folkard M, Prise KM, Vojnovic B, Held KD, Michael BD 2003. Low-dose studies of bystander cell killing with targeted soft X rays. Radiation Research 160(5): 505-511.

Schollnberger H, Mitchel REJ, Azzam EI, Crawford-Brown DJ, Hofmann W 2002. Explanation of protective effects of low doses of gamma-radiation with a mechanistic radiobiological model. International Journal of Radiation Biology 78(12): 1159-73.

Schollnberger H, Mitchel REJ, Redpath JL, Crawford-Brown DJ, Hofmann W 2007. Detrimental and protective bystander effects: A model approach. Radiation Research 168(5): 614-26.

Schöllnberger H, Scott B, Hanson T 2001. Application of Bayesian inference to characterize risks associated with low doses of low-LET radiation. Bulletin of Mathematical Biology 63(5): 865-84.

Schulte-Hermann R, Grasl-Kraupp B, Bursch W 2000. Dose-response and threshold effects in cytotoxicity and apoptosis. Mutation Research-Genetic Toxicology and Environmental Mutagenesis 464(1): 13-18.

Schwartz JL 2004. Abandon hope all ye target theory modelers: on the effects of low dose exposures to ionizing radiation and other carcinogens. Mutation Research 568(1): 3-4.

Schwartz JL 2007. Variability: The common factor linking low dose-induced genomic instability, adaptation and bystander effects. Mutation Research-Fundamental and Molecular Mechanisms of Mutagenesis 616(1-2): 196-200.

Schwartz JL, Jordan R, Evans HH 2001. Characteristics of chromosome instability in the human lymphoblast cell line WTK1. Cancer Genetics and Cytogenetics 129(2): 124-30.

Schwartz JL, Jordan R, Evans HH, Lenarczyk M, Liber H 2003. The TP53 dependence of radiation-induced chromosome instability in human lymphoblastoid cells. Radiation Research 159(6): 730-36.

Schwartz JL, Jordan R, Sedita BA, Swenningson MJ, Banath JP, Olive PL 1995. Different sensitivity to cell-killing and chromosome mutation-induction by gamma-rays in 2 human lymphoblastoid cell-lines derived from a single-donor - possible role of apoptosis. Mutagenesis 10(3): 227-33.

Schwartz JL, Rotmensch J, Sun J, An J, Xu ZD, Yu YJ, Hsie AW 1994. Multiplex polymerase chain reaction-based deletion analysis of spontaneous, gamm-ray-induced and alpha-induced HPRT mutants of CHO-K1 cells. Mutagenesis 9(6): 537-40.

Scott BR 2005a. Low Dose Radiation Risk Extrapolation Fallacy Associated with the Linear-No-Threshold Model. Belle Newsletter. Pp. 22-27.

Scott BR 2005b. Stochastic Thresholds: A Novel Explanation of Nonlinear Dose-Response Relationships for Stochastic Radiobiological Effects. Dose Response. Pp. 547-67.

Scott BR 2007. Low Dose Radiation Induced Protective Process and Implications for Risk Assessment, Cancer Prevention, and Cancer Therapy. Dose Response - an International Journal. Pp. 131-49.

Scott BR 2008. Low-dose radiation risk extrapolation fallacy associated with the linear-no-threshold model. Human & Experimental Toxicology 27(2): 163-68.

Scott BR, Schollnberger H 2000. Introducing biological microdosimetry for ionising radiation. Radiation Protection Dosimetry 91(4): 377-84.

Sedelnikova OA, Nakamura A, Kovalchuk O, Koturbash I, Mitchell SA, Marino SA, Brenner DJ, Bonner WM 2007. DNA double-strand breaks form in bystander cells after microbeam irradiation of three-dimensional human tissue models. Cancer Research 67(9): 4295-302.

Semenenko VA, Stewart RD 2004. A fast Monte Carlo algorithm to simulate the spectrum of DNA damages formed by ionizing radiation. Radiation Research 161(4): 451-57.

Semenenko VA, Stewart RD 2005. Monte Carlo simulation of base and nucleotide excision repair of clustered DNA damage sites. II. Comparisons of model predictions to measured data. Radiation Research 164(2): 194-201.

Semenenko VA, Stewart RD, Ackerman EJ 2005. Monte Carlo simulation of base and nucleotide excision repair of clustered DNA damage sites. I. Model properties and predicted trends. Radiation Research 164(2): 180-93.

Sgouros G, Knox SJ, Joiner MC, Morgan WF, Kassis AI 2007. MIRD continuing education: Bystander and low-dose-rate effects: Are these relevant to radionuclide therapy? Journal of Nuclear Medicine 48(10): 1683-91.

Shadley JD, Afzal V, Wolff S 1987. Characterization of the adaptive response to ionizing radiation induced by low doses of X-rays to human lymphocytes. Radiation Research 111(3).

Shankar B, Pandey R, Sainis K 2006. Radiation-induced bystander effects and adaptive response in murine lymphocytes. International Journal of Radiation Biology 82(8): 537-48.

Shannan B, Seifert M, Boothman DA, Tilgen W, Reichrath J 2007. Clusterin overexpression modulates proapoptotic and antiproliferative effects of 1,25(OH)(2)D-3 in prostate cancer cells *in vitro*. Journal of Steroid Biochemistry and Molecular Biology 103(3-5): 721-25.

Shannan B, Seifert M, Leskov K, Willis J, Boothman D, Tilgen W, Reichrath J 2006. Challenge and promise: roles for clusterin in pathogenesis, progression and therapy of cancer. Cell Death and Differentiation 13(1): 12-19.

Shao C, Folkard M, Prise KM 2008a. Role of TGF-beta 1 and nitric oxide in the bystander response of irradiated glioma cells. Oncogene 27(4): 434-40.

Shao C, Prise KM, Folkard M 2008b. Signaling factors for irradiated glioma cells induced bystander responses in fibroblasts. Mutation Research-Fundamental and Molecular Mechanisms of Mutagenesis 638(1-2): 139-45.

Shao C, Lyng FM, Folkard M, Prise KM 2006. Calcium fluxes modulate the radiation-induced bystander responses in targeted glioma and fibroblast cells. Radiation Research 166(3): 479-87.

Shao C, Stewart V, Folkard M, Michael BA, Prise KM 2003a. Nitric oxide-mediated signaling in the bystander response of individually targeted glioma cells. Cancer Research 63(23): 8437-42.

Shao C, Furusawa Y, Matsumoto Y, Pan Y, Xu P, Chen H 2007. Effect of gap junctional intercellular communication on radiation responses in neoplastic human cells. Radiation Research 167(3): 283-88.

Shao CL, Furusawa Y, Aoki M, Ando K 2003b. Role of gap junctional intercellular communication in radiation-induced bystander effects in human fibroblasts. Radiation Research 160(3): 318-23.

Shao CL, Folkard M, Michael BD, Prise KM 2004. Targeted cytoplasmic irradiation induces bystander responses. Proceedings of the National Academy of Sciences of the United States of America 101(37): 13495-500.

Sheikh MS, Fornace AJ 1999. Regulation of translation initiation following stress. Oncogene 18(45): 6121-28.

Shen NR, Zdzienicka MZ, Mohrenweiser H, Thompson LH, Thelen MP 1998. Mutations in hamster single-strand break repair gene XRCC1 causing defective DNA repair. Nucleic Acids Research 26(4): 1032-37.

Shore RE 2009. Low-Dose Radiation Epidemiology Studies: Status and Issues. Health Physics 97(5): 481-86.

Short SC, Woodcock M, Marples B, Joiner MC 2003. Effects of cell cycle phase on low-dose hyper-radiosensitivity. International Journal of Radiation Biology 79(2): 99-105.

Short SC, Mitchell SA, Boulton P, Woodcock M, Joiner MC 1999. The response of human glioma cell lines to low-dose radiation exposure. International Journal of Radiation Biology 75(11): 1341-48.

Short SC, Kelly J, Mayes CR, Woodcock M, Joiner MC 2001. Low-dose hypersensitivity after fractionated low-dose irradiation *in vitro*. International Journal of Radiation Biology 77(6): 655-64.

Slane BG, Aykin-Burns N, Smith BJ, Kalen AL, Goswami PC, Domann FE, Spitz DR 2006. Mutation of succinate dehydrogenase subunit C results in increased O-2(center dot-), oxidative stress, and genomic instability. Cancer Research 66(15): 7615-20.

Slee EA, O'Connor DJ, Lu X 2004. To die or not to die: how does p53 decide? Oncogene 23(16): 2809-18.

Slovic P 1996. Perception of Risk from Radiation. Radiation protection dosimetry. 68(3-4): 165.

Smilenov LB, Brenner DJ, Hall EJ 2001. Modest increased sensitivity to radiation oncogenesis in ATM heterozygous versus wild-type mammalian cells. Cancer Research 61(15): 5710-13.

Smilenov LB, Hall EJ, Bonner WM, Sedelnikova OA 2006. A microbeam study of DNA double-strand breaks in bystander primary human fibroblasts. Radiation Protection Dosimetry 122(1-4): 256-59.

Smilenov LB, Lieberman HB, Mitchell SA, Baker RA, Hopkins KM, Hall EJ 2005. Combined haploinsufficiency for ATM and RAD9 as a factor in cell transformation, apoptosis, and DNA lesion repair dynamics. Cancer Research 65(3): 933-38.

Smith LE, Nagar S, Kim GJ, Morgan WF 2003. Radiation-induced genomic instability: Radiation quality and dose response. Health Physics 85(1): 23-29.

Snyder AR, Morgan WF 2004a. Radiation-induced chromosomal instability and gene expression profiling: searching for clues to initiation and perpetuation. Mutation Research-Fundamental and Molecular Mechanisms of Mutagenesis 568(1): 89-96.

Snyder AR, Morgan WF 2004b. Gene expression profiling after irradiation: Clues to understanding acute and persistent responses? Cancer and Metastasis Reviews 23(3-4): 259-68.

Snyder AR, Morgan WF 2005a. Differential induction and activation of NF-kappa B transcription complexes in radiation-induced chromosomally unstable cell lines. Environmental and Molecular Mutagenesis 45(2-3): 177-87.

Snyder AR, Morgan WF 2005b. Lack of consensus gene expression changes associated with radiation-induced chromosomal instability. DNA Repair 4(9): 958-70.

Sokolov MV, Smilenov LB, Hall EJ, Panyutin IG, Bonner WM, Sedelnikova OA 2005. Ionizing radiation induces DNA double-strand breaks in bystander primary human fibroblasts. Oncogene 24(49): 7257-65.

Somodi Z, Zyuzikov NA, Kashino G, Trott KR, Prise KM 2005. Radiation-induced genomic instability in repair deficient mutants of Chinese hamster cells. International Journal of Radiation Biology 81(12): 929-36.

Sondhaus CA, Bond VP, Feinendegen LE 1996. The use of cell-oriented factors and the hit size effectiveness function in radiation protection. Health Physics 70(6): 868-76.

Song JM, Milligan JR, Sutherland BM 2002. Bistranded oxidized purine damage clusters: Induced in DNA by long-wavelength ultraviolet (290-400 nm) radiation? Biochemistry 41(27): 8683-88.

Sowa MB, Murphy MK, Miller JH, McDonald JC, Strom DJ, Kimmel GA 2005. A variable-energy electron microbeam: A unique modality for targeted low-LET radiation. Radiation Research 164(5): 695-700.

Sowa Resat MB, Morgan WF 2004. Radiation-induced genomic instability: a role for secreted soluble factors in communicating the radiation response to non-irradiated cells. Journal of Cellular Biochemistry 92(5): 1013-9.

Spitz DR, Azzam EI, Li JJ, Gius D 2004. Metabolic oxidation/reduction reactions and cellular responses to ionizing radiation: A unifying concept in stress response biology. Cancer and Metastasis Reviews 23(3-4): 311-22.

Stannard JN, Baalman RW, United States. Dept. of Energy. Office of H, Environmental R 1988. Radioactivity and health : a history. [Richland, Wash.]; Springfield, Va., Pacific Northwest Laboratory; Available from National Technical Information Service.

Stewart RD 2001. Two-lesion kinetic model of double-strand break rejoining and cell killing. Radiation Research 156(4): 365-78.

Stewart RD, Ratnayake RK, Jennings K 2006. Microdosimetric model for the induction of cell killing through medium-borne signals. Radiation Research 165(4): 460-69.

Stiff T, O'Driscoll M, Rief N, Iwabuchi K, Lobrich M, Jeggo PA 2004. ATM and DNA-PK function redundantly to phosphorylate H2AX after exposure to ionizing radiation. Cancer Research 64(7): 2390-96.

Straume T, Amundson SA, Blakely WF, Burns FJ, Chen A, Dainiak N, Franklin S, Leary JA, Loftus DJ, Morgan WF and others 2008. NASA Radiation Biomarker Workshop, September 27-28, 2007. Radiation Research 170(3): 393-405.

Strozyk E, Poeppelmann B, Schwarz T, Kulms D 2006. Differential effects of NF-kappa B on apoptosis induced by DNA-damaging agents: the type of DNA damage determines the final outcome. Oncogene 25(47): 6239-51.

Sturbaum B, Brooks AL, McClellan RO 1970. Tissue distribution and dosimetry of ^{144}Ce in Chinese hamsters. Radiation Research 44: 359-67.

Sutherland BM, Bennett PV, Sidorkina O, Laval J 2000a. Clustered damages and total lesions induced in DNA by ionizing radiation: Oxidized bases and strand breaks. Biochemistry 39(27): 8026-31.

Sutherland BM, Bennett PV, Sidorkina O, Laval J 2000b. Clustered DNA damages induced in isolated DNA and in human cells by low doses of ionizing radiation. Proceedings of the National Academy of Sciences of the United States of America 97(1): 103-8.

Sutherland BM, Bennett PV, Sutherland JC, Laval J 2002a. Clustered DNA damages induced by X rays in human cells. Radiation Research 157(6): 611-16.

Sutherland BM, Bennett PV, Georgakilas AG, Sutherland JC 2003a. Evaluation of number average length analysis in quantifying double strand breaks in genomic DNAs. Biochemistry 42(11): 3375-84.

Sutherland BM, Bennett PV, Weinert E, Sidorkina O, Laval J 2001a. Frequencies and relative levels of clustered damages in DNA exposed to gamma rays in radioquenching vs. nonradioquenching conditions. Environmental and Molecular Mutagenesis 38(2-3): 159-65.

Sutherland BM, Bennett PV, Saparbaev M, Sutherland JC, Laval J 2001b. Clustered DNA damages as dosemeters for ionising radiation exposure and biological responses. Radiation Protection Dosimetry 97(1): 33-38.

Sutherland BM, Georgakilas AG, Bennett PV, Laval J, Sutherland JC 2003b. Quantifying clustered DNA damage induction and repair by gel electrophoresis, electronic imaging and number average length analysis. Mutation Research-Fundamental and Molecular Mechanisms of Mutagenesis 531(1-2): 93-107.

Sutherland BM, Bennett PV, Cintron NS, Guida P, Laval J 2003c. Low levels of endogenous oxidative damage cluster levels in unirradiated viral and human DNAs. Free Radical Biology and Medicine 35(5): 495-503.

Sutherland BM, Bennett PV, Schenk H, Sidorkina O, Laval J, Trunk J, Monteleone D, Sutherland J 2001c. Clustered DNA damages induced by high and low LET radiation, including heavy ions. Physica Medica 17: 202-4.

Sutherland BM, Bennett PV, Cintron-Torres N, Hada M, Trunk J, Monteleone D, Sutherland JC, Laval J, Stanislaus M, Gewirtz A 2002b. Clustered DNA damages induced in human hematopoietic cells by low doses of ionizing radiation. Journal of Radiation Research 43: S149-S152.

Sutherland JC, Monteleone DC, Trunk JG, Bennett PV, Sutherland BM 2001d. Quantifying DNA damage by gel electrophoresis, electronic imaging and number-average length analysis. Electrophoresis 22(5): 843-54.

Suzuki M, Boothman DA 2008. Stress-induced premature senescence (SIPS) - Influence of SIPS on radiotherapy. Journal of Radiation Research 49(2): 105-12.

Suzuki M, Zhou HN, Geard CR, Hei TK 2004. Effect of medium on chromatin damage in bystander mammalian cells. Radiation Research 162(3): 264-69.

Sykes PJ, Morley AA, Hooker AM 2006a. The PKZ1 Recombination Mutation Assay: A Sensitive Assay for Low Dose Studies. Dose-Response and International Journal. Pp. 91-105.

Sykes PJ, Day TK, Swinburne SJ, Lane JM, Morley AA, Hooker AM, Bhat M 2006b. *In Vivo* mutagenic effect of very low dose radiation. Dose-Response 4(4): 309-16.

Takahashi A, Matsumoto H, Furusawa Y, Ohnishi K, Ishioka N, Ohnishi T 2005. Apoptosis induced by high-LET radiations is not affected by cellular p53 gene status. International Journal of Radiation Biology 81(8): 581-86.

Takata M, Sasaki MS, Tachiiri S, Fukushima T, Sonoda E, Schild D, Thompson LH, Takeda S 2001. Chromosome instability and defective recombinational repair in knockout mutants of the five Rad51 paralogs. Molecular and Cellular Biology 21(8): 2858-66.

Tanaka T, Kajstura M, Halicka HD, Traganos F, Darzynkiewicz Z 2007. Constitutive histone H2AX phosphorylation and ATM activation are strongly amplified during mitogenic stimulation of lymphocytes. Cell Proliferation 40(1): 1-13.

Tapio S, Jacob V 2007. Radioadaptive response revisited. Radiation and Environmental Biophysics 46(1): 1-12.

Tebbs RS, Hinz JM, Yamada NA, Wilson JB, Salazar EP, Thomas CB, Jones IM, Jones NJ, Thompson LH 2005. New insights into the Fanconi anemia pathway from an isogenic FancG hamster CHO mutant. DNA Repair 4(1): 11-22.

Tenforde TS, Brooks AL 2009. Perspectives of U.S. government agencies on the potential role of greater scientific understanding of low-dose radiation effects in establishing regulatory health protection guidance. Health Phys. Health Physics 97(5): 516-18.

Thompson LH, Schild D 1999. The contribution of homologous recombination in preserving genome integrity in mammalian cells. Biochimie 81(1-2): 87-105.

Thompson LH, West MG 2000. XRCC1 keeps DNA from getting stranded. Mutation Research-DNA Repair 459(1): 1-18.

Thompson LH, Schild D 2002. Recombinational DNA repair and human disease. Mutation Research-Fundamental and Molecular Mechanisms of Mutagenesis 509(1-2): 49-78.

Thompson LH, Hinz JM, Yamada NA, Jones NJ 2005. How Fanconi anemia proteins promote the four Rs: Replication, recombination, repair, and recovery. Environmental and Molecular Mutagenesis 45(2-3): 128-42.

Thompson RC 1989. Life-span effects of ionizing radiation in the beagle dog. [S.l.], Battelle.

Toburen LH, Shinpaugh JL, Justiniano ELB 2002. Modelling interaction cross sections for intermediate and low energy ions. Radiation Protection Dosimetry 99(1-4): 49-51.

Tomascik-Cheeseman LM, Coleman MA, Marchetti F, Nelson DO, Kegelmeyer LM, Nath J, Wyrobek AJ 2004. Differential basal expression of genes associated with stress response, damage control, and DNA repair among mouse tissues. Mutation Research-Genetic Toxicology and Environmental Mutagenesis 561(1-2): 1-14.

Tovar C, Rosinski J, Filipovic Z, Higgins B, Kolinsky K, Hilton H, Zhao XL, Vu BT, Qing WG, Packman K and others 2006. Small-molecule MDM2 antagonists reveal aberrant p53 signaling in cancer: Implications for therapy. Proceedings of the National Academy of Sciences of the United States of America 103(6): 1888-93.

Trosko JE, Chang CC, Madhukar BV 1990. Modulation of intercellular communication during radiation and chemical carcinogenesis. Radiation Research 123(3): 241-51.

Tsai KKC, Chuang EYY, Little JB, Yuan ZM 2005. Cellular mechanisms for low-dose ionizing radiation-induced perturbation of the breast tissue microenvironment. Cancer Research 65(15): 6734-44.

Tsai MH, Chen X, Chandramouli GVR, Chen Y, Yan H, Zhao S, Keng P, Liber HL, Coleman CN, Mitchell JB and others 2006. Transcriptional responses to ionizing radiation reveal that p53R2 protects against radiation-induced mutagenesis in human lymphoblastoid cells. Oncogene 25(4): 622-32.

Tubiana M 2005. Dose-effect relationship and estimation of the carcinogenic effects of low doses of ionizing radiation: The joint report of the Academie des Sciences (Paris) and of the Academie Nationale de Medecine. International Journal of Radiation Oncology Biology Physics 63(2): 317-19.

Tubiana M, Aurengo A, Averbeck D, Masse R 2008. Low-dose risk assessment: The debate continues. Radiation Research 169(2): 246-47.

Tucker JD, Morgan WF, Awa AA, Bauchinger M, Blakey D, Cornforth MN, Littlefield LG, Natarajan AT, Shasserre C 1995. A proposed system for scoring structural aberrations detected by chromosome painting. Cytogenetics and Cell Genetics 68(3-4): 211-21.

Tyburski JB, Patterson AD, Krausz KW, Slavik J, Fornace AJ, Jr., Gonzalez FJ, Idle JR 2008. Radiation metabolomics. 1. Identification of minimally invasive urine biomarkers for gamma-radiation exposure in mice. Radiation Research 170(1): 1-14.

Tyson JJ 2004. Monitoring p53's pulse. Nature Genetics 36(2): 113-14.

Uehara S, Toburen LH, Nikjoo H 2001. Development of a Monte Carlo crack structure code for low-energy protons in water. International Journal of Radiation Biology 77(2): 139-154.

Ullrich RL 2003. Genomic instability, susceptibility genes, and carcinogenesis. Health Physics 85(1): 30-30.

Upton AC 1999. The linear-nonthreshold dose-response model: A critical reappraisal. 9-31 p.

Valentin J 2006. Low-dose extrapolation of radiation-related cancer risk. Oxford, England, Published for the International Commission on Radiological Protection by Elsevier.

Valentin J 2007. The 2007 recommendations of the International Commission on Radiological Protection. Oxford, England, Published for the International Commission on Radiological Protection by Elsevier.

Vance MM, Baulch JE, Raabe OG, Wiley LM, Overstreet JW 2002. Cellular reprogramming in the F-3 mouse with paternal F-0 radiation history. International Journal of Radiation Biology 78(6): 513-26.

Vazquez M, Greulich-Bode KM, Arsuaga J, Cornforth MN, Bruckner M, Sachs RK, Hlatky L, Molls M, Hahnefeldt P 2002. Computer analysis of mFISH chromosome aberration data uncovers an excess of very complicated metaphases. International Journal of Radiation Biology 78(12): 1103-15.

Venkatachalam P, de Toledo SM, Azzam EI 2005. Flavin-containing oxidases regulate progression from G1 to S phase of the cell cycle in normal human diploid fibroblasts. Radiation Physics and Chemistry. 72(2-3): 315-21.

Vines AM, Lyng FM, McClean B, Seymour C, Mothersill CE 2009. Bystander effect induced changes in apoptosis related proteins and terminal differentiation in invitro murine bladder cultures. International Journal of Radiation Biology 85(1): 48-56.

Vit JP, Rosselli F 2003. Role of the ceramide-signaling pathways in ionizing radiation-induced apoptosis. Oncogene 22(54): 8645-52.

Vives S, Loucas B, Vazquez M, Brenner DJ, Sachs RK, Hlatky L, Cornforth M, Arsuaga J 2005. SCHIP: statistics for chromosome interphase positioning based on interchange data. Bioinformatics 21(14): 3181-82.

Walter AE, Brooks AL, Cuttler JM, Feinendegen LE, Gonzalez AJ, Morgan WF 2016. The high price of public fear of low-dose radiation. J. Radiat. Prot. In press 2016.

Wang B, Ohyama H, Haginoya K, Odaka T, Itsukaichi H, Yukawa O, Yamada T, Hayata I 2000. Adaptive response in embryogenesis. III. Relationship to radiation-induced apoptosis and Trp53 gene status. Radiation Research 154(3): 277-82.

Wang DJ, Gao L 2005. Proteomic analysis of neural differentiation of mouse embryonic stem cells. Proteomics 5(17): 4414-26.

Wang TL, Tamae D, LeBon T, Shively JE, Yen Y, Li JJ 2005. The role of peroxiredoxin II in radation-resistant MCF-7 breast cancer cells. Cancer Research 65(22): 10338-10346.

Wang X, Li G, Iliakis G, Y W 2002. Ku affects G_2 checkpoint response following ionizing radiation. Cancer Research 62: 6031–34.

Ward JF 1994. The complexity of DNA damage - relevance to biological consequences. International Journal of Radiation Biology 66(5): 427-32.

Warters RL 2002. Post-translational modification of proteins at low doses: A layer of regulation that may impact the concept of risk. Radiation Research 158(6): 792-92.

Warters RL, Williams DL, Zhuplatov SB, Pond CD, Leachman SA 2007. Protein phosphorylation in irradiated human melanoma cells. Radiation Research 168(5): 535-44.

Waters KM, et al. (2012) Network Analysis of Epidermal Growth Factor Signaling Using Integrated Genomic, Proteomic and Phosphorylation Data. PLoS ONE 7(3): e34515. doi:10.1371/journal.pone.0034515

Weber TJ, Siegel RW, Markille LM, Chrisler WB, Lei XYC, Colburn NH 2005. A paracrine signal mediates the cell transformation response to low dose gamma radiation in JB6 cells. Molecular Carcinogenesis 43(1): 31-37.

Weinfeld M, Xing JZ, Lee J, Leadon SA, Cooper PK, Le XC 2001. Factors influencing the removal of thymine glycol from DNA in gamma-irradiated human cells. Progress in Nucleic Acid Research and Molecular Biology, Vol 68: 139-49.

Wemer T, Haller D 2007. Intestinal epithelial cell signalling and chronic inflammation: From the proteome to specific molecular mechanisms. Mutation Research-Fundamental and Molecular Mechanisms of Mutagenesis 622(1-2): 42-57.

Whalen MK, Gurai SK, Zahed-Kargaran H, Pluth JM 2008. Specific ATM-mediated phosphorylation dependent on radiation quality. Radiation Research 170(3): 353-64.

Wickliffe JK, Rodgers BE, Chesser RK, Phillips CJ, Gaschak SP, Baker RJ 2003a. Mitochondrial DNA heteroplasmy in laboratory mice experimentally enclosed in the radioactive Chernobyl environment. Radiation Research 159(4): 458-64.

Wickliffe JK, Bickham AM, Rodgers BE, Chesser RK, Phillips CJ, Gaschak SP, Goryanaya JA, Chizhevsky I, Baker RJ 2003b. Exposure to chronic, low-dose rate gamma-radiation at chornobyl does not induce point mutations in Big Blue (R) mice. Environmental and Molecular Mutagenesis 42(1): 11-18.

Wiese C, Collins DW, Albala JS, Thompson LH, Kronenberg A, Schild D 2002. Interactions involving the Rad51 paralogs Rad51C and XRCC3 in human cells. Nucleic Acids Research 30(4): 1001-8.

Wiktor-Brown DM, Yoon SN, So PTC, Engelward BP, Kwon HS 2008. Integrated one- and two-photon imaging platform reveals clonal expansion as a major driver of mutation load. Proceedings of the National Academy of Sciences of the United States of America 105(30): 10314-19.

Williams BR, Mirzoeva OK, Morgan WF, Lin JY, Dunnick W, Petrini JHJ 2002. A murine model of Nijmegen breakage syndrome. Current Biology 12(8): 648-53.

Williams JR, Zhang Y, Zhou H, Russell J, Gridley DS, Koch CJ, Little JB 2008a. Genotype-dependent radiosensitivity: Clonogenic survival, apoptosis and cell-cycle redistribution. International Journal of Radiation Biology 84(2): 151-64.

Williams JR, Zhang Y, Zhou H, Gridley DS, Koch CJ, Russell J, Slater JS, Little JB 2008b. A quantitative overview of radiosensitivity of human tumor cells across histological type and TP53 status. International Journal of Radiation Biology 84(4): 253-64.

Wilson GD, Marples B 2007. Flow cytometry in radiation research: Past, present and future. Radiation Research 168(4): 391-403.

Wilson JW, Thibeault SA, Cucinotta FA, Shinn JL, Kim M, Kiefer R, Badavi FF 1995. Issues in protection from galactic cosmic rays. Radiation and Environmental Biophysics 34(4): 217-22.

Wilson PF, Nagasawa H, Warner CL, Fitzek MM, Little JB, Bedford JS 2008. Radiation sensitivity of primary fibroblasts from hereditary retinoblastoma family members and some apparently normal controls: Colony formation ability during continuous low-dose-rate gamma irradiation. Radiation Research 169(5): 483-94.

Wilson WE, Lynch DJ, Wei K, Braby LA 2001. Microdosimetry of a 25 keV electron microbeam. Radiation Research 155(1): 89-94.

Wilson WE, Lynch DJ, Wei K, Miller JH, Ans 2000. Microdosimetry for low-dose low-LET selected cell irradiations. 105-11.

Wilson WE, Miller JH, Lynch DJ, Lewis RR, Batdorf M 2004. Analysis of low-energy electron track structure in liquid water. Radiation Research 161(5): 591-96.

Wimmer K, Thoraval D, Asakawa J, Kuick R, Kodaira M, Lamb B, Fawcett J, Glover T, Cram S, Hanash S 1996. Two-dimensional separation and cloning of chromosome 1 NotI-EcoRV-derived genomic fragments. Genomics 38(2): 124-32.

WHO 2013. Health risk assessment from the nuclear accident after the 2011 Great East Japan earthquake and tsunami, based on a preliminary dose estimation. World Health Organization, WHO Press, Geneva, Switzerland.

Wolff S 1998. The adaptive response in radiobiology: Evolving insights and implications. Environmental Health Perspectives 106: 277-83.

Worgul BV, Smilenov L, Brenner DJ, Vazquez M, Hall EJ 2005. Mice heterozygous for the ATM gene are more sensitive to both X-ray and heavy ion exposure than are wild-types. In: Hei TKMJ ed. Space Life Sciences: Ground-Based Iron-Ion Biology and Physics, Including Shielding. Pp. 254-59.

Wu H, Hada M, Meador J, Hu X, Rusek A, Cucinotta FA 2006. Induction of micronuclei in human fibroblasts across the Bragg curve of energetic heavy ions. Radiation Research 166(4): 583-89.

Wykes SM, Piasentin E, Joiner MC, Wilson GD, Marples B 2006. Low-dose hyper-radiosensitivity is not caused by a failure to recognize DNA double-strand breaks. Radiation Research 165(5): 516-24.

REFERENCES 305

Yamaguchi H, Minopoli G, Demidov ON, Chatterjee DK, Anderson CW, Durell SR, Appella E 2005. Substrate specificity of the human protein phosphatase 2C delta Wip1. Biochemistry 44(14): 5285-94.

Yamamoto K, Ishiai M, Matsushita N, Arakawa H, Lamerdin JE, Buerstedde JM, Tanimoto M, Harada M, Thompson LH, Takata M 2003. Fanconi anemia FANCG protein in mitigating radiation- and enzyme-induced DNA double-strand breaks by homologous recombination in vertebrate cells. Molecular and Cellular Biology 23(15): 5421-30.

Yamamoto K, Hirano S, Ishiai M, Morishima K, Kitao H, Namikoshi K, Kimura M, Matsushita N, Arakawa H, Buerstedde JM and others 2005. Fanconi anemia protein FANCD2 promotes immunoglobulin gene conversion and DNA repair through a mechanism related to homologous recombination. Molecular and Cellular Biology 25(1): 34-43.

Yang CR, Leskov K, Hosley-Eberlein KJ, Criswell TL, Mooney MA, Pink JJ, Boothman DA 2000a. KU70-Binding proteins. Moriaty MEMMCWJFSCFRJM ed. 426-29.

Yang CR, Leskov K, Hosley-Eberlein K, Criswell T, Pink JJ, Kinsella TJ, Boothman DA 2000b. Nuclear clusterin/XIP8, an x-ray-induced Ku70-binding protein that signals cell death. Proceedings of the National Academy of Sciences of the United States of America 97(11): 5907-12.

Yang CR, Wilson-Van Patten C, Planchon SM, Wuerzberger-Davis SM, Davis TW, Cuthill S, Miyamoto S, Boothman DA 2000c. Coordinate modulation of Sp1, NF-kappa B, and p53 in confluent human malignant melanoma cells after ionizing radiation. Faseb Journal 14(2): 379-90.

Yang F, Stenoien DL, Strittmatter EF, Wang JH, Ding LH, Lipton MS, Monroe ME, Nicora CD, Gristenko MA, Tang KQ and others 2006. Phosphoproteome profiling of human skin fibroblast cells in response to low- and high-dose irradiation. Journal of Proteome Research 5(5): 1252-60.

Yang HY, Asaad N, Held KD 2005. Medium-mediated intercellular communication is involved in bystander responses of X-ray-irradiated normal human fibroblasts. Oncogene 24(12): 2096-2103.

Yin E, Nelson DO, Coleman MA, Peterson LE, Wyrobek AJ 2003. Gene expression changes in mouse brain after exposure to low-dose ionizing radiation. International Journal of Radiation Biology 79(10): 759-75.

Zeng G, Day TK, Hooker AM, Blyth BJ, Bhat M, Tilley WD, Sykes PJ 2006. Non-linear chromosomal inversion response in prostate after low dose X-radiation exposure. Mutation Research-Fundamental and Molecular Mechanisms of Mutagenesis 602(1-2): 65-73.

Zhang HG, Wang JH, Yang XW, Hsu HC, Mountz JD 2004. Regulation of apoptosis proteins in cancer cells by ubiquitin. Oncogene 23(11): 2009-15.

Zhang N, Chen P, Khanna KK, Scott S, Gatei M, Kozlov S, Watters D, Spring K, Yen T, Lavin MF 1997. Isolation of full-length ATM cDNA and correction of the ataxia-telangiectasia cellular phenotype. Proceedings of the National Academy of Sciences of the United States of America 94(15): 8021-26.

Zhang QM, Williams ES, Askin KE, Peng YL, Bedford JS, Liber HL, Bailey SM 2005. Suppression of DNA-PK by RNAi has different quantitative effects on telomere dysfunction and mutagenesis in human lymphoblasts treated with gamma rays or HZE particles. Radiation Research 164(4): 497-504.

Zhang Y, Zhou J, Held KD, Redmond RW, Prise KM, Liber HL 2008. Deficiencies of double-strand break repair factors and effects on mutagenesis in directly gamma-irradiated and medium-mediated bystander human lymphoblastoid cells. Radiation Research 169(2): 197-206.

Zheng L, Flesken-Nikitin A, Chen PL, Lee WH 2002. Deficiency of Retinoblastoma gene in mouse embryonic stem cells leads to genetic instability. Cancer Research 62(9): 2498-2502.

Zhou G, Bennett PV, Cutter NC, Sutherland BM 2006. Proton-HZE-particle sequential dual-beam exposures increase anchorage-independent growth frequencies in primary human fibroblasts. Radiation Research 166(3): 488-94.

Zhou H, Ivanov VN, Lien Y-C, Davidson M, Hei TK 2008. Mitochondrial function and nuclear factor-kappa B-mediated signaling in radiation-induced bystander effects. Cancer Research 68(7): 2233-40.

Zhou HN, Randers-Pehrson G, Geard CR, Brenner DJ, Hall EJ, Hei TK 2003. Interaction between radiation-induced adaptive response and bystander mutagenesis in mammalian cells. Radiation Research 160(5): 512-16.

Zhou HN, Ivanov VN, Gillespie J, Geard CR, Amundson SA, Brenner DJ, Yu ZL, Lieberman HB, Hei TK 2005. Mechanism of radiation-induced bystander effect: Role of the cyclooxygenase-2 signaling pathway. Proceedings of the National Academy of Sciences of the United States of America 102(41): 14641-46.

Zhu AP, Zhou HN, Leloup C, Marino SA, Geard CR, Hei TK, Lieberman HB 2005. Differential impact of mouse Rad9 deletion on ionizing radiation-induced bystander effects. Radiation Research 164(5): 655-61.

Ziemer PL 2009. Federal programs to reimburse the public for environmental and occupational exposures. Health Physics 97(5): 528-36.

Zimmer KG 1961. Generalized Formal Hit Theory. In: Studies on Quantitative Radiation, Oliver & Boyd.

Zou Y, Gryaznov SM, Shay JW, Wright WE, Cornforth MN 2004. Asynchronous replication timing of telomeres at opposite arms of mammalian chromosomes. Proceedings of the National Academy of Sciences of the United States of America 101(35): 12928-33.

INDEX

A-bomb survivors, 35, 134
advanced light source, 49
adaptive response, 3, 14, 61, 86, 87, 88, 89, 90, 91, 92, 93, 94, 97, 99, 100, 101, 114, 115, 116, 117, 132, 141, 143, 147, 150, 151, 152, 154, 157, 171, 172, 177, 178, 179, 181, 182, 188, 190, 194, 196, 205, 214, 215, 219, 234, 237, 243, 245, 246
   bystander effects, 115
   cancer formation, 99
   cell cycle, 90
   cell killing, 90, 94
   cell transformation, 90, 91, 171
   chemical agents induced, 87
   chromosome aberrations, 89; responders and non-responders, 89
   clinical applications, 116
   dose and dose-rate, 92, 93
   gene expression, 90, 96, 97, 150, 152, 194
   genetic background, 90, 114, 189, 234, 245, 246
   genomic instability, 116, 141, 151
   glutathione, 143, 152; cell killing, 152; chromosome aberrations, 152
   hormesis, 87, 88, 178, 179, 188
   HZE particles, 94, 181, 182
   mechanisms of action, 94, 95, 96, 97, 98, 99, 101, 157, 171, 172
      amifostine (WR-1065), 148, 157; MnSOD, 148
      apigenin, 99
      apoptosis, 96, 98, 99; superoxide dismutase (SOD2), 148; TNFα, 148
      cell cycle genes, 96
      cell signaling genes, 96
      DNA damage, 96
      gene expression, 94, 95
      gluthione, 219
      heat shock proteins, 96

adaptive response (*continued*)
      human studies, 97, 98
      immune response, 96
      mitochondrial response, 157
      MnSOD, 157
      NF-κB, 157
      protein synthesis, 96, 98
      ROS status, 102, 138, 145, 154, 157, 158, 168, 171, 216, 234, 242
      transcription factor p53, 96
   micronuclei response, 90, 93
   MnSOD, 143, 152, 157, 219
   mutations, 90, 99
   non-linear responses from, 88, 101, 214
   prostate, 90
   protective adaptive responses, 88, 101
   radiation risk, 91, 101
   spleen, 90
   threshold model, 88, 100, 101, 237
   types, 100; priming dose (tickle dose) and challenge dose, 89, 100
advances in medical physics, 207
advisory committee, 204
aerosol particle separator, 25
Albuquerque, NM, 22
angiogenesis, 158, 216
apoptosis (programmed cell death), 69, 71, 72, 73, 78, 85, 96, 98, 99, 108, 122, 126, 138, 141, 148, 158, 163, 164, 165, 166, 167, 168, 169, 172, 218, 219, 246
   cancer development and apoptosis, 72, 167
   cancer risk, 168
   cell transformation, 167
   clustrin, 71, 122
   DNA damage removal, 165
   DNA repair genes, 168
   DNA repair proteins, 166
   during fetal development, 71, 164, 165

apoptosis (*continued*)
  endothelial cells, 168
  genomic instability, 167
  mutations, 167
  NF-κB, 166, 169
  p53 role in apoptosis, 72, 165
  ROS status, 168
  terminal deoxynucleotide transferase dUTP Nick End Labeling (TUNEL assay), 71
  TNF-related, apoptosis-inducing ligand gene (TRAIL), 166
  ubiquitin, 168
Argonne National Laboratory, 33, 36
as low as reasonably achievable (ALARA), 44
ATM, 136, 137, 138, 145, 153, 170, 171; interaction with p53, 138
Averbeck, Dr. Dietrich, 208

Biological Effects of Low Dose and Dose Rate Radiation (BERAC Report Program Plan), 233, 250
biomarkers of exposure, dose and disease, 13, 39, 45, 61, 119, 120, 121, 122, 123, 124, 125, 126, 127, 128, 129, 137, 141, 154
  Center for High Throughput Minimally-Invasive Radiation Biodosimetry, 120
  chromosome aberrations, 39, 119, 122, 123, 129, 141; gold standard, 39, 119; fluorescence in situ hybridization (FISH), 123; genomic instability, 123, 141
  dose effects, 125, 129
  dose rate effects, 126, 129
  DNA damage, 120, 128, 129, 137; γH2AX, 120, 128, 137; phosphorylation of proteins at damage site, 137
  gene expression, 13, 45, 61, 121, 122; function of dose and dose-rate, 121; measure of cell sensitivity, 122
  high Z particles, 127

biomarkers of exposure (*continued*)
  micronuclei, 124, 125; and cancer induction, 125; human populations monitored with, 125
  protein and metabolites, 122
  proteomics, 122
  protein microarrays, 122
  radiation type, 126, 129
  symptomatic biomarkers, 119
Biological Effects of Ionizing Radiation VII (BEIR VII), 132, 180, 205, 207, 210, 223
bio-nanotechnology, 217
Bowman Birk protein, 157
Brenner, Dr. David J., 46, 208
Brookhaven National Laboratory, 27, 33, 36, 105
Brooks, Andrew H., 2
Brooks, Dr. Antone L., 3, 4, 23, 26, 31, 46, 203
Brooks, Evan S., 29
Brooks, Jana L., 23
Brooks, Kriston P., 20
Brooks, Lara K., 28
Brooks, Dr. Mark L., 19
Bull, Dr. Richard J., 46
bystander effects, 3, 13, 31, 43, 50, 53, 55, 57, 58, 61, 71, 75, 77, 78, 79, 80, 81, 82, 83, 84, 85, 86, 87, 90, 114, 115, 116, 117, 121, 122, 132, 137, 139, 141, 142, 143, 144, 154, 172, 188, 189, 196, 205, 214, 244, 245
  adaptive response, 115
  animals, 85, 86; ascopal effects in, 86; chromosome aberration in, 85; clastogenic factors in, 85, 86; and liver cancer, 85
  apoptosis induced by, 71, 78, 85
  bystander effects from direct contact of cells, 78, 87, 115
  bystander effects induced by release of soluble factors, 78, 81, 87, 115
  bystander produce point mutations, base substitutions, and base changes, 81

bystander effects (*continued*)
 calcium flux, 143; MAP kinase signaling pathway, 143
 cancer role in, 86, 87
 cell cycle role in, 80, 85
 cell killing induced by, 77
 cell transformation induced by, 78, 79, 86, 142, 172
 cell type limit bystander communication, 83
 chromosome aberrations, 81; bystander effects produce chromatid type aberrations, 81; chromosome type by direct hits, 81
 clinical applications, 116
 connexin, 43, 142, 143, 189
 distance for bystander responses, 84, 85
 DNA damage induced by, 82, 141, 188; γH2AX induced by, 82, 137
 gap junctions, 142, 143, 189
 gene expression changes induced by, 121, 122
 genetic background, 114
 inflammatory disease role in bystander, 83
 media transfer experiments, 144; epidermal growth factor, 144
 micronuclei induced by, 77, 142
 mutations induced by, 80
 nitric oxide, 143; adaptive response, 143; bystander effects, 143
 nutritional status role in, 83
 oxidative status role in, 80, 83, 87
 protein kinase C pathway, 144
 radiation type, 144
 target for induction nucleus, 80, 83
 TGFβ1, 143, 149
 time course of bystander, 84

cancer, 20, 25, 26, 30, 39, 56, 65, 72, 81, 85, 86, 87, 99, 100, 102, 109, 111, 113, 125, 139, 145, 146, 149, 153m, 158, 161, 168, 169, 171, 178, 187, 189, 226-27

cancer (*continued*)
 baseline level, 100
 breast, 169; BRACA1 and BRACA2, 169, 189; CHK2, 169
 liver, 25, 26, 85, 187
 lung, 30
 lymphoma, 109
 mammary, 109
 risk, 100
 thyroid, 227
 trachea, 30
 treatment, 153; beta-lapachone, 153; quinine oxidoreductase, 153
carcinogen, 29
Casarett, Alison P., 20
cataracts, 138; role of ATM, 138; role of Mrad9, 138
Case Western Reserve University, 34
cell killing, 37, 69, 70, 77, 90, 94, 101, 138, 152, 162, 163, 164, 172, 216, 218
 apoptosis, 164
 colony forming assay, 37
 dye exclusion assay, 37
 high LET, 37, 69
 hyper-radiosensitivity (HRS), 70, 162, 163, 172, 218; dose fractionation, 162; G_2 cell cycle, 163; oxygen tension, 162; p53 dependent apoptosis, 163
 increased radioresistance (IRR), 70, 162, 163, 172, 218
 interphase death, 164
 low LET, 37, 69
 mitotic death, 164
 necrosis, 164
 threshold or shoulder, 69
cell-cell communication, 13, 45, 52, 55, 140, 142, 153, 195
cell cycle, 54, 161, 162; CHK1 dependent, 161; G_2 Block, 161; Ku, 161; Cyclin D1, 162
cell-matrix communication, 13, 45, 140, 195

cell transformation, 78, 79, 86, 90, 91, 142, 156, 167, 171, 172
centers of excellence, 224, 225
characteristic X-rays, 56
Chernobyl, 93, 127, 128, 212, 226, 227
chief scientist, 3, 9, 211, 221
Chinese hamsters, 21, 25
chromosome aberrations, 20, 21, 25, 39, 59, 60, 81, 89, 101, 103, 119, 122, 123, 124, 129, 141, 152, 159, 172, 178, 240, 242
   cancer risk, 39
   lymphocytes, 39
   banding, 123
   complex aberrations, 60
   duplications, 159
   loss or change in number, 159
   rogue cells, 123
   translocations, 159
chromosome painting, 59; Co-FISH, 61; FISH, 59, 123; interchromosome inversions, 124; mFISH, 124
Church, Bruce W., 17
Church of Jesus Christ of Latter Day Saints, 231
clastogenic factors, 78
clusterin, 122, 148–149
   apoptosis, 148
   cell survival, 148
   cell proliferation, 148
   nuclear clusterin effects, 148
   secretory clusterin effects, 149
cobalt-60 gamma rays, 26
contaminated environments, 127, 128, 129, 227; Chernobyl, 127–29, 227
Comar, Dr. Cyril, 20
communication, 27, 199, 200, 203, 204, 211, 219, 220, 221, 222, 228, 229, 235, 247, 248, 249, 250, 251
   advisory committee, 199, 204, 211
   bureaucrats, 220
   contractors meeting, 204, 250
   internet, 222
   media communication, 200, 222

communication (*continued*)
   monitoring progress, 199, 250
   news media, 220
   public communication, 200, 222, 247, 250
   regulatory agencies, 220, 222
   stakeholders, 247, 248, 250
   workshop on risk communication, 199
Colorado State University, 34
Columbia University, 34, 52, 120
Comparative Radiation Ecology of Six Utah Dairy Farms, 19
computer power, 217
Cornell University, 20, 34
Couch, Lezlie A., 32
Cross, Dr. Fred T., 30

Decision Research Institute, 199
dirty bomb, 9, 226, 228
"Dirty Harry" fallout, 15
Dixie Junior College, 16
DNA damage, 39, 40, 41, 66, 67, 73, 82, 96, 101, 120, 121, 126, 128, 129, 135, 136, 137, 138, 141, 154, 155, 160, 165, 166, 168, 170, 171, 184, 185, 188, 190, 193, 195, 216, 235, 236, 237, 238, 239, 242, 244, 246
   clustered DNA damage, 66, 121, 135, 238
   damage detection γH2AX, 67, 73, 137, 138
   damage from microbeam, 67
   double strand breaks, 67, 137, 155, 237, 246
   locally multiply damaged sites (LMDS), 66, 73, 135
   repair normal endogenous oxidative metabolism, 40, 66, 135, 136, 235, 238, 239, 242, 244
   single strand breaks, 137
   induced by radiation, 40, 66
   induction of signaling pathways, 40, 135, 136, 138, 195, 216
      ATM gene, 136, 138
      ATM protein, 136

DNA damage (*continued*)
    ATM signaling, 136, 137, 170
    genomic instability, 137
    ROS status, 138, 154
DNA discovery by Watson and Crick, 40
DNA repair, 68, 110, 136, 137, 154, 160, 169, 183, 234, 239, 240, 241
    diseases with defective human repair genes, 68
        ataxia telangiectasia, 136, 169, 240
        Cockayne syndrome, 68
        Fanconi anemia, 68
        Nijmegen breakage syndrome, 68, 169, 183, 240
        retinoblastoma, 110, 169
        trichotriodystrophy, 68
        xeroderma pigmentosum, 68, 110
    DNA repair genes, 68
Domenici, Sen. Peter, 42, 43, 45
dose response relationships, 12, 38, 42, 44, 76, 80, 88, 101, 105, 140, 161, 106, 171, 215, 234, 245, 248
    hormetic response, 76
    linear no threshold responses, 12, 38, 42, 44, 76, 80, 101, 105, 106, 161, 215, 234, 248; DNA damage, 171; mutations, 38
    on/off and all or none responses, 80
    non-linear dose responses, 106, 171
    threshold, 140
dose-dose rate effective factors (DDREF), 43, 44, 73, 206
dose rate, 60
dose ranges chart, 11, 12, 256, 257
down winders, 209
drosophila, 38, 247
dynamic microscopic image processing scanner (DMIPS), 70

ecology, 36
energy barrier, 134
energy production, 9, 28, 29; fossil fuels, 28, 29; nuclear, 9, 28
environmental cleanup, 42

Environmental Protection Agency (EPA), 210, 221, 251
epigenetics, 111, 112, 170, 171, 172, 197, 212, 217
    acethlylation, 217
    diet and cancer, 111, 171
    dose response, 171
    methylation, 217
    stromal radiation, 111
    tandem repeat DNA loci, 112
    transgenerational changes, 112
epidemiological studies, 47, 99, 125, 128, 131, 176, 179, 205, 214, 235, 236, 247

fallen angel (mutated p53 gene), 138; role in cancer, 139
fallout nuclear weapons, 15, 41, 232
fear of radiation, 9, 18, 218, 219, 226, 227, 232
Flint, Dr. Alex, 44
flow cytometry, 59, 121
Fluorescent In Situ Hybridization (FISH), 60, 241
Flynn, Dr. James, 46, 200
Frazier Dr. Marvin, 1, 29, 43, 44, 45, 46, 47, 131
French National Academy, 35, 179, 180, 206, 210
Frenchman Flat, 15
Friedman, Dr. Sharon, 199
Fukushima, 212, 226, 227
funding, 222-23, 224, 226
future directions, 211

gene expression, 13, 45, 61, 82, 90, 94, 95, 96, 97, 121, 122, 129, 138, 139, 140, 142, 150, 152, 158, 194, 197
gene therapy, 153
genetic background, 35, 90, 108, 114, 169, 170, 172, 235, 237, 246, 247; genetic polymorphisms, 236, 237, 246, 247; genomic instability, 169
genetic risk, 39, 245
genomic instability, 13, 14, 61, 86, 90, 101, 102, 103, 104, 105, 107, 108,

genomic instability (*continued*)
109, 111, 113, 114, 115, 116, 117, 123, 132, 137, 139, 141, 143, 146, 147, 148, 151, 154, 155, 156, 157, 158, 160, 161, 169, 170, 171, 195, 196, 205, 215, 216, 219, 239, 240, 242, 245
  adaptive response decreased frequency, 104, 113, 116
  amifostine WR-1065, 148
  apoptosis, 108
  cancer and genomic instability, 113, 161
  cell killing, 101
  cell transformation, 156
  cell types involved in, 103
  chromosome aberrations, 101, 103
  comet assay, 103
  death inducing effects, 108
  DNA breakage, 101
  DNA base changes, 101
  DNA copy number, 104
  DNA repair, 102, 154, 239; base excision repair, 102, 239; homologous recombination repair, 102, 170, 239; non-homologous end-rejoining, 239
  dose, dose-rate, and dose distribution, 101, 105
  genetic background, 108; BALB/c mice, 108; BRCA 1, BRCA 2 gene, 110, 240; C57BL/6 mice, 108; Gadd 45 gene, 110; Rad 51 gene and Rad 51C gene, 110; XRCC3 gene, 110
  hallmark of cancer, 102
  high Z particles, 105
  human cancer, 111, 113
  human populations failed to induce, 104
  interleukin, 116, 155
  micronuclei, 103
  mitochondria, 107, 113
  mutations, 101, 155, 216
  normal cells failed to induce, 104
  nucleus target for induction, 104, 113
  ROS status of cells, 102, 107, 113, 143, 155, 157, 158, 171, 195
  senescence, 156

genomic instability (*continued*)
  telomere dysfunction, 161
guardian of the genome (TP53), 109, 110, 137, 138, 139, 140, 243
  apoptosis, 138
  cancer, 139
  genomic instability, 139
  interaction with MDM2, 140; DNA Repair, 138; gene expression, 138; mutated p53, 138
  radiosensitivity, 139
gene chip technology, 61, 139, 212, 245
  genomics, 61, 212
  radiation induced changes in gene expression, 61, 139; high and low dose genes, 61; modified by exposure variables, 61
  radiation induced stress genes, 61, 245
genetic cells, 20
Gordon Research Conference, 45
Gray Cancer Institute, 56

Hall, Dr. Eric J., 46, 53
Hallmarks of Cancer chart, 216
Hanford advisory board, 202, 220
Harvard University, 34
high Z particles, 32, 47, 60, 105
Hiroshima, 33, 41
hit cells, 25, 31
hit theory, 13, 52, 59, 135, 214
Holmes, Jeffery, 2
hormesis, 78, 88, 178, 179; and birth defects, 178; and cancer, 178; and chronic myeloid leukemia, 179
hot particle hypothesis, 24, 25
human genome project, 45, 49, 59, 61, 212, 247; development of gene chip technology, 61; sequencing genome, 45

ICRU, 133
Indian tribes, 202, 220
International Council on Radiation Protection (ICRP), 35, 132, 206, 210, 221, 229
investigator workshops, 201

INDEX 313

Katz, Dr. Arthur, 47

latent period, 216
Lawrence Berkley National Laboratory, 27, 33, 57
Lawrence Livermore National Laboratory, 27, 33
lead scientist, 201; Barcellos-Hoff, Dr. Mary Helen, 201
Lengermann, Dr. Fred, 20, 21
lessons learned, 211
Los Alamos National Laboratory, 27, 38, 59
Lovelace Fission Product Inhalation Program, 22
Lovelace Inhalation Toxicology Research Institute (ITRI), 22, 27, 28, 31, 33, 36
Lurie, Dr. Robert H., 4
Lyons, Dr. Peter B., 44

management, 222
McClellan, Dr. Roger O., 22, 23, 26, 232
mechanisms of action, 13
mega mouse genetics studies, 38
metabolomics, 65, 74, 141; use in biodosimetry, 65
metabolic pathways, 144, 149, 150
    transforming growth factor β (TGFβ), 145, 154, 214, 219
    glucose metabolism, 150
    glucose transporters, (ECAR) 150
Metting, Dr. Noelle, 1, 11, 12, 47, 231
microbeam, 31, 45, 49, 50, 51, 54, 56, 57, 73, 133, 212
    alpha, 50, 51, 73
    advanced light source, 57
    electron gun, 54, 73
    focused X-ray, 56, 73
microenvironment, 65, 219, 243, 244
microfluidics, 217
modeling, 12, 42, 72, 75, 76, 80, 88, 90, 100, 101, 105, 161, 173, 174, 175, 176, 177, 178, 179, 180, 181, 182, 183, 184, 185, 186, 187, 188, 189, 190, 191, 196, 197, 205, 206, 239
    animal models, 183, 190, 191; dog models, 183; rodent models, 183

modeling (*continued*)
    biologically based models, 174, 189, 190
    bystander models, 188; and direct effects models (BaD), 188
    cell communication models, 188
    cell killing models, 187, 191; two-lesion kinetic models, 187
    cell transformation models, 178
    chromosome aberration models, 185, 190, 197; chromosome domain, 186; complex chromosome aberrations, 186; interaction distance, 186; random breakage and reunion, 186
    DNA damage models, 184, 185, 190
    genetic background, 189, 191
    hit size effectiveness models, 174
    hormesis models, 178, 179
    impact on standards, 179
    linear-quadratic models, 177, 190
    LNT Models, 175, 176, 190, 205, 206, 239
    microdosimetry models, 175, 187
    NASA Models, 181
    non-LNT models, 177, 190
    risk assessment models, 189
    threshold models, 177; endpoint, 179; organ, 179; species, 179; tissue, 179
    two stage clonal expansion models, 182
    two step mutation models, 173, 174
Morgan, Dr. William F., 1, 46, 203
Mutations, 13, 29, 38, 80, 81, 90, 99, 101, 138, 155, 167, 172, 216, 242

Nagasaki, 33
National Academy of Sciences, 35, 132, 179-80, 206, 229
National Aeronautics and Space Administration (NASA), 47, 105
National Council on Radiation Protection (NCRP), 35, 181, 189, 229; board of directors, 231; 44th Annual Meeting, "Low Dose and Low Dose-Rate Radiation Effects and Models," 207, 210
Nevada test site, 15

nuclear factor κB (NF-κB), 146, 147, 216
    activated by low doses, 146
    ATM, 146
    cancer induction, 146
    genomic stability, 146
    inflammatory disease, 146, 216
    tumor necrosis factor alpha (TNAα), 147
nuclear weapons and war, 9, 122, 125, 226, 228, 229
nuclear waste clean-up, 9, 226
NRC, 51, 210, 221, 251

Oak Ridge National Laboratory, 27, 33, 38
Office of Biological and Environmental Research, 43
OSHA, 221

Pacific North West National Laboratory, 1, 29, 30, 31, 32, 33, 36, 47
paradigm shifts in radiation biology, 75, 76, 131, 210, 215
    adaptive responses vs Linear No Threshold, 75, 76
    bystander effects vs hit theory, 75; cell-cell and cell-tissue communication, 76
    hormetic response, 76
    mutation theory vs genomic instability, 76, 215
Partial Test Ban Treaty, 36
Patrinos, Dr. Ari, 44, 47
Pendleton, Dr. Robert C., 17, 19
Pitman, Andrew, 2
Poulsen, Janet L., 18
presentations, 202, 203, 206, 207, 221, 223, 231, 246, 249
    American Chemical Society, 202
    American Pharmacists Association, 202
    American Statistical Association, 202
    DOE National Laboratories, 202
    government agencies, 202; EPA, 202, 207; EPA Radiation Advisory Council (RAC), 207, 231; NIEHS, 221, 246; NIH, 202, 221, 222, 246, 249

presentations (*continued*)
    International Consortium for Research on Health Effects of Radiation (ICERHER), 202
    public presentations, 203
    standard regulators, 202; BEIR VII, 202; Electric Power Research Institute, 202, 206; ICRP, 221; ICRU, 203; NCRP, 203, 221
    universities, 202
    Washington State Department of Ecology, 202
Preston, Dr. R. Julian, 46
proteomics, 14, 62, 63, 64, 65, 74, 141, 194, 217
    biodosimetry, 64
    liquid chromatography/tandem mass spectrometry (LC-MS/MS), 63, 194
    mitrochondrial protein changes, 65
    NEDylation, 217
    phosphoproteome, 63, 217
    protein changes and cancer, 65
    shed proteins, 63
    SUMOylation, 217
    ubiquitination, 217
prodromal radiation syndrome, 34
protein expression, 14
PubMed, 139; p53 research and, 139

Raabe, Dr. Otto, 25
radiation, 9, 10, 14, 20, 24, 42, 43, 59, 72, 73, 81, 125, 180, 209, 214, 219, 223, 226, 232, 241, 245, 247
    low dose risks, 10
    medicine, 9, 226
    nuclear industry, 9, 125, 226, 232
    Radiation Exposure Compensation Act (RECA), 180, 209
    radiation risk, 20; cancer, 20, 81; genetic, 20, 81; teratogenic (zebrafish as model), 72
    paradigms, 59, 214, 223, 241, 245, 247; hit theory, 59
    Relative Biological Effectiveness (RBE), 24

radiation (*continued*)
    standards, 14, 42
    tissue weighting factors, 43, 73
ROS status, 102, 107, 113, 138, 143, 145, 154, 155, 157, 158, 164, 168, 171, 195
radiation ecology, 19, 35
radionuclides, 17, 18, 19, 23, 24, 26, 36
    Americium-241, 24
    Californium-252, 24
    Cerium-144, 17, 24
    Cesium-137, 17, 18, 19, 24
    internally deposited, 36
    Iodine-131, 17
    Plutonium, 239, 17, 24
    Strontium-90, 17, 23
    Tritium, 24, 26
Radon, 30; The Health Effects of Radon (BEIR VI), 30
RNA, 217; circular RNA, 217; long non-coding RNA, 217; micro RNA, 217
Raman spectrometry, 217
Rocky Flats, mayor of, 202, 220

St. George, Utah, 15, 16
science advisory committee, 223
secretomics, 65, 74
Sedan atomic test, 18
shape of dose-response relationship, 14
sister chromatid exchanges, 77, 141, 142, 170, 243
Slovic, Dr. Paul, 200
Smith, Dr. David, 29, 43
somatic cells, 20
standards radiation, 205
stochastic effects, 178
stress, 158, 235
    aging, 158
    angiogenesis, 158
    cancer risk, 158, 235
    gene expression, 158
    selective apoptosis, 158
    senescence, 158

systems biology approach, 14, 189, 193, 194, 195, 196, 197, 224
    astronauts, 196
    cell-cell interactions, 195, 197
    cell-matrix interactions, 195
    data sets require, 193
    functional endpoints, 197
    gene expression, 194
    integrates responses, 193
    levels of biological organization, 193, 197
    protein expression, 194, 195, 197
    protein modification, 194
    risk assessment, 197
    signaling pathways, 195, 196

Taylor Lecture, 232
telomeres, 61, 159, 160, 161, 172
    DNA double strand breaks, 160
    DNA-PK kinase, 160
    DNA repair, 160
    genomic instability, 160
    telomerase, 160
teratogenic effects of radiation, 72, 74
terrorist attack, 122, 125, 147, 209, 226
Texas A & M University, 50, 51, 54
Thomassen, Dr. David, 45, 46, 47, 199
tickle dose, 3
total energy deposited, 134

Ullrich, Robert, 46
United Nations Scientific Committee on the Effects of Atomic Radiation, 206, 210
University of California-Berkeley, 34
University of California-Davis, 36
University of Rochester, 34
University of Tennessee, 34
University of Texas, 34
University of Utah, 16, 34, 36
University of Wisconsin-Madison, 34

Venethum, Gary, 44
Vernal, Utah, 18
vitamin E analogues, 158

Waters, Dr. Katrina, 1
Wagner, Dr. Henry N., 46
Wallace, Dr. Susan S., 46
Washington, D.C., 26, 28
Washington State University, 31, 32
Washington State University Press, 2
Wassam, John, 203
website, 203, 221
Wiley, Julie, 2
Woloschak, Dr. Gayle, 4, 46

X-ray microprobe, 49
X-ray microscopy, 217